The Carbon Balance of Forest Biomes

EXPERIMENTAL BIOLOGY REVIEWS

The Carbon Balance of Forest Biomes

Edited by

HOWARD GRIFFITHS
Department of Plant Sciences, University of Cambridge, Cambridge, UK

PAUL G JARVIS
School of GeoSciences, University of Edinburgh, Edinburgh, UK

Taylor & Francis
Taylor & Francis Group

A CIP catalogue record for this book is available from the British Library.

ISBN 1 85996 214 9

Taylor & Francis Group
4 Park Square, Milton Park, Abingdon, Oxon OX14 4RN, UK and
270 Madison Avenue, New York, NY 10016, USA
World Wide Web home page: www.garlandscience.com

Distributed in the USA by
Taylor & Francis Group
Customer Service
7625 Empire Drive
Florence, KY41042, USA

Distributed in Canada by
Taylor & Francis
74 Rolark Drive
Scarborough, Ontario M1R 4G2, Canada
Toll Free Tel.: +1 877 226 2237; E-mail: tal_fran@istar.ca

Distributed in the rest of the world by
Thomson Publishing Services
Cheriton House
North Way
Andover, Hampshire SP10 5BE, UK
Tel.: +44 (0)1264 332424; E-mail: salesorder.tandf@thomsonpublishingservices.co.uk

Library of Congress Cataloging-in-Publication Data

The carbon balance of forest biomes / edited by H. Griffiths, P. Jarvis.
    p. cm.
    Includes bibliographical references and index.
    1. Global environmental change. 2. Carbon--Environmental aspects. 3. Forests and forestry--Environmental aspects. 4. Plants--Effect of atmospheric carbon dioxide on. 5. Carbon dioxide mitigation. I. Griffiths, Howard, 1953- II. Jarvis, P. G. (Paul Gordon)
    GE149.C35 2005
    577.3'144--dc22
                                                    2004025748

Editor: Nigel Farrar
Editorial Assistant: Dominic Holdsworth
Production Editor: Erika Pennington
Typeset by Phoenix Photosetting, Chatham, Kent, UK
Printed by Cromwell Press, Trowbridge, UK

Contents

3.   Carbon sequestration in European croplands   47
Pete Smith and Pete Falloon

4.   Estimating forest and other terrestrial carbon fluxes at a national scale: the UK experience   57
Ronnie Milne and Melvin G.R. Cannell

Contributors

Amiro, B.D., Department of Soil Science, University of Manitoba Winnipeg, Winnipeg, Canada

Arrouays, D., INRA Unité de Science du Sol, 45166, Ardon, France

Aurela, M., Climate and Global Change Research, Finnish Meteorological Institute, Helsinki, Finland

Bakwin, P.S., Climate Monitoring and Diagnostics Lab, National Oceanic and Atmospheric Administration, Boulder, USA

Baldocchi, D., Ecosystem Science Division, Department of Environmental Science, University of California, Berkeley, California, USA

Berry, J.A., Department of Global Ecology, Carnegie Institution of Washington, Stanford, USA

Betts, A.K., Atmospheric Research, Pittsford, USA

Bhupinderpal-Singh, Department of Forest Ecology, SLU, Umeå, Sweden

Black, T.A., Department of Agricultural Sciences, University of British Columbia, Vancouver, Canada

Brumme, R., Institute of Soil Science and Forest Nutrition, Göttingen, Germany

Butler, M.P., Department of Meteorology, Pennsylvania State University, Pennsylvania, USA

Cannell, M.G.R., Centre for Ecology & Hydrology (Edinburgh), Penicuik, UK

Ciais, P., LSCE, Unité mixte CEA/CNRS, France

Conen, F., Institute of Environmental Geosciences, Universtiy of Basel, Basel, Switzerland

Czimczik, C.I., Department of Earth System Science, University of California, Irvine, USA

Davis, K.J., Department of Meteorology, Pennsylvania State University, Pennsylvania, USA

Dolman, A.J.H., Department of Hydrology and Geo-environmental sciences, Vrije Universiteit, Amsterdam, Netherlands

Dunn, A., Department of Earth and Planetry Sciences, Harvard University, Cambridge, USA

Ehleringer, J.R., Department of Biology, University of Utah, Utah, USA

Falloon, P., Agriculture and the Environment Division, Rothamsted Research, Harpenden, UK

Gaumont-Guay, D., Department of Agriculture, University of British Columbia, Vancouver, Canada

Gower, S.T., Department of Forestry, University of Wisconsin-Madison, Wisconsin, USA

Grace, J., School of GeoSciences, University of Edinburgh, Edinburgh, UK

Heimann, M., Max-Planck-Institut für Biogeochemie, Jena, Germany

Helliker, B.R., Department of Global Ecology, Carnegie Institution of Washington, Stanford, USA

Högberg, M.N., Department of Forest Ecology, SLU, Umeå, Sweden

Högberg, P., Department of Forest Ecology, SLU, Umeå, Sweden

Ibrom, A., Plant Research Department, Risø National Laboratory, Roskilde, Denmark

Janssens, I., Department of Biology, University of Antwerpen, Belgium

Jarvis, P.G., School of GeoSciences, University of Edinburgh, Edinburgh, UK

Jassal, R.S., Department of Agriculture, University of British Columbia, Vancouver, Canada

Jolivet, C., INRA, Unité de Science du Sol, 45166, Ardon, France

Kelliher, F.M., Department of Agriculture, University of British Columbia, Vancouver, Canada

Laurila, T., Climate and Global Change Research, Finnish Meteorological Institute, Helsinki, Finland

Linder, S., Southern Swedish Forest Research Centre, SLU, Alnarp, Sweden

Lohila, A., Climate and Global Change Research, Finnish Meteorological Institute, Helsinki, Finland

Malhi, Y., School of Geography and the Environment, University of Oxford, UK

Martikainen, P.J., Department of Environmental Sciences, University of Kuopio, Kuopio, Finland

Mencuccini, M., School of GeoSciences, University of Edinburgh, Edinburgh, UK

Miglietta, F., IBIMET, Instituto de Biometeorologia, Firenze, Italy

Milne, R., Centre for Ecology & Hydrology (Edinburgh), Penicuik, UK

Mund, M., Max-Planck-Institut für Biogeochemie, Jena, Germany

Noble, I.R., Environmentally and Socially Sustainable Development Network, World Bank, Washington DC, USA

Nordgren, A., Department of Forest Ecology, SLU, Umeå, Sweden

Olsson, P., Department of Forest Ecology, SLU, Umeå, Sweden

Ottosson-Löfvenius, M., Department of Forest Ecology, SLU, Umeå, Sweden

Phillips, O., Earth and Biosphere Institute, School of Geography, University of Leeds, Leeds, UK

Potter, C.S., Ecosystem Science and Technology Branch, NASA Ames Research Center, USA

Ricciuto, D.M., Department of Meteorology, Pennsylvania State University, Pennsylvania, USA

Ronda, R., Department of Hydrology and Geo-environmental sciences, Vrije Universiteit, Amsterdam, Netherlands

Schulze, E.-D., Max-Planck-Institut für Biogeochemie, Jena, Germany

Shvidenko, A., International Institute for Applied Systems Analysis, Laxenburg, Austria

Smith, P., School of Biological Sciences, University of Aberdeen, Aberdeen, UK

Tuovinen, J.-P., Climate and Global Change Research, Finnish Meteorological Institute, Helsinki, Finland

Verchot, L.V., International Center for Research in Agroforestry, Nairobi, Kenya

Watson, R.T., Environmentally and Socially Sustainable Development Network, World Bank, Washington DC, USA

Wirth, C., Department of Ecology and Evolutionary Biology, Princeton University, Princeton, USA

Wofsy, S.C., Department of Earth and Planetry Sciences, Harvard University, Cambridge, USA

Xu, L., Ecosystem Science Division, Department of Environmental Science, University of California, Berkeley, California, USA

Zerva, A., School of GeoSciences, University of Edinburgh, Edinburgh, UK

Abbreviations

$\delta^{13}C$	carbon isotope ratio
A&R	afforestation and reforestation
ABL	atmospheric boundary layer
ARCA	actual rate of carbon accumulation
a.s.l.	above sea level
AVHRR	advanced very high resolution radiometer
BCI	Barro Colorado Island, Panama
BEP	background emission pattern
BERMS	Boreal Ecosystems Research and Monitoring Sites
BOREAS	Boreal Ecosystem–Atmosphere Study
BP	before present
CAI	current annual increment
CBL	convective boundary layer
CDM	clean development mechanism
CER	certified emission reduction
CLASS	Canadian Land Surface Scheme
COBRA	CO_2 Budget and Regional Airborne study
CoP	Conference of the Parties
CP	Commitment Period
CV	coefficient of variance
DIC	dissolved inorganic carbon
DOC	dissolved organic carbon
EC	eddy covariance
ECMWF	European Center for Medium-Range Weather Forecasting
EG	early girdling
EUROFLUX	EU flux measurement project 1997–2001
FACE	free air carbon dioxide enrichment
FAO	Food and Agriculture Organization of the United Nations
FC	Forestry Commission (UK)
FLUXNET	global network of eddy flux sites
GCM	general circulation model
GEP	gross ecosystem productivity
GHG	greenhouse gas
GIS	geographic information system
GPP	gross primary productivity
GPS	global positioning system
GWP	global warming potential
ha	hectare; 1 ha = 10^4 m^2
InTEC	Integrated Terrestrial Ecosystem C-budget model
IPCC	Intergovernmental Panel on Climate Change
JI	joint implementation

LAI	leaf area index
LG	late girdling
LORCA	long-term rate of carbon accumulation
LUCF	Land-Use Change and Forestry (IPCC)
LULUCF	Land Use, Land-Use Change and Forestry (IPCC)
MAI	mean annual increment
NBP	net biome productivity
NCEP	National Centers for Environmental Prediction
NCAR	National Center for Atmospheric Research
NDVI	normalized difference vegetation index
NEE	net ecosystem exchange
NEP	net ecosystem productivity
NGO	non-governmental organization
NIR	National Inventory Report (UK)
NIWT	National Inventory of Woodlands and Trees (UK)
NOBS	northern BOREAS old black spruce site
NOAA-CMDL	National Ocean and Atmospheric Administration-Climate Monitoring and Diagnostics Lab.
NOPEX	Northern Hemisphere climate processes land-surface experiment
NPP	net primary productivity
PAR	photosynthetically active radiation (400–700 nm)
p.p.b.	parts per billion (10^9)
PPFD	photosynthetic photon flux density (400–700 nm)
p.p.m.v.	parts per million by volume
ppt	precipitation
QELRC	quantified emission limitation or reduction commitment
RAMS	Regional Atmospheric Modelling System
RECAB	CarboEurope Regional Carbon Balance project
RH	relative humidity
RUBISCO	ribulose 1,5-bisphosphate carboxylase-oxygenase
RUBP	ribulose 1,5-bisphosphate
SD	standard deviation
SEP	seasonal emission pattern
Sib	simple biosphere model
SOA	southern BOREAS aspen site
SOBS	southern BOREAS old black spruce site
SOC	soil organic carbon
SOJP	southern BOREAS old jack pine site
SOM	soil organic matter
SRES	IPCC Special Report on Emission Scenarios
STILT	stochastic time inverted Langrangian transport model
SVAT(S)	soil–vegetation–atmosphere transfer (scheme)
UNFCCC	United Nations Framework Convention on Climate Change
VOC	volatile organic compound
WFPS	water-filled pore space

WLEF	national public broadcast station call letters (Wisconsin, USA)
YC	yield class
ZERO	Zackenberg Ecological Research Operations

Preface

Climate change is now increasingly recognized by the public at large to be a real and present problem. Recent authoritative reviews, co-ordinated by the IPCC (Intergovernmental Panel on Climate Change), had also reached such a compelling conclusion after critically evaluating the scientific evidence. Thus, the 0.6 °C increase in global temperatures during the past 50 years has been driven, we think, by the annual release of carbon from fossil fuel emissions and land-use change of around 8 to 9 Pg carbon (where a petagram, Pg, is equal to 10^{15} grammes or one gigatonne, Gt). Around two-thirds of this emitted CO_2 is removed from the atmosphere each year, with terrestrial vegetation and physical plus biological processes in oceans each acting as equivalent sinks. However, such changes are superimposed upon enormous inter-annual carbon fluxes: of the total 750 Pg (C) present as CO_2 in the atmosphere, each year some 90 Pg (C) is taken up and released by oceans, and 55 Pg (C) taken up by terrestrial photosynthesis and released by respiration.

In this volume, we focus on the role of terrestrial vegetation in contributing to global carbon uptake and sequestration. From a biological perspective, the carbon pools in terrestrial vegetation and soils (respectively, some 650 and 3300 Pg carbon) outweigh the 2 Pg (C) of the marine biota, even considering the rapid turnover and carbon sequestration potential of phytoplankton biomass. Additionally, the dynamics of gaseous diffusion enable perhaps 30% of the atmospheric CO_2 pool to exchange across leaf surfaces, via stomata, on an annual basis. In considering the *Kyoto Protocol*, we need to address both theoretical and practical requirements for the monitoring and assessment of carbon sequestration by terrestrial biomes. The uncertainties associated with the magnitude of carbon fluxes between the atmosphere, terrestrial vegetation and soils need to be assessed, as do constraints to scaling from leaf, via canopy, to ecosystem. Another priority is to assess above- and below-ground carbon stocks, and their responses to increasing CO_2 and nitrogen fertilization. The interaction between global temperatures and accelerating greenhouse gas emissions from peat-based forest biomes remains another important way that terrestrial vegetation may indirectly contribute to climate change processes. Ultimately, we need to understand for how long terrestrial sinks will endure, particularly in the context of increasing disturbance and occurrence of fire, so that their potential can be managed and exploited for mitigating climate change.

This volume represents the proceedings of a symposium organized and supported by the Plant Environmental Physiology Group of the Society for Experimental Biology (SEB) and the British Ecological Society (BES), held at the SEB annual general meeting in Southampton, April 2003. In order to address the issues regarding the carbon balance across such a range of scales, we have organized a progression in the sequence of chapters. Initially, global issues are considered in terms of the practicalities for implementing the *Kyoto Protocol* from the perspective of the World Bank (Bob Watson and Ian Noble), including international accounting for Land Use and Land-Use Change and Forestry (LULUCF) and the eligibility for credit using the Clean Development Mechanism (CDM). Having considered the size, magnitude and endurance of potential sinks in an overview of global carbon budgets by John Grace, the regional potential for carbon sequestration is provided for both agricultural landscapes (Chapter 3) and for UK forestry (Chapter 4).

The latest developments in methods and models for integrating gaseous fluxes across the planetary boundary layer show how variations within the 'global gas exchange cuvette' can be addressed at regional scales (Chapters 5 and 6). The interrelationship between net primary productivity (NPP) and the longer-term potential for carbon storage (net biome productivity (NBP)) is compared in a meta-analysis at regional scales by Philippe Ciais and colleagues (Chapter 7). Here the uncertainties relate to the extent that carbon sequestration is driven by increased photosynthetic drawdown, or reduced losses from respiration or disturbance, and these issues are subsequently addressed for specific biomes. In considering Boreal, temperate/ Mediterranean and tropical forest systems, (Chapters 8, 9 and 10), we scale from individual leaf and canopy gas exchanges and eddy flux measurements, as compared to the longer-term dynamics derived from traditional standing-crop mensuration techniques. The confirmation that tropical old-growth forest has been stimulated, representing a potential sink of 0.54 Pg (C) per year in the Neotropics and 1.2 ± 0.4 Pg (C) per year worldwide, is given by Yadvinder Malhi (Chapter 10).

Turning to address components of carbon balance within forest biomes, the errors and uncertainties in assessing carbon sequestration and storage in soils are considered, firstly, in terms of forest management and disturbance (Chapter 11). Subsequently, the practical difficulties in distinguishing between autotrophic and heterotrophic respiration, framed by the complexities of carbon transfer through the canopy-root-mycorrhizal-soil continuum, are addressed in a summary of the elegantly destructive 'tree-strangling' approaches undertaken by Peter Högberg and colleagues (Chapter 12). Soils are important both for regulating fluxes of greenhouse gases (other than CO_2) with relatively higher warming potential, such as CH_4 and N_2O, and revised estimates of global fluxes for contrasting forest and peatland biomes are given in Chapters 13 and 14. Finally, the future potential for carbon sequestration is considered (Chapters 15 and 16). This has enormous practical relevance: firstly, for the management of forests to maximize carbon storage and sink potential for the future, and secondly, for assessing the impact of mitigation, CDM credits and BioCarbon Forestry management of forests. For instance, following re-afforestation it may take 25–40 years for a newly-planted forest to become a net sink for carbon (Chapters 11 and 15), whereas disturbance as windthrow can enhance carbon storage as regeneration occurs in naturally-forested systems (Chapter 16).

We are grateful for the irrepressible efforts of Chris Trimmer at the SEB during the organisation of the symposium, together with generous financial support from both the SEB and the BES, and an additional input from the CarboEurope programme for associated delegates. The meeting was notable for the number of participants and their spirited discussions, for which we are also grateful. A number of equipment manufacturers were also generous in their support, including ADC Bioscientific, Campbell Scientific, Glen Spectra, Li-Cor, PDZ-Europa, PP Systems and Skye Instruments. Finally, the support and cheerfully-constructive promptings of Dominic Holdsworth, Fran Kingston, Erika Pennington and Lisa Blake at Taylor and Francis, were essential for keeping the editorial process on schedule.

In conclusion, we are unable to reassure readers that terrestrial vegetation, both in forests and agriculture, will sequester enough carbon to mitigate climate change. Whilst forests currently remove around one third of the anthropogenic CO_2 released to the atmosphere, this must decline as emissions go up and forests come down! In his rousing opening to the symposium, Bob Watson made it absolutely clear that 'we have exceeded the operating range of the atmospheric system'. He also concluded that 'sinks matter too much to be excluded', since a straight-jacket has been imposed by the *Kyoto Protocol* that allows for only 0.033 Pg (C) per year to be sequestered. He estimated that up to 2.3 Pg (C) per year could potentially be sequestered by terrestrial vegetation and soils, from a combination of remaining natural forests, and careful forest management via LULUCF or BioCarbon Forestry (whereby payments are offered of $3 to $4 per Mg (C) sequestered). However, it is clear that we need a globally inclusive programme in which Russia, China, India and the developing world participate, as well, of course, as the USA. Ultimately, we must reduce emissions and develop alternative energy sources – plants and soils at best can give us a partial 'quick fix' for the next 20 to 30 years. For many of us, that sinking feeling could well then be associated with the likely continued warming and sea-level rise.

Howard Griffiths, Cambridge   Paul Jarvis, Edinburgh

[NB: 1 Pg = 10^{15} g = 10^9 t (tonne) = 1 Gt; 1 Mg = 10^6 g = 1 tonne; 1 ha (hectare) = 100 m × 100 m = 10^4 m^2 = 0.01 km^2]

The global imperative and policy for carbon sequestration

Robert T. Watson and Ian R. Noble

1. Introduction

Human activities are significantly altering the carbon cycle through energy and land-use practices and policies. In 1992, governments acknowledged that this human perturbation to the carbon cycle is changing the Earth's climate, and most have signed and ratified the United Nations Framework Convention on Climate Change (UNFCCC)[1]. In 1997, parties to the UNFCCC negotiated the Kyoto Protocol, which required all parties to accept their differentiated responsibilities to combat climate change and to set specific emission reduction targets for industrialized countries. The rules for the Kyoto Protocol were further elaborated in the Marrakech Accords in 2001.

Throughout these negotiations, governments have recognized that a range of agricultural, rangeland, and forestry activities can play a vital role in decreasing the net emissions of carbon into the atmosphere. Given that many of the rules under the Kyoto Protocol relating to these Land Use, Land-Use Change and Forestry (LULUCF) activities apply only for the first commitment period, an interesting question is how they will be addressed during the second and subsequent commitment periods. The first commitment period being 2008–2012, the second commitment period is not yet established but is likely to be 2013–2017. Key questions include: (i) whether 'wall to wall' accounting of all LULUCF uptakes and emissions in industrialized countries will be adopted or whether only site-specific activities will be eligible for credit; (ii) whether certain factors influencing carbon uptake and release, i.e. increasing atmospheric CO_2 concentrations, nitrogen and sulphur deposition rates, climate variability, and forest age structure, can or should be factored out; and (iii) which, if any, additional activities in developing countries will be eligible for credit under the clean development mechanism (CDM), e.g., agricultural and rangeland activities, and avoided deforestation.

---

1 The text of the UNFCCC, its Kyoto Protocol, and the Marrakech Accords are all available at http://unfccc.int/; the reports of the Intergovernmental Panel on Climate Change (IPCC) are available at http://www.ipcc.ch/.

The Carbon Balance of Forest Biomes, edited by H. Griffiths and P.G. Jarvis.
© 2005 Taylor & Francis Group

2. Climate change

There is little doubt that the Earth's climate has warmed, on average by about 0.6 °C, over the past 100 years in response to human activities; precipitation patterns have changed, sea levels have risen, and most non-polar glaciers are retreating. These changes can be primarily attributed to increasing atmospheric concentrations of greenhouse gases (GHGs) resulting from the combustion of fossil fuels and land-use changes. The question is not whether climate will change further in the future in response to human activities, but rather by how much, where, and when. Based on plausible future demographic, economic, socio-political, technological, and behavioural changes, the Intergovernmental Panel on Climate Change (IPCC) projected that the atmospheric concentration of CO_2 would increase from the current level of about 370 parts per million (p.p.m.), to between 540 and 970 p.p.m. by 2100, without taking into account possible climate-induced additional releases from the biosphere in a warmer world. The IPCC also projected that the Earth's climate would warm an additional 1.4–5.8 °C between 1990 and 2100, assuming that there are no coordinated international policies to seriously address the issue of climate change (Watson, 2001).

Climate change will, in many parts of the world, adversely effect socio-economic sectors, including water resources, agriculture, forestry, fisheries, human settlements, ecological systems, and human health. The IPCC concluded that developing countries, and especially poor people within them, are the most vulnerable to climate change, that more people will be adversely affected by climate change than will benefit from it, and that the greater the rate and magnitude of change the more adverse the consequences.

The near-term challenge for most industrialized countries is to achieve their Kyoto targets, whereas the longer-term challenge is to meet the objectives of Article 2 of the UN Framework Convention on Climate Change, i.e., stabilization of greenhouse concentrations in the atmosphere at a level that would prevent dangerous anthropogenic interference with the climate system, with specific attention being paid to food security, ecological systems, and sustainable economic development. To stabilize the atmospheric concentrations of CO_2 will require that emissions are eventually reduced to only a small fraction of current levels, i.e., 5–10%. For example, to stabilize at 550 p.p.m. by volume of CO_2, emissions globally will have to peak between 2020 and 2030, then be reduced below current emissions between 2030 and 2100, followed by significant decreases thereafter (*Table 1*).

Table 1. *Timescales associated with achieving stabilization of the atmospheric concentrations of CO_2 at different levels based on scenarios developed for the IPCC Third Assessment (Watson, 2001).*

Stabilization level (p.p.m.)	Date for global emissions to peak	Date for global emissions to fall below current levels
450	2005–2015	Before 2040
550	2020–2030	2030–2100
650	2030–2045	2055–2145
750	2050–2060	2080–2180
1 000	2065–2090	2135–2270

The IPCC concluded that significant reductions in net GHG emissions are technically feasible by the use of an extensive array of technologies in the energy supply, energy demand, and agricultural and forestry sectors, many at little or no cost to society. However, realizing these emission reductions involves the development and implementation of supporting policies to overcome barriers to the diffusion of these technologies into the market-place, increased funding for research and development, and effective technology transfer.

3. The carbon cycle

The atmospheric concentration of CO_2 has historically oscillated between about 180 p.p.m. during glacial periods and 280 p.p.m. during interglacial periods. However, since the industrial revolution, human activities, primarily through the combustion of fossil fuels and land-use changes, have and are continuing to perturb the carbon cycle, increasing the atmospheric concentration of CO_2 to the current level of about 370 p.p.m. Human activities have caused the atmospheric concentration of CO_2 to exceed the previous operating range of the atmospheric system, without knowing the consequences (Falkowski *et al.*, 2000).

Whereas the terrestrial biosphere has historically (from 1800 until about 1930) been a net source of carbon to the atmosphere, it has in the past several decades become a net sink. There was a net release of CO_2 into the atmosphere by the terrestrial biosphere until about 1930, but since then there has been an ever-increasing uptake by the terrestrial biosphere, with the gross terrestrial uptake exceeding emissions from land-use changes (Watson and Noble, 2002). Estimates of the annual average CO_2 budget (*Table 2*) are still subject to large uncertainty. *Table 2* shows the budget as estimated in the IPCC Third Assessment Report (Watson, 2001) and by Houghton (2003), who takes more recent results (and in particular, results on the net ocean uptake) into account. The points to note are that during the 1980s and 1990s the global increase in storage of carbon in the terrestrial biosphere was about 2–3 Pg (C) per year. This was partly offset by human-induced land-use change of about 2 Pg (C) per year. Inverse modelling suggests that about half of the global uptake in the 1990s occurred in the tropics, and the other half in the mid- and high-latitudes of the Northern Hemisphere (Prentice *et al.*, 2002; Watson and Noble, 2002). The primary cause of the current uptake of about 1–1.5 Pg (C) per year in North America, Europe, and northern Asia is thought to be re-growth resulting from management practices on abandoned agricultural land, and climate change (largely increased length of growing season), with CO_2 and nitrogen fertilization also contributing, but possibly to a smaller extent (Hicke *et al.*, 2002; Houghton, 2002).

Analyses of the year-to-year variation in the major fluxes suggest that the terrestrial sink may vary in magnitude by as much as ± 2 Pg (C), i.e., in some years the accumulation of carbon in terrestrial biomes may be as high as 4 Pg (C) whereas in others there may be no accumulation (Bousquet *et al.*, 2000). Accumulation in the oceans also varies, but by a lesser amount. The most likely cause of the temporal variability in terrestrial biomes is the effect of climate on carbon pools with short lifetimes, largely through variations in respiration and fires (Cao *et al.*, 2002; Dargaville *et al.*, 2002; Langenfelds *et al.*, 2002; Vukicevic *et al.*, 2001).

Table 2. *Global carbon budgets from IPCC (2001) and Houghton (2003). (Fluxes are in Pg (C) per year (equivalent to Gt (C) per year), with positive terms representing release to the atmosphere, plus or minus one standard error; e.g. for the atmospheric increase in the 1980s there is a 67% probability that the rate of increase was between 3.2 and 3.4.).*

| | IPCC (2001) | | Houghton (2003) | |
	1980s	1990s	1980s	1990s
Atmospheric increase	+3.3 ± 0.1	+3.2 ± 0.1	+3.3 ± 0.1	+3.2 ± 0.2
Fossil emissions	+5.4 ± 0.3	+6.3 ± 0.4	+5.4 ± 0.3	+6.3 ± 0.4
Ocean–atmosphere flux	−1.9 ± 0.6	−1.7 ± 0.5	−1.7 ± 0.6	−2.4 ± 0.7
Net land–atmosphere flux	−0.2 ± 0.7	−1.4 ± 0.7	−0.4 ± 0.7	−0.7 ± 0.8
Land-use change	1.7 ± ?	No estimate	2.0 ± 0.8	2.2 ± 0.8 ?
Residual terrestrial sink	−1.9 ± ?	No estimate	−2.4 ± 1.1	−2.9 ± 1.1

There is some indication that the rate of accumulation in terrestrial ecosystems is increasing, possibly by about 0.5 Pg (C) per year according to Houghton's most recent estimates. Whether this is part of a long-term trend or simply a chance difference between the two decades is unclear. However, in a modelling comparison, most models suggest that current uptake by the terrestrial biosphere is likely to increase until the middle of the 21st Century, peak at about 5 Pg (C) per year and then decline, possibly even becoming a source by the end of the century (Cramer *et al.*, 2001).

A key question is how might this accumulation of carbon into terrestrial biomes and its variability of about ± 2 Pg (C) around the mean accumulation affect targets and compliance with the Kyoto Protocol.

4. The Kyoto Protocol

The Kyoto Protocol recognizes that land use, land-use change, and forestry activities can play an important role in meeting the ultimate objective of the UNFCCC, i.e., stabilization of GHG concentrations at a level that does not cause dangerous anthropogenic perturbation to the climate system, by reducing the net emissions of GHGs to the atmosphere. Biological mitigation of GHGs through LULUCF activities can occur by three strategies: (i) conservation of existing carbon pools, e.g., avoiding deforestation; (ii) sequestration by increasing the size of carbon pools, e.g., through afforestation and reforestation; and (iii) substitution of fossil fuel energy by use of modern biomass.

The Marrakech Accords included several key principles: LULUCF activities must contribute to the conservation of biodiversity and sustainable use of natural resources; any later reversals of uptake must be accounted; and the accounting system must exclude removals of GHGs from the atmosphere resulting from elevated CO_2 concentration ([CO_2]) above pre-industrial levels, indirect nitrogen deposition, and the dynamic effects of age structure resulting from activities and practices before the reference year (1990). As discussed later in the chapter, reversals of uptake can be accounted for, but removing the indirect effects (elevated [CO_2] above pre-industrial levels, indirect nitrogen deposition, and the dynamic effects of age structure) will be very difficult with current knowledge.

The Marrakech Accords to the Kyoto Protocol included several key definitions. Each party can define a forest consistent with their current definition providing that it is based on a canopy cover between 10 and 30%, a minimum tree height between 2 and 5 m, and a spatial extent between 0.05 and 1 ha (1 hectare (ha) = 10^4 m^2).

The definitions of afforestation, reforestation, and deforestation all use land-use change as a criterion, with afforestation being the establishment of a forest on land that has not been forested for at least 50 years, and reforestation on land that has not been forested since 1990. These definitions of afforestation and reforestation remove the possible scam of clearing forest to gain credits for subsequent reforestation.

The Marrakech Accords also stated that all carbon pools must be accounted, i.e., above-ground biomass, below-ground biomass, litter, dead wood, and soil organic carbon. However, a pragmatic approach to monitoring is allowed in that 'a Party may choose not to account for a given pool in a commitment period, if transparent and verifiable information is provided that the pool is not a source'.

Under Article 3.3 of the Kyoto Protocol, Annex 1 countries (i.e., industrialized countries) must account for all afforestation, reforestation, and deforestation activities. The Protocol also allowed additional activities to be accounted under Article 3.4 but these activities were not determined at the Kyoto meeting. There was a major debate over the implementation of Article 3.4 at the 6th Conference of the Parties (CoP6) in The Hague in 2000, which was resumed in Bonn the following year. The USA had put forward a suggestion that sinks should be accounted for far more comprehensively than some other Parties wished to see. The USA proposal suggested that all uptakes and emissions from all forest, agricultural, and grazing lands should be included (often called the 'wall-to-wall' accounting system). This proposal was contentious because many parties and non-governmental organizations (NGOs) viewed the use of sinks as a way to avoid reducing GHG emissions from the energy sector and because it would potentially allow the USA to claim sink credits for regrowth on lands cleared before 1990. This was not resolved before the USA announced its intention not to ratify the Kyoto Protocol. Negotiations at CoP7 in Marrakech led to an outcome similar to the USA proposal in that during the first commitment period, Parties may elect to account for sinks from forest management, cropland management, grazing land management, or revegetation, but a cap was placed on the maximum credits allowable from forest management activities. The activities must have occurred since 1990, and the emission reductions or uptakes must be human induced. Rules for the second commitment period are yet to be decided.

The Marrakech Accords limited the eligibility of LULUCF activities under the CDM in the first commitment period to afforestation and reforestation projects. Annual use of carbon credits from such projects during the first commitment period cannot exceed 1% of base year emissions of an Annex 1 Party. This is equivalent to about 50 Tg (C) per year for all Annex 1 countries. This could be achieved by about 5–10 million ha of new plantings in agroforestry or reforestation before 2008. The current rate of establishment of plantations throughout the developing world is about 4.5 million ha per year but a large proportion of these plantings are not additional; i.e., they would have occurred without the incentives of the Kyoto Protocol and are, thus, not acceptable for credit under the CDM. With strict enforcement of the additionality rule and the lack of a significant market for credits from sink projects in the CDM, it is likely that there will be a very limited use of sinks in the CDM in the first commitment period (CP).

Negotiators agreed that there should be no credits through the CDM in the first commitment period for better forest management, reduced impact logging, forest protection (avoided deforestation), reduced tillage agriculture, or grazing management for which there is, in principle, significant potential and which could contribute to the sustainable development goals of many developing countries.

5. Quantitative potential for LULUCF activities

A key question is what is the potential for LULUCF activities to sequester carbon under Articles 3.3, 3.4, and 12 during the first commitment period and beyond (*Table 3*)? In the list of 'best' estimates that follows, there are significant uncertainties, primarily because of assumptions made about areas affected, carbon sequestration rates, and the rate of adoption of improved management techniques by foresters and farmers.

1. The potential carbon credits for industrialized countries through afforestation, reforestation and deforestation activities during the first commitment period according to current national communications will contribute little net credit because the projected activities will be limited. If Australia were to ratify the Kyoto Protocol and achieve its targets for reducing land clearing it will achieve a small credit for what amounts to avoided deforestation.
2. The potential afforestation and reforestation, including agroforestry, in non-Annex 1 countries during the first commitment period is as much as 700 Tg (C) per year. However, as noted in the previous section, credits are limited to about 50 Tg (C) in the first commitment period, and current market indications are that this cap will not be reached.
3. The potential carbon credits for avoided deforestation in non-Annex 1 countries is theoretically equal to the rate of deforestation, i.e., about 1.6 Pg (C) per year, although only a fraction of this could be achieved over the next few decades. This huge potential caused considerable concerns during the negotiations and the activity was excluded from the CDM at least for the first commitment period.
4. The potential carbon credit for forest, crop, and grazing land management through Article 3.4 activities during the first commitment period is about 250 Tg (C) per year for industrialized countries, but it is likely that only a small portion will be claimed because of a cap of about 100 Tg (C) on forest management. The potential for equivalent activities in developing countries is about 300 Tg (C) per year, but none of this is eligible.
5. There is also significant potential for using biomass fuels to displace fossil fuels as a source of energy, but these activities are not accounted for under the LULUCF Articles, except for the standing biomass in plantations.
6. Despite these huge potentials, the application of caps and decisions not to elect to use certain Article 3.4 activities by Annex 1 countries are likely to lead to a limited use of LULUCF credits in the first commitment period. Our estimates are that they will contribute less than 100 Tg (C) per year from ratifying Annex 1 (including CDM) sinks or about 10–15% of the emission reductions required compared with business-as-usual projects for the 1990–2010 period.

Table 3. *Comparison of the estimated potential contribution of sinks with what appears likely to be achieved in the first commitment period after the application of caps applied by the negotiations and with the non-ratification of Australia and USA. (All units are Tg (C) per year.)*

	Estimated potential[1]	Indicated use at CoP6[2]	Caps	Estimated use in first CP without the USA and Australia	Estimated use if USA and Australia ratified
Annex 1					
ARD[13]	−50 to −40[3]	−4[4]	—	3	−4
Avoided deforestation	20[5]	15[6]	—	0	15
Forest management	100	720[7]	98[8]	70[9]	98
Crop and grazing-land management	150	18	—	10	18
Total		754	—	83	127
Non-Annex 1					
Afforestation and reforestation	up to 700[10]	<300[11]	50[8]	32[12]	50
Avoided deforestation	1 600	≤1 600	0	0	0
Forest management	70		0	0	0
Crop and grazing-land management	240		0	0	0
Total			—	<32	50
Total sinks in first CP[14]				*ca.* 100	*ca.* 180
Emission reductions required below 1990 levels				140	250
Emission reductions compared with 15% BAU[15] increase over 1990 levels				640	1 000

[1]These data are from IPCC (2000), which preceded the agreements in Marrakech and the revised estimates prepared by parties of carbon gains and losses from forestry activities. On the whole, these estimates do not include many of the factors contributing to the 'free-ride'.

[2]Based on national submissions before CoP6 and Food and Agriculture Organization of the United Nations (FAO) data in use at the time of those negotiations.

[3]This is based on an IPCC estimate of 20–30 Tg (C) per year from uptake from afforestation and reforestation (A&R) and 90 Tg (C) per year emissions from deforestation. Deforestation of 20 Tg (C) per year falls under Article 3.7, as shown in the row below.

[4]Net gain from A&R activities and losses from deforestation under Article 3.3 as reported by Annex 1 countries at CoP6.

[5]Eligible under Article 3.7; see Noble and Scholes (2000) for a detailed explanation.

[6]Based on Australia reducing land-clearing activities and the use of Article 3.7.

[7]The IPCC estimates are for increased uptake over a 1990 baseline from activities performed since 1990, whereas the Marrakech Accords allow forest management to be measured simply as the net uptake in managed forests. This leads to a far higher potential credit from managed forests, but the Marrakech Accords limited total credit by applying a cap for each Annex 1 country.

[8]Including the USA cap as allocated in the Marrakech Accords.

[9]This assumes that all ratifying Annex 1 countries will use their full cap. A portion of these sinks is derived from countries that are likely to achieve compliance without the need to use sinks (e.g., Russia). Nevertheless, these sinks could enter the market either through Article 6 (JI) or Article 17 activities.

[10]Including agroforestry.

[11]Based on A&R activities in tropical countries, but many plantations are not eligible as they are not on land that was clear of forest in 1990 and many others would fail a strict additionality test.

[12]Current market indications are that this cap will not be reached.

[13]Afforestation, Reforestation and Deforestation.

[14]Commitment Period.

[15]Business As Usual.

The IPCC also estimated the potential global carbon uptake using LULUCF activities over the next 50 years for activities started after 1990, not including any contribution from the current terrestrial uptake. The estimated global potential is of the order of 100 Pg (C) (cumulative) by 2050, equivalent to about 10–20% of projected fossil-fuel emissions during that period, although there are substantial uncertainties associated with this estimate. Realization of this potential depends upon land and water availability as well as the rates of adoption of land-management practices. The largest biological potential for atmospheric carbon mitigation is in subtropical and tropical regions.

6. The technical debate

In the negotiations leading up to the Kyoto Protocol in 1997 there was relatively little attention given to sinks. Many of the scientific and technical issues were poorly understood, and there was uncertainty until the final stages as to whether and how sinks might be included. Since then, a major scientific and technical debate has continued, including the following issues.

6.1 *Technical detail of processes and quantities*

One example is the IPCC Special Report on LULUCF, which assessed: (i) options for definitions of forests, afforestation, reforestation, and deforestation; (ii) the potential advantages and disadvantages of different accounting systems; (iii) the potential of different LULUCF activities in Annex 1 and non-Annex 1 countries; (iv) the ability to measure carbon in all pools; (v) the economic costs associated with different LULUCF activities; and (vi) the feasibility of project-based accounting.

6.2 *Methodological issues*

An example is the IPCC Good Practice Guidance, which was published in 2003. This established approaches for quantifying net emissions from a range of LULUCF activities, including: (i) the basis for a consistent representation of land areas to keep track of the land and its carbon; (ii) good practice guidelines on how to measure all carbon pools in a range of ecosystems (forests, grasslands, managed lands, soils, others); (iii) how to deal with uncertainties; and (iv) how to account for revegetation and forest degradation.

In addition, the IPCC has been requested to develop methodologies to factor out the impact of: (i) increases in the atmospheric concentrations of CO_2 since the pre-industrial era; (ii) increased deposition of nitrogen and sulphur; (iii) changes in the age structure of the forests; and (iv) changes in climate, on the rate of carbon sequestration. A key question for the scientific community is whether the 'direct' effects of human-induced activities, e.g. the growth increment resulting from 'normal' forest growth after afforestation or reforestation, can be separated out from the growth increment resulting from 'indirect' human-induced activities, e.g. that resulting from CO_2 and nitrogen fertilization and climate change. The IPCC Special Report on LULUCF (IPCC, 2000) concluded that for activities that involve land-use changes (e.g., conversion of grassland/pasture to forest) it may be very difficult, if not impossible, to distinguish with present scientific tools that portion of the observed stock change that is directly human-induced from that portion that is caused by indirect or natural factors. For activities that involve land-management changes (e.g., tillage to no-till agriculture), it should be feasible to distinguish partly between the direct and indirect human-induced components through control plots and modelling, but not to separate out climate variability.

The IPCC has also been requested to develop methodologies to factor out the impact of natural climate variability. Emissions and removals from natural causes such as El Niño may be large compared with commitments, and the climate is predicted to become more El Niño-like because of human-induced climate change (IPCC, 2001). As noted earlier, year-to-year global carbon accumulation varies by as much as 0–4 Tg (C) per year because terrestrial systems do not sequester as efficiently during El Niño events.

6.3 *Implementation issues*

Throughout the negotiations there has been concern whether sink-based projects and accounting could be effectively implemented, especially through the CDM. The World Bank BioCarbon Fund is developing a 'learning by doing' market mechanism that includes LULUCF projects performed in developing countries' CDM. There will be two windows: the first, where projects are Kyoto compliant (i.e., afforestation and reforestation); and the second, where credits are currently excluded from Kyoto compliance (e.g., revegetation, agricultural, and forest management, and forest conservation projects). Each project must demonstrate: (i) atmospheric benefits, i.e., the project must contribute to reducing GHG concentrations in the atmosphere, and must be additional – i.e., the reduction would not have happened without the incentives from the Kyoto Protocol; (ii) environmental benefits, e.g., conservation of biodiversity, reduction in soil losses, or the rehabilitation of degraded lands; and (iii) social benefits, including additional income, income stability, education, capacity building, technology transfer, or health benefits.

6.4 *Criticisms of supposed failings and potential scams*

The use of LULUCF activities presents an opportunity for some of the poorest nations to benefit from carbon mitigation activities, but it is vigorously opposed by some (mostly European-based) NGOs. The critics suggest several reasons:

1. Sinks cannot be measured. However, several scientific assessments, including the IPCC, have concluded that sinks can be measured with sufficient accuracy and at a cost that is compatible with compliance and trading regimes. The parties agreed that all carbon stocks that are decreasing, and all carbon stocks for which credit is being claimed, must be measured. However, if it can be shown that a specific stock is increasing, but is difficult or costly to measure, it does not need to be measured. Technical methods sufficient to serve the requirements of the Protocol exist for above-ground stocks (above-ground biomass, including litter) and most likely for below-ground stocks (below-ground biomass and soil carbon). Industrialized countries generally have the technologies available, but few currently apply them

routinely for monitoring, whereas developing countries may require assistance to develop the necessary capacities and cover costs. Given that methods and research results are highly transferable between countries, rapid improvements in monitoring capabilities are occurring. Sinks are not permanent. This is a misconception of mistaking the trees for the forest. The issue of the permanence of sinks, which are reversible through human activities, disturbances, or environmental change, including climate change, is more critical than for activities in other sectors, e.g., the energy sector. However, for Annex 1 countries, the pragmatic solution is to ensure that any credit for enhanced carbon stocks on a particular plot of land resulting from Article 3.3 or 3.4 activities is balanced by accounting for any subsequent reductions in those carbon stocks on the same plot of land, regardless of the cause. In addition, the risk of loss is low, probably amounting to only a few per cent in properly designed projects, and temporary crediting and insurance systems are under discussion to address the permanency issue.

2. One major 'permanence' issue is how to deal with the carbon emitted from fires. The IPCC suggested that if the fire is not used as an opportunity to change the use of the land, the party should not be debited with the carbon emitted due to the fire, but neither should it be credited with the regrowth carbon following the fire. It should be noted that many fires burn only the brush, litter, and foliage, and do not destroy the trees; hence the carbon emitted during many forest fires is often restored within 5–10 years. For example, we estimate that the fires near Sydney, Australia, in 2002 released about 20 Mt (C) into the atmosphere, but this will be restored within a few years as most of the *Eucalyptus* trees survived the fire and will quickly re-sprout. In contrast, if the fires burn the peat layer, as in the Indonesian fires in 1997–98, the carbon emitted will not be quickly re-sequestered.

3. Sinks in the CDM will swamp the original intent of the Kyoto Protocol. This concern is totally unfounded given that the use of sink credits derived from CDM projects is limited to no more than 1% of emissions reductions in each party, and few countries are likely to reach their cap.

4. Afforestation or reforestation in the Boreal system will increase the magnitude of climate change through decreases in albedo in winter, i.e., through reduced 'snow cover' (Betts, 2000). Although this is true, these areas are not likely to be priority areas for afforestation and reforestation activities because of slow growth rates. However, current models indicate that large-scale land-use changes will affect the global surface energy budget and lead to changes in local, regional, and global climates. Some of these changes will reduce the mitigation effects arising purely from changes in GHGs in the atmosphere, such as the albedo effects in the Boreal zone, whereas others will enhance the mitigation effect; e.g., reforestation or afforestation may increase transpiration leading to cooler local climates (Marland *et al.*, 2003).

5. Climate change will eventually lead to the release of all of the carbon stored through afforestation or reforestation activities. Vague fears that we will somehow be worse off because of planting additional forests are unfounded. Although there may be some release of carbon from afforested or reforested lands as climate changes, it can be accounted for, and the amount of carbon remaining sequestered will still be much larger than if the land had not been afforested or reforested.

There is an indirect effect of the use of sinks that may be the basis for a better-founded concern. Under the Kyoto compliance system, all forms of emission reductions and uptakes are fully fungible, i.e., a tonne of sinks can be used to offset a tonne of emissions from fossil fuels. In such a trade off, a tonne of carbon is moved from the stable, geological pool of fossil fuel into the much more dynamic atmosphere–ocean–land carbon cycle. Thus, as this system equilibrates over the very long-term, atmospheric concentrations should increase. The core issue here is the importance of long-term effects (many centuries) versus short-term impacts of climate change. If it were possible to cease the release of fossil carbon immediately, or to quickly re-sequester that fossil carbon already released into the atmosphere into long-term geological stores, then biological sinks need not be a compliance issue. However, technological change and institutional change of the scale required to reduce our dependence on fossil fuels cannot be achieved other than over several decades. Sinks provide an opportunity to counteract some of the increase in GHGs during that transition stage and to lessen the impacts of climate change on the Earth's ecosystems and human society.

7. Accounting options for sinks in later commitment periods

Negotiations for the second commitment period will start formally in 2005, although informal discussions about targets within the Kyoto mechanisms and for non-ratifying Annex 1 nations are ongoing. The role of sinks is likely to remain controversial, although the scientific, social, and commercial understanding of sinks is far better than in the lead up to the Kyoto Protocol and even the Marrakech Accords. Some parties and players in the negotiations focus on potential negative impacts of badly planned sink activities and the supposed difficulties in measuring and accounting for sinks. However, there is increasing recognition, especially among developing countries, that well-planned LULUCF activities and projects can have a broad range of positive environmental, social, and economic impacts on biodiversity, forests, soils, water resources, food, fibre, fuel, employment, health, poverty, and equity. For example, the BioCarbon Fund[2] of the World Bank is considering sink projects that promote sustainability goals and adaptation to climate change, or the slowing of deforestation to achieve multiple environmental and social benefits, especially in threatened or vulnerable forests that are unusually species-rich, globally rare, or unique to that region.

7.1 *Indirect anthropogenic effects*

Parties to the UNFCCC have already indicated that they wish to see the 'free-ride' arising from indirect anthropogenic effects (CO_2 fertilization, nitrogen deposition, age structure changes, etc.) factored out of the accounting system. If the accounting system includes only a small component of sink-based credits, the errors and distortions from not factoring out the free-ride indirect effects would be small. This will be the case in the first commitment period. Afforestation, reforestation, and deforestation under Article 3.3 will deliver only a few million tonnes of carbon, and the credits

---

2 Web site for the World Bank BioCarbon Fund is www.biocarbonfund.org.

for forests under Article 3.4 and the CDM are capped. In both cases the capped amount will not be used by some countries whereas in others the cap will be reached irrespective of whether factoring out of the free-ride occurs or not.

The IPCC has been asked to assess options for factoring out the indirect anthropogenic effects (see Section 6, the technical debate, above). There is considerable doubt whether there is the scientific knowledge to do this, taking into account variations across ecosystems, prevailing climate variability, and different management activities. Scientifically determined correction factors taking into account these influences will be costly to prepare, fraught with uncertainties, and controversial. An alternative approach is to use a simple discounting factor or factors and to apply these generically. This would cause inequities between some countries, but these could be taken up in the negotiation of future targets.

7.2 *Wall-to-wall accounting*

As described above, there has been a debate about whether the accounting of sinks should be comprehensive (here called the 'wall-to-wall' approach) or limited to a restricted range of activities, circumstances, and locations (here called the 'project approach'). A major consequence of a wall-to-wall approach is that that most of the free-ride and its year-by-year variation in carbon uptake would be reflected in the accounting system. Below, we quantify the impacts and show that the inclusion of a wall-to-wall approach could lead to significant problems in allocating targets and assessing compliance.

The estimates in Table 2 provide an indication of the approximate magnitude of the major carbon fluxes. From it, we estimate that the free-ride is increasing by about 0.5 Pg (C) per year globally per decade. We also assume that half of the free-ride derives from ecosystems in Annex 1 countries (Watson and Noble, 2002). These assumptions lead to the conclusion that in Annex 1 countries the increase in net uptake by vegetation and soils in the second commitment period (2013–2017) compared with the net uptake in the baseline year of 1990 (i.e., net–net accounting in the jargon of the negotiations) is 0.6 Pg (C) per year (i.e., 2.5 decades of 0.25 Pg (C) per decade increase in Annex 1 countries for each of the five years of the commitment period) or 3 Pg (C) over the commitment period. Note that in the first commitment period there is also an increase in net uptake arising from the free-ride, but the Marrakech Accords have capped the amount of credits that can be claimed from forest management and thus the free-ride component in calculating compliance is greatly reduced.

Thus, Annex 1 countries would receive a bonus of about 3 Pg (C) without any extra actions if wall-to-wall accounting was used. If this additional 0.6 Pg (C) per year were to be factored out of the compliance calculations in the second commitment period, the assigned amounts for Annex 1 countries would need to be reduced by this amount. This would require a reduction in assigned amounts equivalent to about another 15% of 1990 energy emissions on top of whatever reduction target is agreed for the second commitment period. Recall that the first commitment goal was a 5.2% reduction on 1990 emissions. The adjustment of targets would be an enormously risky process given the large uncertainties that remain in trying to estimate the size of the free-ride even at continental scales, let alone at the individual country level. Any over-estimation of the free-ride would be translated into an apparent shortfall in emission reductions and that would need to be made up through energy-based activities. Any underestimate would have the opposite effect.

Natural variation in uptake by terrestrial ecosystems would add another burden of uncertainty, even when averaged over a 5-year accounting period and all Annex 1 countries. If we take the results of Bousquet *et al.* (2000) as indicative, then a 5-year averaging period results in variation in net uptake of over ±0.25 Pg (C) per period (again assuming about half of the effect is in Annex 1 countries). Because banking (i.e., a carry over of credits from one reporting period to the next) is not allowed for credits derived from sinks, this variation would show up in compliance outcomes. This variation is of approximately the same size (6%) as the emission-reduction targets in the first commitment period under the Kyoto Protocol (5.2%). Increasing the averaging period, allowing banking of sink credits and agreements to exchange credits between countries that are affected differently by global climate fluctuations such as El Niño, would reduce this problem. Nevertheless, variations of this size would have major impacts on trading markets, leading to significant price uncertainty

There is still a high degree of uncertainty associated with these estimates. The calculation of the free-ride may be significantly in error, and it remains possible that the size of the free-ride is not increasing. Until we understand the quantities involved in the full global carbon cycle and the variation year-by-year, anticipating the impact of wall-to-wall accounting on compliance targets will be fraught with difficulties and major uncertainties.

7.3 *Other options for sink accounting*

Sinks could be dropped from the compliance regime completely. However, we believe that this would have several undesirable outcomes. First, sinks are a significant component of the global carbon cycle and we must manage and monitor them if we are to succeed in moderating climate change. There are many opportunities through appropriate management to achieve significant sequestration for periods of many decades to centuries, thus allowing more flexibility in our path to achieving a society less dependent on fossil fuels. There is a range of sink activities that would also contribute to other goals such as environmental protection, biodiversity conservation, and adaptation to climate change. Removing sinks from the compliance regime would not necessarily prevent such actions but it would remove a major incentive to implement them.

Another approach is to include sinks, but only through a project approach. This would severely limit the area of land subject to reporting because projects would only cover a small fraction of the land area within a country. Each project would pick up some elements of the free-ride and would be subject to year-by-year climate variability, but the absolute amounts would be significantly reduced. Projects are also more likely to be intensively managed and hence less subject to variability. Much of the year-by-year variability arises from major events such as fires, droughts, or exceptional wet periods; although some events will affect uptakes and emissions despite management actions, others can be avoided or moderated by appropriate management. Components of the free-ride (e.g., CO_2 effects, temperature changes, and nitrogen and sulphur deposition) could be factored out by a discount factor appropriate to the project type, although again these would be difficult and costly to estimate. Early experience in the BioCarbon Fund of the World Bank suggests that most proponents of sink projects plan to maintain large buffers of unsold carbon as a backup against losses from fires and other disturbances. An additional discount may be acceptable in some circumstances, but it will make other projects financially unviable.

The main impact of a project-by-project approach will be on activities under Article 3.4 of the Kyoto Protocol, as the CDM and joint implementation (JI) are already project-based and Article 3.3 activities are much like projects. However, Annex 1 countries would most likely report as projects only those activities that were likely to provide positive carbon credits over time, and there would be little incentive to avoid activities that lead to net GHG emissions. Under Article 3.3 deforestation is recognized as a discrete activity and the opposite to afforestation and reforestation. Parties are required to report any deforestation, and this can be readily verified through remote sensing or ground observations. However, it would be difficult to identify equivalent, discrete opposites to such activities that sequester carbon through forest management, cropland management, or grazing land management. Also, projects would be subject to leakage both within a Party's borders and beyond and much of this may remain unreported. Parties could be legally mandated to include all activities within a certain class of intervention, i.e., all agricultural activities or all forestry activities, thus avoiding the perverse outcome of only taking credit for projects within a certain class that are sequestering carbon, and ignoring those that are losing carbon. However, if all forestry activities were included, it would suffer from the same free-ride problem discussed under wall-to-wall accounting.

There are more radically different ways to retain the benefits of sinks in the wider campaign to combat climate change, but without the problems that they are seen to cause in the compliance regime. One approach is to allocate Parties separate targets for sink activities. These could be subject to longer reporting frameworks, include banking and possibly reduced penalties for non-compliance within a particular period if this can be linked to regional-scale climate anomalies. Thus, Parties to Annex 1 could accept two targets: one similar to the current assigned amount and another based on sink activities. Some Parties with small fluxes and absolute uptakes or emissions from sinks may elect not to have and report sink compliance. Some Parties with demonstrably efficient energy and industrial sectors could accept sink targets in exchange for lesser energy-sector targets, and seek to achieve their sink targets through activities at home or through a modified CDM.

8. Conclusions

About a third of CO_2 emissions to the atmosphere since the industrial revolution were derived from land-cover changes, and currently about a third of anthropogenic CO_2 emissions to the atmosphere are absorbed and accumulate in terrestrial ecosystems. Thus, management and monitoring of the terrestrial carbon cycle is an essential component of tackling climate change.

Deliberate biological mitigation of GHG emissions through LULUCF activities can occur through three strategies: (i) conservation of existing carbon pools, e.g., avoiding deforestation; (ii) sequestration by increasing the size of carbon pools, e.g., through afforestation and reforestation; and (iii) substitution of fossil fuel energy by use of modern biomass. Well designed, biological mitigation activities would also contribute to other goals such as environmental protection, biodiversity conservation, and adaptation to climate change.

Biological mitigation activities provide many opportunities through appropriate management to achieve significant sequestration of carbon for periods of many decades to centuries, thus enabling more flexibility in our path to achieving a society less dependent on fossil fuels. However, to stabilize the atmospheric concentration of CO_2 (Article 2 of the Convention) will require significant emission reductions globally, and this can only be achieved either by reducing energy production emissions or by capture and storage of such emissions.

The magnitude of trading under the CDM for LULUCF activities during the first commitment period will be limited given the absence of the USA from the Kyoto Protocol, and the restrictions made on the use of sinks under Article 3.4 and the CDM. Nevertheless, sinks may account for about 15% of total emission reductions (based on a business-as-usual scenario from 1990 to 2010) in the first commitment period.

Debate about the second commitment period will start in earnest in 2005. There is also likely to be a parallel debate originating in the USA about alternatives to Kyoto mechanisms. The role of biological projects, both for mitigating net GHG emissions and for adaptation, is likely to be contentious. There needs to be a major re-examination of how to address biological mitigation activities within industrialized countries, given there are no simple approaches to factoring out the free-ride (caused by increasing atmospheric CO_2 concentrations, nitrogen and sulphur deposition, climate change, and forest age structure) and the effects of natural climate variability, especially if wall-to-wall accounting is adopted. The site-specific (project-based) approach is likely to suffer from Parties ignoring activities that lead to net emissions.

There also needs to be a re-examination of the eligibility of sink activities within the CDM, especially given the increasing interest among developing countries in the constructive use of sinks in future commitment periods. Avoided deforestation, forest, cropland, and grassland management activities can all contribute to reducing net greenhouse emissions, and if well designed provide significant other environmental (e.g., conservation of biodiversity) and social benefits.

New scientific knowledge is urgently required for informed policy formulation, especially the magnitude and causes of the free-ride by region (or country); learning-by-doing projects are required to demonstrate the environmental and social efficacy of biological mitigation projects.

References

Betts, R.A. (2000) Offset of the potential carbon sink from boreal forestation by decreases in surface albedo. *Nature* **408**: 187–190.

Bousquet, P., Peylin, P., Ciais, P., Le Que're, C., Friedlingstein, P. and Tans, P.P. (2000) Regional changes in carbon dioxide fluxes of land and oceans since 1980. *Science* **290**: 1342–1346.

Cao, M.K., Prince, S.D. and Shugart, H.H. (2002) Increasing terrestrial carbon uptake from the 1980s to the 1990s with changes in climate and atmospheric CO_2. *Global Biogeochemical Cycles* **16**: 1059–1069.

Cramer, W., Bondeau, A., Woodward, F.I., Prentice, I.C., Betts, R.A., Brovkin, V. *et al.* (2001) Global response of terrestrial ecosystem structure and function to CO_2 and climate change: results from six dynamic global vegetation models. *Global Change Biology* **7**: 357–373.

Dargaville, R.J., Heimann, M., McGuire, A.D., Prentice, I.C., Kicklighter, D.W., Joos, F. *et al.* (2002) Evaluation of terrestrial carbon cycle models with atmospheric CO_2 measurements: Results from transient simulations considering increasing CO_2, climate, and land-use effects. *Global Biogeochemical Cycles* **16**: 1078–1092.

Falkowski, P., Scholes, R.J., Boyle, E., Canadell, P., Canfield, D., Elser, J. *et al.* (2000) The global carbon cycle: A test of our knowledge of earth as a system. *Science* **290**: 291–296.

Hicke, J.A., Asner, G.P., Randerson, J.T., Tucker, C., Los, S., Birdsey, R., Jenkins, J.C. and Field, C. (2002) Trends in North American net primary productivity derived from satellite observations, 1982–1998. *Global Biogeochemical Cycles* **16**: 998–1019.

IPCC, 2001: *Climate Change 2001: The Scientific Basis, Contribution of Working Group I to the Third Assessment Report of the Intergovernmental Panel on Climate Change*; **Houghton, J.T., Ding, T., Griggs, D.J., Noguer, M., van der Linden, P.J., Dai, X., Maskall, K. and Johnson, C.A. (eds)**, sections 9.3.5–9.3.6, Cambridge University Press, Cambridge.

Houghton, R.A. (2002) Magnitude, distribution and causes of terrestrial carbon sinks and some implications for policy. *Climate Policy* **2**: 71–88.

Houghton, R.A. (2003) Why are estimates of the terrestrial carbon balance so different? *Global Change Biology* **9**: 500–509.

IPCC (2000) *Special Report on Land Use, Land-Use Change and Forestry.* Watson, R.T., Noble, I.R., Bolin, B., Ravindanath, N.H., Verado, D.J. and Dokken, D.J. (eds), Cambridge Universtity Press, Cambridge.

Langenfelds, R.L., Francey, R.J., Pak, B.C., Steele, L.P., Lloyd, J., Trudinger, C.M. and Allison, C.E. (2002) Interannual growth rate variations of atmospheric CO_2 and its $\delta^{13}C$, H_2, CH_4, and CO between 1992 and 1999 linked to biomass burning. *Global Biogeochemical Cycles* **16**: 1027–1048.

Marland, G., Pielke, R.A. Sr., Apps, M., Avissar, R., Betts, R.A., Davis, K.J. *et al.* (2003) The climatic impacts of land surface change and carbon management, and the implications for climate-change mitigation policy. *Climate Policy* **3**: 149–157.

Noble, I.R. and Scholes, R.J. (2000) Sinks and the Kyoto Protocol. *Climate Policy* **1**: 1–20.

Prentice, I.C., Heimann, M. and Sitch, S. (2002) The carbon balance of the terrestrial biosphere: Ecosystem models and atmospheric observations. *Ecological Applications* **10**: 1553–1573.

Vukicevic, T., Braswell, B.H. and Schimel, D. (2001) A diagnostic study of temperature controls on global terrestrial carbon exchange. *Tellus Series B – Chemical and Physical Meteorology* **53**: 150–170.

Watson, R.T. (ed.) (2001) *Climate Change 2001: Synthesis Report.* Cambridge University Press, Cambridge.

Watson, R.T. and Noble, I.R. (2002) Carbon and the science-policy nexus: The Kyoto Challenge. In: Steffen, W., Jager, J., Carson, D. and Bradshaw, C. (eds) *Challenge of a Changing Earth*, Springer, Berlin.

2

Role of forest biomes in the global carbon balance

John Grace

1. Introduction: defining the questions about carbon sinks

The past decade has seen an unprecedented research effort aimed at understanding the contemporary carbon cycle. The reason for this is the recognition in 1994 by the United Nations Framework Convention on Climate Change (UNFCCC) that anthropogenic emissions of CO_2 and other greenhouse gases (GHGs) are now increasing fast enough to cause 'dangerous interference' to the climate system. This concern is embodied in the 1997 Kyoto Protocol, which gives each of the 38 richer countries of the world a target for reduction of GHG emissions, and enables them to use forests at home and abroad to transfer carbon from the atmosphere to the land, as a contribution towards meeting reduction targets.

Research on the role of forests and other terrestrial ecosystems in the carbon cycle was stimulated especially by the papers of Woodwell *et al.* (1978), Broeker *et al.* (1979), Houghton *et al.* (1983), Tans *et al.* (1990) and Siegenthaler and Sarmiento (1993). Over this time controversy raged, and the pendulum swung from Woodwell's initial view that vegetation is a net source of carbon to the atmosphere (Woodwell *et al.*, 1978), to the view that vegetation is a net carbon sink. Forests are the most likely candidates as carbon sinks, as their wood is an immense store of carbon while their growth rates are known to be increasing.

Various hypotheses have been proposed as reasons for this (Galloway *et al.*, 1995; C.D. Keeling *et al.*, 1996; Nagy *et al.*, 2003; Nemani *et al.*, 2003; Spiecker *et al.*, 1996), as follows.

1. Enhanced deposition of nitrogen from anthropogenic sources, acting as a fertilizer, has stimulated plant growth everywhere.
2. Increases in temperature may have increased the duration of the growing season in temperate and Boreal regions.
3. Elevated CO_2 concentration may have acted as a fertilizer everywhere.
4. There may have been a general amelioration of conditions for plant growth, as a result of combined changes in temperature, radiation, and precipitation.

The important research questions arising from these considerations are:

1. How much carbon is being extracted annually from the atmosphere by forests compared with other terrestrial ecosystems?
2. Where are the terrestrial carbon sinks?
3. What controls their strength?
4. Will the sinks endure?
5. What observing networks are required to monitor the sinks?
6. What are the feedbacks in the carbon cycle?
7. Is it feasible to manage forests as sinks?

Addressing these questions has brought researchers together in large interdisciplinary and international groups on a scale hitherto unknown, and has resulted in the development of entirely new research methodologies and observational strategies. In this chapter, I consider how far the research community has progressed toward answering these seven research questions.

2. How much carbon is extracted annually from the atmosphere by forests and other biomes?

To address this question I draw on data collected at a range of spatial scales using traditional and new methodologies.

2.1 *Net primary productivity*

Tables of net primary productivity (NPP) of the biomes of the world have existed for several decades but are often contradictory (Atjay *et al.*, 1979; Whittaker and Likens, 1975; Woodwell and Whittaker, 1968). A recent re-evaluation of the world's NPP (Roy *et al.*, 2001) finds that the total terrestrial NPP may be 62.6 Pg (C) per year, of which more than half is attributable to forest biomes (*Table 1*). It should be realized that the observational evidence for this or any other figure is quite fragile – given the uncertainty in measuring NPP, particularly when evaluating the below-ground NPP. Classical methods that involve measuring growth increments and litter fall are not adequate. They ignore fine root turnover and emissions of exudates and volatiles, and so they almost certainly provide an underestimate. For now, we will assume global NPP is 63 Pg (C) per year, while noting in passing the suggestion by Nemani *et al.* (2003) that NPP has increased by 3.4 Pg (C) per year over the past 18 years as a result of a more favourable climate for plant growth.

This 63 Pg (C) per year of NPP is made up of about 123 Pg gained from photosynthesis and about 60 Pg lost from respiration of green plants. The massive withdrawal of CO_2 from the atmosphere in the Northern Hemisphere summer produces the well-known seasonal change in the CO_2 concentration of the atmosphere, first detected by Charles Keeling at Mauna Loa in the Pacific Ocean and now observed in all parts of the world (IPCC, 2001). In photosynthesis and aerobic respiration, the exchange of one mole of CO_2 between vegetation and the atmosphere is accompanied by the exchange of one mole of O_2, in the opposite direction. Atmospheric CO_2

Table 1. *Carbon fixed by the Earth's vegetation, as NPP (Saugier et al., 2001), and the 'potential' sink strength by the Taylor and Lloyd (1992) method. The total C pool includes vegetation and soil organic matter. See text for the assumptions.*

Biome	NPP (Mg (C) ha^{-1} per year)	Area (million km^2)	Total carbon pool (Pg (C))	Total NPP (Pg (C) per year)	Estimated sink (Pg (C) per year)	Average sink per hectare (Mg (C) ha^{-1} per year)
Tropical forests	12.5	17.5	553	21.9	0.66	0.37
Temperate forests	7.75	10.4	292	8.1	0.35	0.34
Boreal forests	1.9	13.7	395	2.6	0.47	0.34
Arctic tundra	0.9	5.6	117	0.5	0.14	0.02
Mediterranean shrublands	5	2.8	88	1.4	0.21	0.75
Crops	3.05	13.5	15	4.1	0.20	0.15
Tropical savannah and grasslands	5.4	27.6	326	14.9	0.39	0.14
Temperate grasslands	3.75	15	182	5.6	0.21	0.14
Deserts	1.25	27.7	169	3.5	0.20	0.07
Ice		15.3				
Total		149.3		62.6	2.85	

concentration is rising, and the corresponding expected decrease in oxygen concentration has also been detected (Keeling and Shertz, 1992).

The NPP represents the rate at which dead plant material enters the soil as organic matter. The subsequent breakdown of this organic matter is through a cascade of biological processes involving temperature-sensitive consumption by a myriad of soil organisms, including microfauna, fungi and bacteria. The material is chemically diverse, containing some fractions that are easily broken down. Sugars, for example, decay in hours or days whereas woody material is extremely resistant to decay and thus contributes to the long-lived carbon store. It was thought by most ecologists that if the global climate were to be constant, then the rate of decomposition would match NPP. In this case the pool of carbon in the soil would be constant and the fluxes of carbon into the system would be matched exactly by the total respiratory fluxes from the ecosystem, arising from both heterotrophic and autotrophic respiration. However, if the global climate is changing, or even if it were to remain constant, one can imagine that production and decomposition may often occur at different rates, so that over a period of time an ecosystem can be a net source or a net sink of carbon. We return to this in Section 3.

2.2 Net ecosystem productivity

To understand and measure the carbon flows of ecosystems, new methods were required that measured all the main carbon flows (including photosynthesis, autotrophic respiration, and heterotrophic respiration). One important method, developed for CO_2 flux measurement in the early 1990s, was eddy covariance (EC). A single EC sensor mounted on a tower above the vegetated surface measures the net flux of CO_2 and water vapour from an area of forest over a 'footprint' of 0.1–1 km². By making certain assumptions, the net flux of carbon at ecosystem scale may be resolved into photosynthesis and respiration. By making further assumptions based on experimentation, respiration can be further resolved into autotrophic and heterotrophic components. Details of the theory and practice of EC measurement can be found elsewhere (see, for example, Aubinet *et al.*, 2000; Moncrieff *et al.*, 1997). So, from the early 1990s it has become possible to compare the carbon fluxes to forest ecosystems (see, for example, Grace *et al.*, 1995), and to explore the underlying causes of the variability among ecosystems (see, for example, Malhi *et al.*, 1999). There are still some missing components of the observed flux, for example volatile organic compounds are generally not accounted, and some lateral fluxes into and out of the ecosystem through the soil may be missed (Richey *et al.*, 2002). Nevertheless, major research projects in the 1990s, such as BOREAS, EUROFLUX, LBA, CARBOEUROFLUX and AMERIFLUX, have resulted in comparative data about the carbon balance of ecosystems worldwide, and led to the establishment of a global network of over 200 observing stations, collectively known as FLUXNET. An example of data from FLUXNET is shown in *Figure 1*, where average values of net ecosystem productivity (NEP) and gross primary productivity (GPP) are plotted for a global data set tabulated by Falge *et al.* (2002). GPP varies hugely, largely as a result of the duration of the

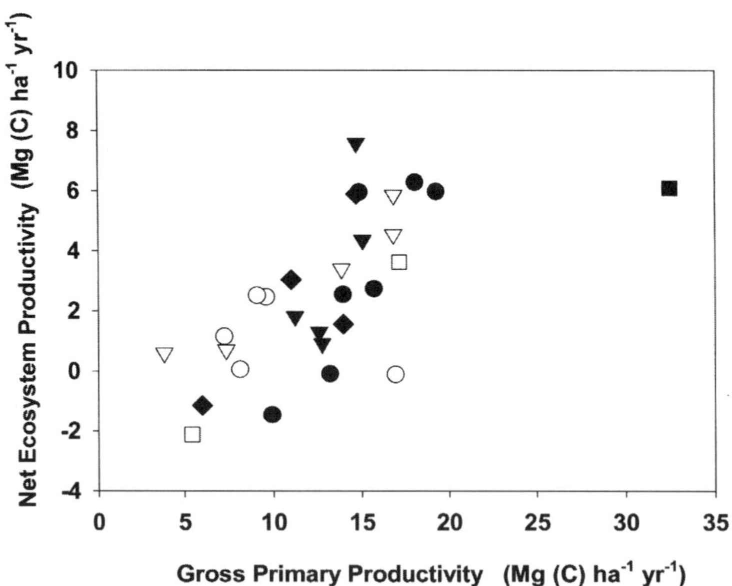

Figure 1. *Magnitudes of carbon fluxes as NEP and GPP for forests, crops, and grasslands (plotted from FLUXNET data tabulated by Falge et al. (2002)), and the relationship between NEP and GPP. Symbols: temperate coniferous forest (●); Boreal coniferous forest (○); temperate deciduous forest (▼); Mediterranean forest (▽); tropical rain forest (■); grassland (□); crops (◆).*

summer (e.g., a very short growing season in the Boreal zone and a long one in the tropics). NEP varies from –1 to 7 Mg (C) ha^{-1} (1 hectare (ha) = 10^4 m^2) per year and shows a rough correlation with GPP. Note that the NPP is not readily estimated from eddy flux data, as the fraction of ecosystem respiration that is attributable to plant respiration is not generally known, although Waring *et al.* (1998) presented data to support the idea that NPP is a constant fraction, 0.47, of GPP.

FLUXNET sites are mostly forests in their middle age, and little disturbed. However, most of the world's forests undergo disturbance, natural and man-made. Additional losses of carbon occur whenever the soil is disturbed, and whenever trees are felled, burned, or are blown down. To understand the impacts on carbon fluxes associated with disturbances, chronosequences (sets of forest stands of differing age in the same area and on the same soil type) are important, and several such sites have been established over the past 10 years. Chronosequences are showing that up to 40% of the carbon uptake observed in middle-aged, managed forests is 'lost' as a result of soil disturbance during the management cycle (see, for example, Law *et al.*, 2001; Schulze *et al.*, 1999). Similar losses may also occur as a result of natural changes in forest structure, even in unmanaged forests where trees have finite lifetimes and are subject to natural destructive agencies during their life cycle. For example, trees in tropical rainforest turn over at a rate of 1–2% per year (Phillips and Gentry, 1994).

2.3 *Net biome productivity*

Steffen *et al.* (1998) introduced a new term, net biome productivity (NBP), to allow for the carbon losses associated with forest management, changes in land use and natural disturbance, such as fire and windthrow (*Figure 2*). This term is intended for carbon accounting at a scale much larger than the forest stand, and may include a number of different ecosystems within the biome. NBP describes the carbon budget at the landscape, national, and continental scales, where it encompasses land-use changes as well as management practices. To be comparable with estimates of sink strength from atmospheric studies (see below), NBP should also encompass product use. The products of forest ecosystems include those that are extremely short-lived, such as newsprint, and those that last for decades and even centuries, such as structural timber, furniture, carvings, and books. To make accounting even more complicated, many countries export a significant fraction of their forest products (e.g., wood, pulp, and packaging) and others import them, so that sequestration in one part of the world can result in emissions elsewhere in a different biome, region, or continent.

3. Where are the sinks?

We can address this question by using a range of techniques, some new, some old, at a range of spatial and temporal scales. Here I summarize the approaches that are contributing to answer this question.

3.1 *Sinks at stand scale using forest inventory*

When woody vegetation is in its establishment phase, as in plantations (and new growth of woody vegetation on the tundra), we have a special case. Here the 'store'

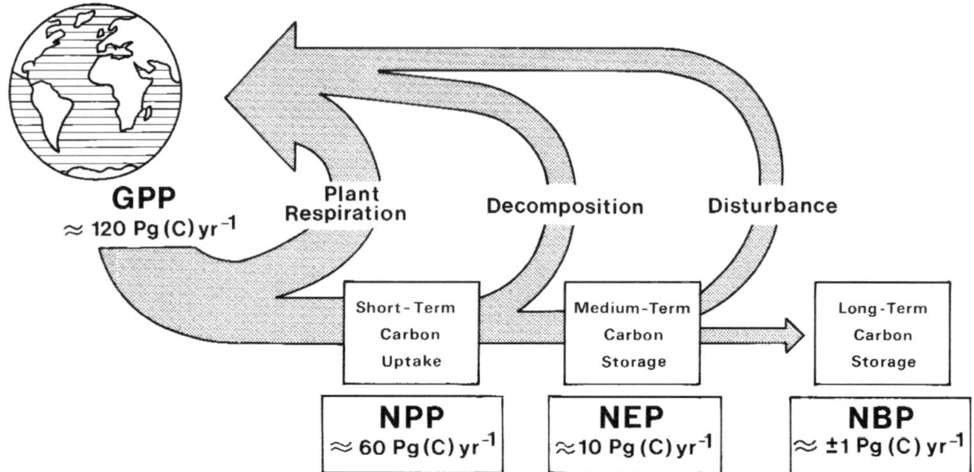

Figure 2. Representation of terrestrial carbon flows as GPP, NPP, NEP, and NBP, and the inter-relationship of these flows.

in the stems of woody plants is growing as a result of the youth of the plant population, and there is very little microbial decomposition of the annual production to balance the gains. The growth rates can be controlled by thinning, to some extent by irrigation, and by fertilization (see Chapter 16, this volume). Foresters express forest growth rate as the average yield in volume of timber over the duration of the tree crop, the 'mean annual increment' or MAI. The 'current annual increment', the CAI, rises as the canopy develops, reaches a maximum, and then declines in old age (see, for example, Cannell, 2003). Trees are traditionally harvested sometime after the maximal CAI, when the MAI reaches its maximal value. Fast-growing conifers in plantations may have a MAI of 14 m^3 ha^{-1} per year, known by foresters in the UK as 'yield-class 14'. Assuming a density of 0.5 Mg (dry mass) m^{-3}, and assuming the carbon content of dry mass to be 50%, the boles alone will accumulate 3.5 Mg (C) ha^{-1} per year. Trees also produce branches, leaves, and coarse and fine roots with much shorter lifetimes than the boles, and all of these may be deposited as litter and accumulate in the stand as debris, and a proportion may add to the soil carbon stock. For estimating carbon accumulation rates (as required for national inventories), some researchers use an 'expansion factor', which is simply the multiplier required to find the total carbon accumulation rate from the rate of bole growth. However, the storage of carbon as woody debris and as soil organic matter is substantial in cold climates, and the long-term behaviour of these carbon stocks is complex and not well understood. There are several models available for the decomposition of soil organic matter that have been very useful for agricultural soils (see, for example, Falloon and Smith, 2002; Chapter 3, this volume), but rather fewer on the breakdown of fine and coarse woody debris in forests (Titus and Malcolm, 1999; Yin, 1999). In general, more work is required to enable us to model the dynamic aspects of the whole-ecosystem carbon stocks and fluxes in forests using the traditional forestry inventory techniques.

3.2 *Sinks at forest scale using eddy covariance*

So far, measurements of carbon fluxes by EC have indicated that most forests are carbon sinks. This is not surprising, as most of the measurements have been made in middle-aged forests that have been managed for timber production and are therefore growing quite fast. For example, European forests were found to have a sink strength that decreased with latitude from 6 to 7 Mg (C) ha^{-1} per year in southern Europe and along the Atlantic fringe to a weak source in central Sweden (Valentini *et al.*, 2000). In the Boreal region, pine forests in Siberia were also found to be a weak sink (Ciais *et al.*, 1995; Schulze *et al.*, 1999; Chapters 7 & 8, this volume), and pine, spruce, and aspen forests in northern Canada were also weak sinks (Falge *et al.*, 2002). The existence of high-latitude sinks seems unsurprising as climate warming has been more pronounced in these cold regions, and quite profound changes have been reported at the tree line, where trees are growing faster than before (Paulsen *et al.*, 2000; *Figure 3*).

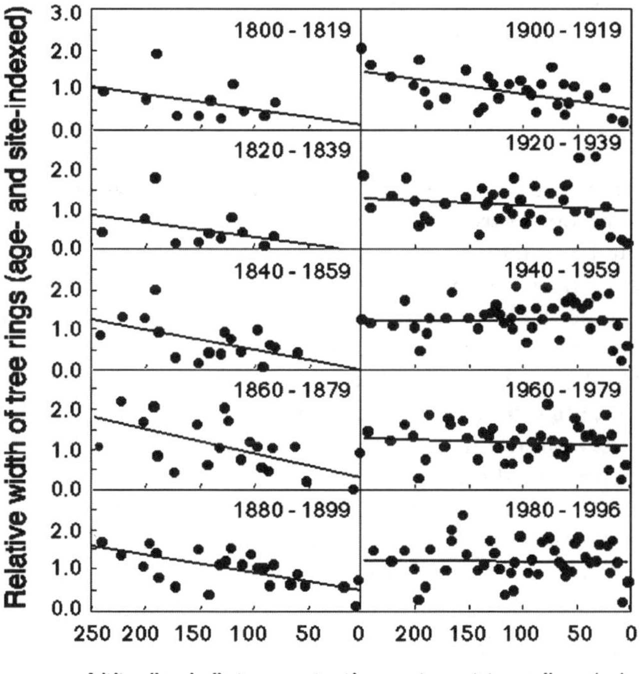

Figure 3. Evidence for increased growth of trees at the alpine tree line, where temperatures have increased faster than the global average, and where there has been significant deposition of active nitrogen (reproduced with permission from Paulsen et al. (2000)). Until 1900 the growth declined sharply as the tree line was approached; now it does not decline at all. Tree growth in such locations is usually controlled by summer temperature and the length of the growth period. Copyright (2000) originally published in Arctic, Antarctic and Alpine Research. Used with permissions.

Most relatively undisturbed rain forests in Brazil appear to be substantial carbon sinks (Grace *et al.*, 1995; Kruijt *et al.*, 2004; Malhi and Grace, 2000; Malhi *et al.*, 2002). One site, with rather a large amount of coarse woody debris, has been shown to be a carbon source (Saleska *et al.*, 2003). However, independent estimates of carbon fluxes to forest stands in the Amazon basin, made from measurements on trees in many permanent forest sample plots, show most of them to be accumulating carbon (Phillips *et al.*, 1998; Chapter 10, this volume).

The existence of a measurable sink implies that the biomass and/or the soil carbon stock is increasing. In the case of the Amazonian rain forest, tree biomass is increasing at a detectable rate, as noted above (Phillips *et al.*, 1998), whereas in the Boreal forest it may well be that it is the soil organic matter that is increasing. The stock of soil carbon in Boreal forest soils has increased to be twice that in tropical forest soils since the last glaciation (Dixon *et al.*, 1994). The recalcitrant part of soil organic matter is lignin, consisting of polymers of three very similar phenolic alcohols. Lignin is resistant to attack by bacteria but vulnerable to a relatively small group of fungi (Gleixner *et al.*, 2001). Another component of the sink is, ironically, black carbon derived from combustion of biomass. Until recently, black carbon was considered to be inert but recent research suggests that it may also decay, albeit slowly (Bird *et al.*, 1999).

3.3 *Sinks resulting from land-use change*

Land-use changes involving forests have the potential to absorb carbon (through planting) or emit carbon (through deforestation). Currently, new planting of forests amounts to about 4.5 million ha per year, although some of that is achieved by harvesting and restocking existing forests (FAO, 2000), and not all new plantings of forests survive. None the less, there are about 190 million ha of plantations globally (mostly in the tropics), much of it growing quite rapidly, perhaps absorbing an average of 4 Mg (C) ha^{-1} per year and thus creating a man-made sink of up to 0.76 Pg (C) per year. However, the source from deforestation (also mostly in the tropics) is larger than that. About 6.4 million ha per year of forest (FAO, 2000), assumed to have an average carbon density of about 150 Mg (C) ha^{-1}, is destroyed each year and replaced by agriculture, with an assumed carbon density of about 5 Mg (C) ha^{-1}, thus giving rise to a net source flux of 0.93 Pg (C) per year. On a global scale, changes in land use are dominated by tropical deforestation (*Figure 4*).

3.4 *Sinks at regional and global scales*

At global and regional scales, the whereabouts of the carbon sinks may be investigated by analysing the small variations (a few parts per million (p.p.m.)) in the atmospheric concentrations of CO_2 in the atmosphere. This depends on the existence of an adequate number of sampling stations in the world (there are currently too few, and hardly any in the tropics). The sinks are inferred from the draw-down in CO_2 concentration, using an atmospheric transport model to correct for the 'smearing' that occurs because of atmospheric motion. Uptake of CO_2 by terrestrial ecosystems and the oceans may be distinguished by means of its isotopic composition: terrestrial photosynthesis discriminates against $^{13}CO_2$ but dissolution in the

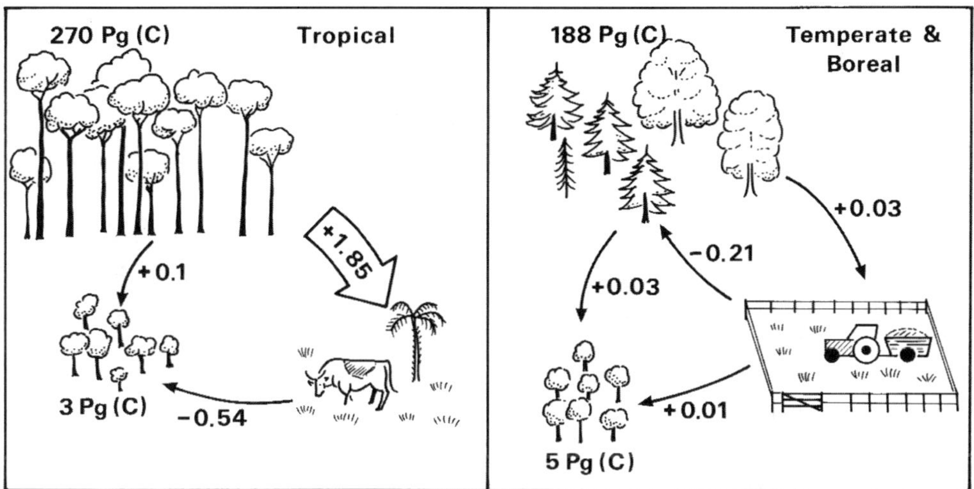

Figure 4. Carbon sources and sinks resulting from land-use changes (mature forest in the top left, plantations in the lower left, and agriculture in the lower right). Arrows show fluxes of carbon (Pg (C) per year) resulting from afforestation and deforesation. Drawn from FAO data.

ocean does not. In assigning regions of the world as carbon sinks by this method, the anthropogenic emissions of CO_2 from burning fossil fuels need to be taken into account as they are large in relation to the biogenic sinks. Anthropogenic emissions are known from governmental statistics or may be estimated from the concentrations of carbon monoxide, CO, or $^{14}CO_2$ in the atmosphere. Using this approach, sinks have been detected in North America and Europe, whereas the tropical region, where deforestation is considerable, is often near to equilibrium (Ciais *et al.*, 1995; Gurney *et al.*, 2002). After losses of CO_2 from tropical deforestation are taken into account, temperate, Boreal and tropical zones all reveal themselves as biogenic carbon sinks. The most thorough recent study, using 26 sample stations and averaging over a 20 year period, suggests a Northern Hemisphere land sink of 0.4 Pg (C) per year and a tropical land sink of 0.8 Pg (C) per year (Rödenbeck *et al.*, 2003). Modelling and observational studies suggest that the tropical rain forest is the main cause of the inter-annual variability, a consequence of the large stocks that are held there and the strong effect of El Niño, which influences the balance between photosynthesis and respiration (Lloyd, 1999; Prentice and Lloyd, 1998).

Attempts have been made to reconcile regional estimates of the sink inferred from atmospheric studies with those from scaled-up, ground-based studies based on EC and inventories (Janssens *et al.*, 2003; Pacala *et al.*, 2001). Land-based estimates are 'consistent' with atmospheric estimates, but both have large uncertainties. The USA has an overall annual sink strength of 0.30–0.58 Pg (C) (Pacala *et al.*, 2001), whereas Europe's is between 0.13 and 0.20 Pg (C) (Janssens *et al.*, 2003).

Viewed globally, the observational data outlined above lead to the estimates of the main carbon fluxes shown in *Figure 5*.

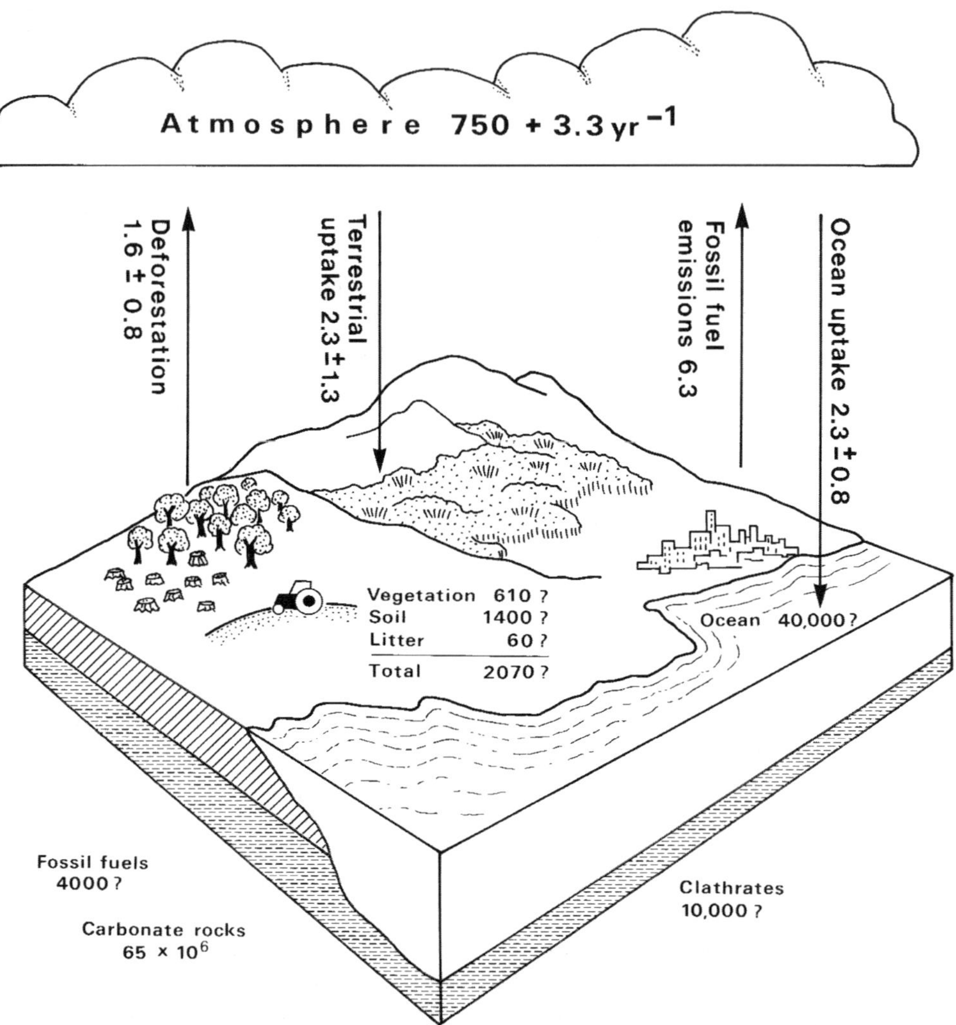

Figure 5. *Carbon stocks and net carbon fluxes to land and ocean. Fluxes to land by photosynthesis are about 120 Pg (C) per year, but about 60 Pg (C) is returned annually by autotrophic respiration, and about another 60 Pg (C) by heterotrophic respiration. The deforestation flux is shown separately in this diagram. Similarly, for the ocean, about 90 Pg (C) per year is dissolved in the colder regions of the ocean and approximately the same is lost from the tropical zone.*

4. What controls the strength of sinks?

It is widely considered that an ecosystem in a constant environment would reach an equilibrium state in which the gains by photosynthesis would be balanced by the losses by respiration, and births and deaths of plants would be equal. In this case the net exchange of carbon would be zero. One can think of a few ecosystems where this evidently does not apply: for example, peat bogs in the Boreal zone have accumulated carbon over many thousands of years because low temperatures and waterlogging have restricted decomposition of soil organic matter. For the case of virgin forests and grasslands the steady-state assumption seems a good *first* approximation. Thereafter, any perturbation in climate might differentially affect the rates at which carbon is added or subtracted from the pool of biomass and soil organic matter.

4.1 *Atmospheric CO_2 concentration*

Photosynthesis and production of biomass in trees is in most cases sensitive to CO_2 concentration, although less so in old than in young trees (Idso, 1999; Medlyn *et al.*, 1999; Wullschleger *et al.*, 1995). These reviews of experimental data suggest that a doubling of atmospheric CO_2 concentration will increase production rates of tree stands by 10–40%, consistent with some recent data from a mesocosm containing a forest ecosystem (Lin *et al.*, 1999).

4.2 *Temperature*

Other climatological variables, particularly temperature, may be just as important as CO_2 concentration. Tree production is extremely sensitive to temperature in the range 0–15 °C, but not very sensitive to temperature above 15–20°C. In contrast, respiration is demonstrably sensitive to temperature above 15°C, showing a more or less exponential relationship with temperature (Lloyd and Taylor, 1994). In the Boreal region, temperature has increased much more rapidly than in the world as a whole, and there are several lines of evidence for the 'greening up' of tundra, and its consequent increased rate of photosynthesis (R.F. Keeling *et al.*, 1996; Myneni *et al.*, 1997; Nagy *et al.*, 2003).

4.3 *Nitrogen*

In addition to CO_2 and temperature, it is to be expected that nitrogen deposition has some stimulatory effect on global photosynthesis, because current deposition rates of active anthropogenic nitrogen are much increased over background rates in the developed countries and regions (Galloway *et al.*, 1995). It is well established that temperate and Boreal ecosystems respond very strongly to fertilization with ammonium nitrate when other nutrients are not limiting (see Chapter 16, this volume). Moreover, it is known that tree growth has increased in the temperate forest zone (Lebourgeois and Becker, 1996; Spiecker *et al.*, 1996). Even where phosphorus rather than nitrogen is limiting, as in many parts of the tropics, an increase in organic acid production and root-surface phosphatase activity, resulting from the increase in atmospheric CO_2 concentration, may release soil phosphorus and cause an increase in plant production (Lloyd *et al.*, 2001a).

4.4 *A simple model to estimate sink strength*

Thus, NPP is likely to be increasing appreciably at the present time because of three factors: the rise in atmospheric CO_2 concentration, the warming over those parts of the world that are too cold for plant growth for much of the year, and the deposition of nitrogen in places where N is limiting. As a result, the rate at which carbon enters the terrestrial pools is increasing. The availability of this increased production to the decomposers will, however, be delayed because of the residence time of carbon in woody tissues. Decomposition of the trunk of a tree, for example, will not begin until the tree dies, and thereafter it is likely to take decades.

These considerations lead to the following simple method for estimating the sink strength (proposed and first used by Taylor and Lloyd (1992), and later used by Saugier *et al.* (2001)). Let us assume that heterotrophic respiration at time t is equal to the NPP (N_p) at time $(t - t_r)$, where t_r is the residence time (estimated as the carbon in vegetation plus soil divided by NPP), then:

$$R_h = N_p(t - t_r).$$

Assume now that NPP is increasing linearly in response to rising CO_2 concentration at a rate a so that:

$$N_p(t) = N_p(t_o)\,(1 + at).$$

Based on experiments on woody plants grown at twice the present CO_2 concentration, it is reasonable to propose that a doubling of the atmospheric CO_2 concentration would increase NPP by 30% (Idso, 1999; Medlyn *et al.*, 1999; Wullschleger *et al.*, 1995). Consequently, the observed annual increase of atmospheric CO_2 concentration of 0.4% per year might cause a proportional annual increase in NPP of $a = 0.0012$ per year.

By definition, NEP (N_e) is

$$N_e = N_p - R_h,$$

and thus

$$N_e(t) = N_p(t) - N_p(t - t_r) = N_p(t_o)at_r.$$

We can see from this simple model that rising CO_2 concentration causes an increase in sink strength that is proportional to the residence time of carbon in the plant–soil store. Applying the above equation to the global productivity data suggests that the sinks are predominantly in the tropical forests, the Boreal forests and the savannahs (*Table 1*). Using this method, we arrive at a 'potential' global terrestrial sink of 2.8 Pg (C) per year. This is a somewhat larger estimate than the estimate obtained from atmospheric methods, perhaps because of disturbance (Gurney *et al.*, 2002; IPCC, 2001; Rödenbeck *et al.*, 2003; Royal Society, 2001), although still within the range of uncertainty.

5. Will the sinks endure?

In the previous section we saw how sinks arise, and why they occur especially in forest ecosystems. Clearly, there are limits to the duration of sinks; they are vulnerable to fire, flood, disease, disturbance as a part of management, and accelerated decomposition caused by global warming. The first four of these are more or less self-explanatory but the last requires some explanation.

5.1 *Decomposition and global warming*

Decomposition of debris and soil organic matter (heterotrophic respiration) is expected to increase as global temperature rises, as long as there is respiratory substrate in the form of dead organic matter. If we assume a Q_{10} of 2.0 for heterotrophic respiration and no positive impact on net photosynthesis and NPP, we are forced to conclude that the sink is unlikely to endure for more than a few decades longer, as heterotrophic respiration will be increased by warming and will soon overtake NPP. For example, Raich *et al.* (2002) fitted a regression model to 25 data sets on soil respiration and obtained a simple equation that accounted for 62% of the total variation using only two variables, temperature and precipitation. They used the model to estimate the mean annual global heterotrophic respiration as 80.4 Pg (C), with a temperature sensitivity of 3.3 Pg (C) per degree Celsius. This is, however, a controversial approach (Jones *et al.*, 2003) because long-term and short-term observations provide conflicting indications of the temperature sensitivity of soil respiration. In the short term, decomposition of organic material, like other enzymatic reactions, increases exponentially with temperature, unless water is restricting the growth and activity of micro-organisms, or unless the substrate is depleted (Lloyd and Taylor, 1994). However, longer-term observations provide evidence that the process is relatively insensitive to temperature, and more dependent on the supply of suitable organic matter (Giardina and Ryan, 2000; Liski *et al.*, 1999).

5.2 *Photosynthesis, CO$_2$ concentration, and nutrition*

We do not expect photosynthesis to continue indefinitely to respond to elevated CO_2 concentration. Almost all the relevant long-term experiments on growth at elevated CO_2 have investigated the response only up to 'twice-recent' CO_2 concentration (about 700 p.p.m.). However, in short-term photosynthesis experiments where the measured rate of photosynthesis is plotted against the internal CO_2 concentration, the rate often does not increase much above a CO_2 concentration of 750 p.p.m. This is because photosynthesis is rate-limited by the enzyme ribulose 1,5-bisphosphate carboxylase oxygenase (RUBISCO), and this in turn is usually limited by the nitrogen supply (see, for example, Schulze *et al.*, 1994). Most forests are chronically nitrogen deficient, and deposition of atmospheric nitrogen from anthropogenic sources, application of nitrogen fertilizers, and enhanced nitrogen turnover may contribute to increase leaf area and to maintain stand photosynthesis. In addition, the carbon-dioxide-saturated rate of photosynthesis is also to some extent limited by temperature; the increase in temperature will also considerably increase the length of the growing season in the northern temperate and Boreal regions. For these reasons, some forest-scale models incorporating linked carbon and nitrogen cycles have led to the conclusion that NEP will continue to increase in parallel with heterotrophic respiration for some considerable time to come (see, for example, McMurtrie *et al.*, 2001).

5.3 *Lack of water*

Notwithstanding, it is likely that other resources, in particular water, may ultimately limit the overall rate of the process, particularly in the tropics. Recently, elaborate global-scale models have suggested that the terrestrial sink will weaken and may not endure (Cox *et al.*, 2000; Cramer *et al.*, 2001). Cox's model is a general circulation model (GCM) containing a simple representation of the carbon cycle. In this model, carbon is released as CO_2 from the decomposition of organic matter, forming a positive feedback, so that more release of CO_2 leads to a faster rate of warming. Another GCM, based on similar principles, shows a similar but weaker effect (Friedlingstein *et al.*, 2003). The reason for the difference may be that the GCM of Cox *et al.* (2000) generates severe droughts in the tropical zone, more so than other GCMs. Critics of the Cox *et al.* model believe that their representation of the carbon cycle is too simplistic and that the temperature sensitivity of decomposition in the model is too large, pointing out that the rate should decline when the easily consumed fraction of the substrate is diminished (see, for example, Grace and Rayment, 2000; Liski, 1999). Other evidence, for example from the tundra (Oechel *et al.*, 2000), shows signs of long-term acclimation of ecosystems to temperature which are difficult to reconcile with the view that the rise in temperature will increase respiration. The question as to whether the present carbon sinks will endure and for how long is currently unresolved.

6.   What observation networks are required to monitor the sinks?

As a result of international discussions on the management of the carbon cycle in the framework of the Kyoto Protocol, it has become important to track the strength of the terrestrial carbon sinks. Several approaches are being developed.

6.1 *Land-based: eddy covariance with forest inventory*

As we have seen, this method provides the NEP over an area of 0.1–1 km^2, and provides excellent information on the sensitivity of the processes of photosynthesis and respiration to climatological variables: solar radiation, temperature, vapour pressure deficit of the air. It enables comparison of different types of land use cover. Most EC sites are already supported by ground-based measurements of soil respiration and forest and soil inventory, and through the FLUXNET it is possible to continue the process of refining the techniques and certifying the standards to which a particular site is operating. New techniques are continually becoming available to make these measurements more effective, accurate, and perhaps cheaper. Automation will enable sites to run with relatively little attention by researchers: EC stations may soon automatically relay data by telemetry, performing the necessary corrections and gap-filling using agreed protocols. Aircraft-borne sensors enable 'snapshots' of the entire landscape: EC sensors may be flown using suitable light aircraft; aircraft can also carry biomass sensors based on LIDAR (Lefsky *et al.*, 2002).

6.2 *Atmospheric methods*

Inference of the sink strength from analysis of many atmospheric concentrations is effective at a global or continental scale. In the latter case, tall towers with well-calibrated infra-red gas analysers (accurate to better than 0.1 p.p.m.) form a 'box' around a continent, and offer the possibility of reporting hour-by-hour the carbon uptake of the entire continent. However, in developed countries, the biogenic fluxes are small in relation to the anthropogenic fluxes, and the common 'marker' for anthropogenic fluxes, CO, is not necessarily a good marker as the ratio $CO:CO_2$ depends on the fuel source. This poses a difficulty as the anthropogenic signal may mask the biogenic one. In some countries it may be possible soon to monitor anthropogenic CO_2 emissions in real time, for example by traffic sensors on main roads and by data supplied automatically from power stations.

Another atmospheric method is to measure the mass balance of gases by aircraft flights through the planetary boundary layer (Lloyd *et al.*, 2001b; Chapter 5 this volume). By repeated integration of gas concentrations from the vegetation surface to the top of the boundary layer, it is possible to estimate the exchange of gases between the vegetation and atmosphere from the changes in mass. In future, a monitoring program might 'piggy-back' on civil aircraft, recording concentration profiles as they take off and land at thousands of sites each day, thus providing independent estimates of the global carbon fluxes.

6.3 *Remote sensing of CO_2*

Some satellite sensors provide measurements of CO_2 concentrations of the atmosphere, and these will soon become widely available. However, the resolution is not likely to be very good (a few parts per million), and data capture will be confined to 'fair weather' periods. It is too early to say how far this methodology is likely to be used, as the technique is still under development.

6.4 *Remote sensing of the land surface*

Optical remote sensing using the long-established sensors continues to be valuable to track the rate of land-use change, particularly in the tropics. Brazil's PRODES project has used LANDSAT to follow the conversion of forest to agriculture, and from such data it is possible to estimate the land-use-related component of carbon balance. Other satellites give higher spatial and spectral resolution, although LANDSAT has the advantage of having been in use since the 1970s, so that there is a long enough record to facilitate studies of land-use change. Recently, sensors with high spectral resolution have been introduced; these offer the possibilities of probing photosynthetic efficiency (Gamon *et al.*, 2002; Nichol *et al.*, 2000).

6.5 *Remote sensing of biomass*

Optical techniques using satellite sensors may be used to estimate biomass in the low to medium mass range, but are not suitable for forests. Essentially, optical methods see only the surface and as soon as the canopy is closed the signal saturates. More useful in principle is radar remote sensing. In this case, the satellite produces an active output of electromagnetic radiation with a wavelength from a few centimeters to several meters. The sensor on the satellite detects the back-scattered signal, which arises from the interaction of the vegetation elements with the radar waves. One positive feature is that radar frequencies are not attenuated by cloud cover, so that data capture can be complete. Early satellites such as ERS and JERS used short wavebands

(C-band, wavelength 4–8 cm), which 'see' only the short elements in the canopy (leaves, twigs). C-band radar is useful as it can, in principle, measure the height of the forest, which scales approximately with biomass. However, use of much longer wavebands is possible in future. For example, P-band (wavelength about 70 cm) produces a back-scattered signal that relates to the longer elements (branches, stems), and therefore enables moderately dense woodland to be measured (Lucas *et al.*, 2000, Melon *et al.*, 2001). Hitherto, P-band was not allowed, as it was reserved for communications. However, during September 2003 an important frequency 432–438 MHz (wavelength 68.5–69.4 cm) was made available to the remote sensing community by the World Radio Conference, paving the way for a P-band, biomass-measuring, satellite. Tests using aircraft-borne sensors over forests suggest that P-band should allow estimation of biomass in the range 0–120 Mt (biomass) ha^{-1}. Although still not sufficient for estimation of biomass of rain forest, this would be adequate for biomass estimation of many plantations and savannah woodlands.

6.6 *An observation network*

Such a network would use the best sources of data from all of the above sources (land-based forest inventories, EC towers, CO_2 measurements on tall towers, and satellite data) and merge the results to tune models. Later, systems like these would include data from biomass-sensing satellites (*Figure 6*). In the scheme proposed, the multiple-constraint principle would apply, whereby several completely independent observing systems would provide a continuous stream of data to be assimilated by a model. The model would be acting as a learning machine, and would return the best estimates of the national, regional and global carbon sinks.

    Running costs of such an observing system may be of similar magnitude to that currently used for weather forecasting.

7. What are the feedbacks to the carbon cycle?

The biogeochemical cycles (especially carbon, nitrogen, phosphorus, and water) interact strongly with each other, and it is impossible to consider any one of them in isolation. For example, the current high rate of deposition of nitrogen as ammonium and nitrate inevitably stimulates photosynthesis and therefore increases the capacity of the vegetation to act as a carbon sink. Moreover, global warming stimulates the mineralization of dead organic matter, especially in northern latitudes, and thus releases those forms of nitrogen that can be used for photosynthesis and so the sink strength increases still further.

7.1 *Emergent properties of forest ecosystems*

Such interacting systems may result in unexpected emergent properties, homeostasis in some cases, instability in others. For example, forests reduce the shortwave reflectance of energy relative to grasslands and agriculture, and tend to have higher rates of water use. Thus, some of the good they do by absorbing CO_2 may be offset by a warming effect by absorbing a larger fraction of solar energy. Betts (2000) presents calculations on the extent of this phenomenon, showing

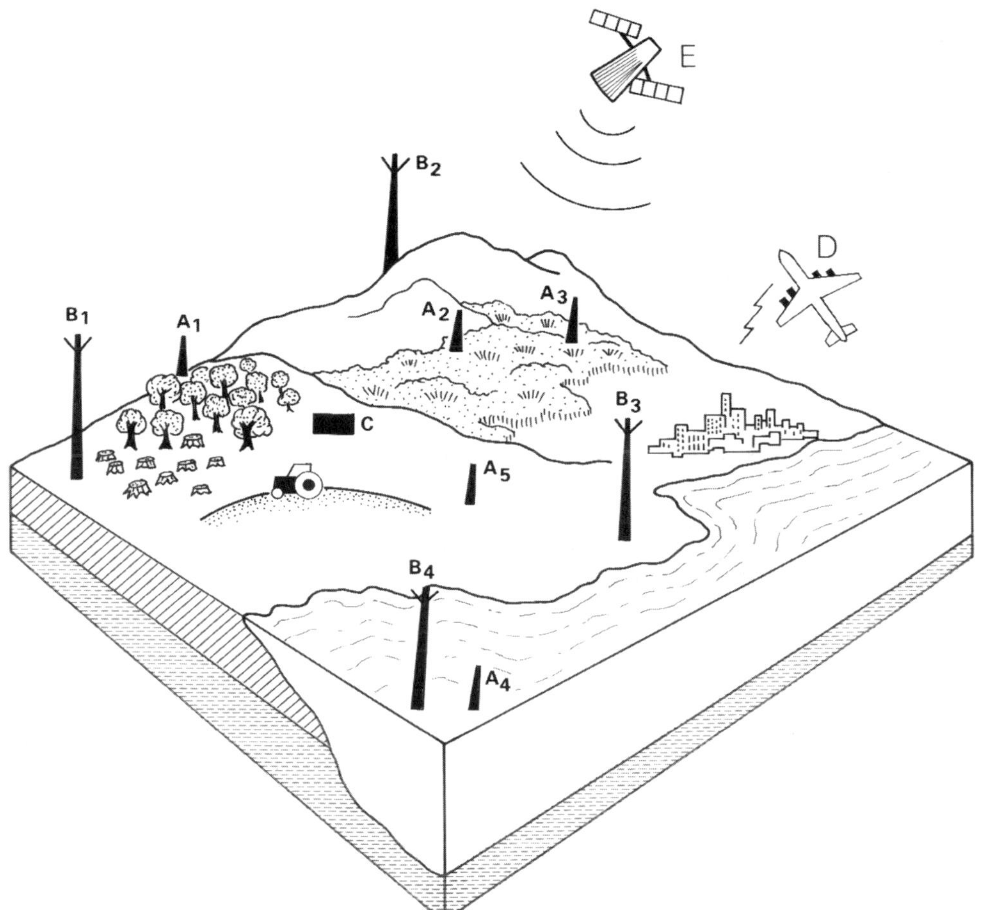

Figure 6. *An observational system to determine the global carbon balance. A1–A5, several hundred flux towers placed in representative vegetation; B1–B4, a set of tall towers situated around continents; C represents a central station receiving data by satellite links; D, civil aircraft used as 'ships of convenience' to make observations on the gaseous concentrations in the planetary boundary layer (during take off and landing); E, satellites that provide additional measurements of changes in gas concentrations in the boundary layer, and also biomass surveillance using optical and radar sensing methods.*

that it is especially problematic in the Boreal region, where snow lies for a long period on the non-forested ground, increasing albedo and therefore having a cooling effect.

Predicting the behaviour of the carbon cycle requires an understanding of these essential feedbacks. We may speculate about the nature and extent of the feedbacks but evidence is so far scant. In general, two extreme views are possible. According to one (a Gaia view, *sensu* Lovelock (1988)), the carbon cycle contains negative feedbacks tending to prevent catastrophic change and maintaining a hospitable planet in the face of interference by humans. According to the other, positive feedbacks will dominate, causing run-away warming. So far, global warming has been progressing for insufficient time to enable rigorous testing of these competing hypotheses of negative and positive feedbacks. Moreover, neither GCM models (Cox *et al.*, 2000) nor dynamic vegetation models (Cramer *et al.*, 2001) are mature enough to deal very well with interactions and feedbacks among some of the processes that are thought to be important in Gaia (e.g., cloud formation).

7.2 *Thinking about feedbacks*

Here I simply list some of the feedbacks that have been mooted but (usually) scarcely investigated. The list is not exhaustive.

Positive feedbacks
1. Warming will cause release of CO_2 from decomposition, primarily in the forest regions of the world but also the tundra, thus accelerating warming.
2. Warming will melt snow and ice, decreasing solar reflectance and thus increasing warming, melting even more snow and ice, and so on.
3. Tropical deforestation will cause warming and drying, which will itself cause a decline in the rain forests of the world.
4. Increased cover of woody vegetation in the Boreal zone, caused by warming in the Boreal zone, will decrease the reflectance of the land surface, and thus accelerate warming.
5. Warming will increase the decomposition rate of clathrates (methyl hydrates), leading to a release of methane, a potent GHG that will increase warming.

Negative feedbacks
1. Deforestation will lead to an increase of soil erosion, atmospheric aerosols will increase and solar radiation at the surface will decline, causing cooling.
2. Replacement of coniferous forest by warmth-loving broadleaves and by agriculture will increase planetary reflectance, causing cooling.
3. Warming will accelerate turnover of soil organic matter and release nutrients that will accelerate uptake of CO_2 and increase growth of forests in the Boreal region.
4. Increased transpiration in a warmer world will lead to more clouds, cooling the planet.

Attempts have been made to evaluate some of these influences as single effects, at forest-stand scale and biome scales, but we are very far from being able to build a reliable model to explore all such interactions together at the whole-world scale. Earth-system modelling is in its infancy, and the research community is still rather fragmented. On the one hand, we have the GCMs that take months to run on super-computers, but do not model the vegetation and the feedbacks very well (see for example, Cox *et al.*, 2000). Their model was the first to couple a representation of the terrestrial carbon cycle to a GCM. On the other hand, dynamic vegetation models have been developed by a different research community to simulate vegetation changes over periods of decades, in response to scenarios of climate change. In these models the land cover responds to changes in temperature and CO_2 concentration

(Cramer *et al.*, 2001), but these are 'given' as outputs from GCMs. At a longer time-scale, there are Gaia theorists who bring rich ideas that emphasize biogeochemical processes over immense time-scales (see, for example, Lenton and van Oijen, 2002). All of these approaches are valuable but none of them claims to be able to model more than a few of the feedbacks, most of which remain poorly understood at the global scale.

8. Is it feasible to manage forests as sinks?

It is likely that the carbon sequestration capacity of the land surface is much more sensitive to management than it is to climate change. If this is true, is it possible to manage the world's vegetation for carbon sequestration? The context of this question is the Kyoto Protocol, which encourages us to think of forests as a means of extracting carbon from the atmosphere and storing it as timber and soil organic matter, at least until such time as we can develop non-carbon fossil energy economies.

8.1 *The Kyoto charge*

It is worth reiterating the wording of Article 3.3, which covers land-use change and forestry in the 38 richer countries of the world, known as the 'Annex I countries'. Collectively, they emit 58% of the CO_2 from fossil fuel burning, and each has an emission reduction target, relative to CO_2 emissions in 1990. If the Kyoto Protocol were to be successful in its objectives, these countries would see an overall reduction by 2008–2012 of 5.2% relative to their emissions in 1990. Article 3.3 allows human-induced sinks:

> The net changes in greenhouse gas emissions by sources and removals by sinks resulting from direct human-induced land-use change and forestry activities, limited to afforestation, reforestation and deforestation since 1990, measured as verifiable changes in carbon stocks in each commitment period, shall be used to meet commitments under this Article of each Party included in Annex I.

Objections are often raised against the use of human-induced sinks, because forests are not always permanent, and there is a feeling that over-reliance on forests may inhibit development of alternative (renewable) energy sources such as wind, wave, tide, and solar. There are other objections too. For example, few, if any of the positive feedbacks relating to forests, spelt out in the previous section, were taken into account by those who formulated the Protocol (See Chapter 1, this volume).

A cause of much confusion has been that different international agencies adopted different definitions of terms such as 'forest', 'afforestation' and 'reforestation'. For the purposes of forestry and forest activities in the Kyoto Protocol, it was agreed that forest should be defined as an area of at least 0.05–1.0 ha of trees, with a canopy cover of at least 10–30%, and with trees capable of reaching 5 m in height. 'Afforestation' should mean that the site has not been forested for at least 50 years and 'reforestation' refers to the planting of trees on sites that were not forested on 31 December 1989 (but not including areas cleared of trees as part of the normal management cycle) (IPCC, 2000).

8.2 *Clean development mechanism projects*

In the terms of the Kyoto Protocol, Annex I countries may also be credited for carbon that has been sequestered in certain development projects, including some forestry projects, which they have established in non-Annex I countries, under the provision of Article 12:

> The purpose of the clean development mechanism shall be to assist Parties not included in Annex I in achieving sustainable development and in contributing to the ultimate objective of the Convention, and to assist Parties included in Annex I in achieving compliance with their quantified emission limitation and reduction commitments under Article 3.

Such projects range from those that directly reduce the emissions of GHGs (for example through the provision of more efficient power stations) to those that create carbon sinks by afforestation or reforestation. Annex I countries will be credited for any carbon sequestered in this way towards their national emission reduction targets. It has been agreed that in the first commitment period (2008–2012), forestry activities will be limited to afforestation and reforestation, and that the carbon absorbed in these projects will attract certified emission reduction (CER) units to be traded with other countries (Grace *et al.*, 2003).

The quantity of carbon that might be involved world-wide is potentially significant in relation to national emission reduction targets, although there is a limit on how much 'foreign' carbon can be counted in this way: only 1% of each country's emissions reduction target. Countries with difficulties in meeting their targets directly by emission reductions are therefore likely to take advantage of the forestry options within the clean development mechanism (CDM).

9. Theoretical and possible sink capacities, and what cannot be counted

It is difficult to assess the size of the carbon sink that could be created by forestry and land-use change. Most countries have 'marginal land' and also 'degraded land' which can be acquired cheaply and on which appropriately selected tree species are capable of rapid growth. An important point here is that trees will grow in very poor soils that are not suitable for herbaceous crops. Cannell (2003) has attempted to estimate the global, European, and UK capacity (*Table 2*). It is clear that a considerable sink capacity could be created. One of the primary limitations is the wording of the Protocol itself and the subsequent Marrakech Accords, which place stringent limitations on the sort of projects that are allowed under the CDM. Nevertheless, a substantial sink could be created through the CDM. The current rate of establishment of plantations in the tropics is 4.3 million ha per annum, according to FAO data (*Figure 4*). Assuming each hectare becomes a sink of 5 Mg (C) ha^{-1} per year, and assuming afforestation at this rate over a 20 year period, we may expect to establish a sink of 0.4 Pg (C) per year, which could absorb as much as 20 Pg (C) over 50 years.

Global tropical deforestation amounts to 1–2 Pg (C) per year (Achard *et al.*, 2002; DeFries *et al.*, 2002) and preventing it would make a substantial contribution to reduc-

Table 2. *Theoretical, potential and achievable sink capacities, over the next 50 years, compared with fossil fuel emissions. Units: Pg (C) per year. Source: Cannell (2003).*

	Theoretical	Realistic potential	Achievable capacity	Current fossil fuel emissions
World	2–4	1–2	0.2–1.0	6.4
Europe (EU 15)	0.2–0.5	0.05–0.1	0.02–0.05	0.94
UK	0.03–0.07	0.003–0.005	0.001–0.002	0.147

ing the CO_2 addition to the atmosphere, as well as providing some useful other benefits, such as conservation of biodiversity (Swingland, 2002). Currently, forest protection may not be counted under the CDM. If preventing deforestation could be a part of the CDM, there would be a real flow of money from rich countries to poor countries and the magnitude of carbon 'saved' could be very large. Large-scale deforestation of Brazil's Amazon might be prevented through payments made by the richer countries (Laurance *et al.*, 2001a, b). A fundamental obstacle is that people need timber, fibre, and food, which they generally get from replacing forest with croplands. A political obstacle is that for many tropical countries there are issues of sovereignty, and not all of the Annex 1 countries have ratified the Protocol. At the time of the Symposium (April 2003) those countries that have already ratified account for 42% of the emissions, still short of the 50% required to trigger the Protocol to come into effect. In 2001 the USA announced that it would not ratify the Protocol. On 30th September 2004 the Russian government approved the Kyoto Protocol after much prevarication, and despite the fact that it is one of the few countries that would benefit from a warmer climate. It is now expected that the Protocol will come into force in February 2005.

10. Concluding remarks

Several independent lines of evidence show that the forested regions of the world are currently acting as sinks for carbon. These are: (i) measurements of stocks and fluxes of carbon; (ii) inferences of regional sink distributions from atmospheric measurements; and (iii) model calculations based upon physiological experimentation. The reason for the existence of these sinks is well understood. The rate of carbon assimilation is increasing as a result of gradual improvement of conditions for plant growth (elevated atmospheric CO_2 concentration may be only one of several factors), but the substantial storage of carbon in forest ecosystems means that the release of carbon by decomposition lags behind.

Research on the role of forests in the carbon cycle has made much progress, but it is difficult to predict what may happen to the sink over the next 50 years. This is because of uncertainty in the behaviour of carbon in the soil under climate change, and because the biophysical and biogeochemical feedbacks between the vegetation and atmosphere require much more attention and elucidation, and may yet produce surprises.

Forest sinks may be manipulated by appropriate silviculture. Indeed, management of the forest is likely to have a larger impact than climate change, at least in the short term (20 years). Useful increases in sink capacity are possible by a global increase in forest cover, requiring urgent action by governments. The prospect of carbon sequestration through management of forests and other landscapes offers real opportunity, and may also benefit nature conservation.

Success of carbon sequestration does depend on the development of cost-effective ways of monitoring and reporting stocks and fluxes of carbon. The research effort to achieve this requires further development of Earth observation: new sensors, new platforms (satellites, aircraft, tall towers), and an enhancement of political resolve at an international level.

Acknowledgements

The author wishes to acknowledge the support of the Natural Environment Research Council (NERC) through the Centre of Terrestrial Carbon Dynamics (CTCD), and the European Commission through CARBOEUROPE-IP.

References

Achard, F., Eva, H.D., Stibig, H.J., Mayaux, P., Gallego, J., Richards, T. and Malingeau, J.P. (2002) Determination of the deforestation rate of the world's humid tropical forests. *Science* **297**: 999–1002.

Atjay, G.L., Ketner, P. and Duvigneaud P. (1979) Terrestrial Primary Productivity and phytomass. In: Bolin, B., Degens, E.T. and Ketner, P. (eds) *The Global Carbon Cycle, SCOPE 13*, pp. 129–182. Wiley, New York.

Aubinet, M., Grelle, A., Ibrom, A., Rannik, U., Moncrieff, J., Foken T. *et al.* (2000) Estimates of the net carbon and water exchange of forests: the EUROFLUX methodology. *Advances in Ecological Research* **30**: 113–175.

Betts, R.A. (2000) Offset of the potential carbon sink from boreal forestation by decreases in surface albedo. *Nature* **408**: 187–190.

Bird, M.I., Moyo, C., Veenendaal, E.M., Lloyd, J. and Frost, P. (1999) Stability of elemental carbon in a savanna soil. *Global Biogeochemical Cycles* **13**: 923–932.

Broeker, W.S., Takahashi, T., Simpson, H.J. and Peng, T.-H. (1979) Fate of fossil fuel carbon dioxide and the global carbon budget. *Science* **206**: 409–410.

Cannell, M.G.R. (2003) Carbon sequestration and biomass energy offset: theoretical, potential and achievable capacities globally, in Europe and the UK. *Biomass and Bioenergy* **24**: 97–116.

Ciais, P., Tans, P.P., White, J.W.C., Trolier, M., Francey, R.J., Berry, J.A. *et al.* (1995) Partitioning of ocean and land uptake of CO_2 as inferred by $\delta^{13}C$ measurements from the NOAA Climate Monitoring Diagnostics Laboratory global air sampling network. *Journal of Geophysical Research* **100**: 5051–5070.

Cox, P.M., Betts, R.A., Jones, C.D., Spall, S.A. and Totterdell, I.J. (2000) Acceleration of global warming due to carbon-cycle feedbacks in a coupled climate model. *Nature* **408**: 184–187.

Cramer, W., Bondeau, A. Woodward, F.I., Prentice, I.C., Betts, R.A., Brovkin, V. *et al.* (2001) Global response of terrestrial ecosystem structure and function to CO_2 and climate change: results from six dynamic global vegetation models. *Global Change Biology* **7**, 357–373.

DeFries, R.S, Houghton, R.A., Hansen, M., Field, C.B., Skole, D. and Townsend J. (2002) Carbon emissions from tropical deforestation and regrowth based on satellite observations for the 1980s and 1990s. *Proceedings of the National Academy of Sciences of the USA* **99**: 14256–14261.

Dixon, R.K., Brown, S., Houghton, R.A. (1994) Carbon pools and fluxes of global forest ecosystems. *Science* **263**: 185–190.

Falge, E., Baldocchi, D., Tenhunen, J., Aubinet, M., Bakwin, P., Berbegier, P. *et al.* (2002) Seasonality of ecosystem respiration and gross primary productivity as derived from FLUXNET measurements. *Agricultural and Forest Meteorology* **113**: 53–74.

Falloon, P. and Smith, P. (2002) Simulating SOC changes in long-term experiments with RothC and CENTURY: model evaluation for a regional scale application. *Soil Use and Management* **18**: 101–111.

FAO (2000) Forest Resources Assessment 2000 (FRA 2000). FAO, Rome.

Friedlingstein, P., Dufresne, J.L., Cox, P.M. and Rayner, P. (2003). How positive is the feedback between climate change and the carbon cycle. *Tellus Series B – Chemical and Physical Meteorology* **55**: 692–700.

Galloway, J.N., Schlesinger, W.H., Levy-II, H., Michaels, A. and Schnoor, L.J. (1995) Nitrogen fixation: anthropogenic enhancement and environmental response. *Global Biogeochemical Cycles* **9**: 235–252.

Gamon, J.A., Serrano, L. and Surfus, J.S. (2002) The photochemical reflectance index as an optical indicator of photosynthetic radiation use efficiency across species, functional types and nutrient levels. *Oecologia* **112**: 492–501.

Giardina, C.P. and Ryan, M.G. (2000) Evidence that decomposition rates of organic carbon in mineral soil do not vary with temperature. *Nature* **404**: 858–861.

Gleixner, G., Czimczik, C.J., Kramer, C., Lühker, B. and Schmidt, M.W.I. (2001) Plant compounds and their turnover and stabilization as soil organic matter. In: Schulze, E.-D., Heimann, M., Harrison, S., Holland E., Lloyd, J., Prentice. I.C. and Schimel, D. (eds) *Global Biogeochemical Cycles in the Climate System*, pp. 210–215. Academic Press, London.

Grace, J. and Rayment, M. (2000). Respiration in the balance. *Nature* 404: 819–820.

Grace, J., Lloyd, J., McIntyre, J., Miranda, A.C., Meir, P., Miranda, H. *et al.* (1995). Carbon dioxide uptake by an undisturbed tropical rain forest in South-West Amazonia 1992–1993. *Science* **270**: 778–780.

Grace, J., Kruijt, B., Freibauer, A., Benndorf, R., Carr, R., Dutschke, M. *et al.* (2003) *Scientific and Technical Issues in the Clean Development Mechanism.* CarboEurope Office, Max Planck Institute for Biogeochemistry, Jena.

Gurney, K.R., Law, R.M., Denning, A.S., Rayner, P.J., Baker, D., Bousquet, P. *et al.* (2002) Towards robust regional estimates of CO_2 sources and sinks using atmospheric transport models. *Nature* **415**: 626–630.

Houghton, R.A., Hobbie, J.E., Mellilo, J.M., Moore, B., Peterson, B.J., Shaver, G.R. *et al.* (1983) Changes in the carbon content of the terrestrial biota and soils between 1860 and 1980 – a net release of CO_2 into the atmosphere. *Ecological Monographs* **53**: 235–262.

Idso, S.B. (1999) The long-term response of trees to atmospheric CO_2 enrichment. *Global Change Biology* **5**: 493–495.

IPCC (2000) *Land Use, Land-use Change and Forestry.* Watson, R.T. *et al.* (eds). Cambridge University Press, Cambridge.

IPCC (2001) *Climate Change 2001: The Scientific Basis. Contribution of Working Group I to the Third Assessment Report of the Intergovernmental Panel on Climate Change.* Houghton, J.T. *et al.* Houghton, J.T., Ding, T., Griggs, D.J., Noguer, M., van der Linden, P.J., Dai, X., Maskall, K. and Johnson, C.A. (eds). Cambridge University Press, Cambridge, UK, and New York, USA.

Janssens, I.A., Freibauer, A., Ciais, P., Smith, P., Nabuurs, G.J., Folberth, G. *et al.* (2003) Europe's terrestrial Biosphere Absorbs 7 to 12 % of European anthropogenic CO_2 emissions. *Science,* **300**: 1438–1541.

Jones, C.D., Cox, P. and Huntingford, C. (2003) Uncertainty in climate-carbon-cycle projections associated with the sensitivity of soil respiration to temperature. *Tellus Series B – Chemical and Physical Meteorology* **55**: 642–648.

Keeling, C.D., Chin, J.F.S. and Whorf, T.P. (1996) Increased activity of northern vegetation inferred from atmospheric measurements. *Nature* **382**: 146–149.

Keeling, R.F. and Shertz, S.R. (1992) Seasonal and interannual variations in atmospheric oxygen and implications for the global carbon cycle. *Nature* **358**: 723–727.

Keeling, R.F., Piper, S.C. and Heimann, M. (1996) Global and hemispheric CO_2 sinks deduced from changes in atmospheric O_2 concentration. *Nature* **381**: 218–221.

Kruijt, B., Elbers, J.A., Von Randow, C., Aravjo, A.C., Olweira, P.J., Culf, A. *et al.* (2004) The Robustness of eddy correlation fluxes for Amazon rainforest conditions. *Ecological Applications* **14**: S101–S113.

Laurance, W.F., Cochrane, M.A., Bergen, S., Fearnside, P.M., Delamonica, P., Barber, C., D'Angelo, S. and Fernandes, T. (2001a). Environment – the future of the Brazilian Amazon. *Science* **291**: 438–439.

Laurance, W.F., Fearnside, P.M., Cochrane, M.A., D'Angelo, S., Bergen, S. and Delamonica, P. (2001b). Development of the Brazilian Amazon: Response. *Science* **292**: 1652–1654.

Law, B.E., Thornton, P.W., Irvine, J., Anthoni, P.M. and Van Tuyl, S. (2001) Carbon storage and fluxes in ponderosa pine at different development stages. *Global Change Biology* **7**: 755–777.

Lebourgeois, K. and Becker, M. (1996) Dendroecological study of Corsican pine in West France. Growth potential evolution during the last decades. *Annals des Sciences Forestieres* **53**: 931–946.

Lefsky, M.A., Cohen, W.B., Harding, D.J., Parker, G.G., Acker, S.A. and Gower, S.T. (2002) Lidar remote sensing of above-ground biomass in three biomes. *Global Ecology and Biogeography* **11**: 393–399.

Lenton, T.M. and van Oijen, M. (2002) Gaia as a complex adaptive system. *Philosophical Transactions of the Royal Society Series B – Biological Sciences* **357**: 683–695.

Lin, G.H., Adams, J., Farnsworth, B., Wei, Y.D., Marino, B.D.V. and Berry, J.A. (1999) Ecosystem carbon exchange in two terrestrial ecosystem mesocosms under changing atmopsheric CO_2 concentrations. *Oecologia* **119**: 87–108.

Liski, J., Ilvesniemi, H., Makela, A. and Westman, C.J. (1999) CO_2 emissions from soil in response to climatic warming are overestimated. *Ambio* **28**: 171–174.

Lloyd, J. (1999) The CO_2 dependence of photosynthesis, plant growth responses to elevated CO_2 concentrations and their interaction with soil nutrient status. II. Temperate and boreal forest productivity and the combined effects of increasing CO_2 concentrations and increased nitrogen deposition at a global scale. *Functional Ecology* **13**: 439–459.

Lloyd, J. and Taylor, J.A. (1994) On the temperature dependence of soil respiration. *Functional Ecology* **8**: 315–323.

Lloyd, J., Bird, M.I., Veenendaal, E.M. and Kruijt, B. (2001a) Should phosphorus availability be constraining moist tropical forest responses to increasing CO_2 concentrations. In: Schulze E.-D. (ed) *Global Biogeochemical Cycles in the Climate System*, pp. 95–114. Academic Press, New York.

Lloyd, J., Francey, R.J., Mollicone, D., Raupach, M.R., Sogachev, A., Arneth, A. *et al.* (2001b) Vertical profiles, boundary layer budgets, and regional flux estimates for CO_2 and its $^{13}C/^{12}C$ ratio and for water vapour above a forest/bog mosaic in central Siberia. *Global Biogeochemical Cycles* **15**: 267–284.

Lovelock, J. (1988) *The Ages of Gaia*. WW Norton, New York.

Lucas, R.M., Milne, A.K., Cronin, N., Witte, C. and Denham, R. (2000) The potential of synthetic aperture radar (SAR) for quantifying the biomass of Australia's woodlands. *Rangeland Journal* **22**: 124–140.

Malhi, Y. and Grace, J. (2000) Tropical forests as atmospheric carbon sinks. *Trends in Ecology and Evolution.* **15**: 332–337.

Malhi, Y., Baldocchi, D.D. and Jarvis, P.G. (1999) The carbon balance of tropical, temperate and boreal forests. *Plant Cell and Environment* **22**: 715–740.

Malhi, Y., Pegoraro, E., Nobre, A.D., Pereira, M.G.P., Grace, J. and Culf, A.D. (2002) Energy and water dynamics of a central Amazonian rain forest. *Journal of Geophysicsl Research* **107 (D20)**: 8061.

McMurtrie, R., Medlyn, B.E., and Dewar, R.C. (2001) Increased understanding of nutrient immobilization in soil organic matter is critical for predicting the carbon sink strength of forest ecosystems over the next 100 years. *Tree Physiology* **21**: 831–839.

Medlyn, B.E., Badeck, F-W., de Pury, D.G.G., Barton, C.V.M., Broadmeadow, M., Ceulemans, R., *et al.* (1999) Effects of elevated CO_2 on photosynthesis in European forest species: a meta-analysis of model parameters. *Plant, Cell and Environment* **22**: 1475–1495.

Melon, P., Martinez, J.M., Le Toan, T., Ulander, L.M.H. and Beaudoin, A. (2001) On the retrieving of forest stem volume from VHFSAR data: observing and modelling. *IEEE Transactions on Geoscience and Remote Sensing* **39**: 2364–2372.

Moncrieff, J.B., Massheder, J.M., de Bruin, H., Elbers, J., Friborg, T., Heusinkveld, B., Kabat, P., Scott, S., Soegaard, H and Verhoef, A. (1997). A system to measure surface fluxes of momentum, sensible heat, water vapour and carbon dioxide. *Journal of Hydrology* **189**: 589–611.

Myneni, R.B., Keeling, C.D., Tucker, C.J., Asrar, G. and Nemani, R.R. (1997) Increased plant growth in the northern latitudes from 1981–1991. *Nature* **386**: 698–702.

Nagy, L., Grabherr, G., Körner, Ch. and Thompson, D.B.A. (2003) *Alpine biodiversity in Europe. Ecological Studies 167*. Springer, Berlin.

Nemani, R.R., Keeling, C.D., Hashimoto, H., Jolly, W.M., Piper, S.C., Tucker, C.J., Myneni, R.B. and Running, S.W. (2003) Climate-driven increases in global net primary productivity from 1982 to 1999. *Science* **300**: 1560–1563.

Nichol, C.J., Huemmrich, K.F., Black, T.A., Jarvis, P.G., Walthall, C.L., Grace, J. and Hall, F.G. (2000) Remote sensing of photosynthetic-light-use efficiency of boreal forest. *Agricultural and Forest Meteorology* **101**: 131–142.

Oechel, W.C., Vourlituis, G.L., Hastings, S.J., Zulueta, R.C., Hinzman, L. and Kane, D. (2000) Acclimation of ecosystem CO_2 exchange in the Alaskan tundra in response to decadal climate warming. *Nature* **406**: 978–981.

Pacala, S.W., Hurtt, G.C., Baker, D., Peylin, P., Houghton, R.A., Birdsey, R.A. *et al.* (2001) Consistent land- and atmosphere-based US carbon sink estimates. *Science* **292**: 2316–2328.

Paulsen, J., Weber, U.M. and Korner, C. (2000) Tree growth near the treeline: abrupt or gradual reduction with altitude? *Arctic Antarctic and Alpine Research* **32**: 14–20.

Phillips, O.L. and Gentry, A.H. (1994) Increasing turnover through time in tropical forests. *Science* **263**: 954–958.

Phillips, O.L., Malhi, Y., Higuchi, N., Laurance, W.F., Nunez, P.V., Vasquez, R.M., *et al.* (1998) Changes in the carbon balance of tropical forests: Evidence from long-term plots. *Science* **282**: 439–442.

Prentice, I.C. and Lloyd, J. (1998) C-quest in the Amazonian Basin. *Nature* **396**: 619–620.

Raich, J.W., Potter, C.S. and Bhagawati, D. (2002) Interannual variability in global soil respiration, 1980–1994. *Global Change Biology* **8**: 800–812.

Richey, J.E., Melack, J.M., Aufdenkampe, A.K., Ballester, V.M. and Hess, L.L. (2002) Outgassing from Amazonian rivers and wetlands as a large tropical source of atmospheric CO_2. *Nature* **416**: 617–620.

Rödenbeck, C., Houweling, S., Gloor, M. and Heimann, M. (2003) CO_2 flux history 1982–2001 inferred from atmospheric data using a global inversion of atmospheric data. *Atmospheric Chemistry and Physics Discussions* **3**: 2575–2659.

Roy, J., Saugier, B. and Mooney, H.A. (2001) *Terrestrial Global Productivity*. Academic Press, San Diego.

Royal Society (2001) *The Role of Land Carbon Sinks in Mitigating Global Climate Change*. The Royal Society, London.

Saleska, S.R., Miller, S.D., Matross, D.M., Goulden, M.L., Wofsy, S.C., da Rocha, H.R. *et al.* (2003) Carbon in amazon forests: unexpected seasonal fluxes and disturbance-induced losses. *Science* **302**: 1554–1557.

Saugier, B., Roy, J. and Mooney, H.A. (2001) Estimations of global terrestrial productivity: converging toward a single number? pp. 543–557. In: Roy J., Saugier B. and Mooney, H.A. *Terrestrial global productivity*. Academic Press, San Diego.

Schulze, E.D., Kelliher, F.M., Körner, Ch., Lloyd, J. and Leuning, R. (1994) Relationships among maximum stomatal conductance, ecosystem surface conductance, carbon assimilation rate, and plant nutrition: a global scaling exercise. *Annual Review of Ecology and Systematics* **25**: 629–694.

Schulze, E.D., Lloyd, J., Kelliher, F.M., Wirth, C., Rebmann, C., Luhkcer, B. *et al.* (1999) Productivity of forests in the Eurosiberian boreal regions and their potential to act as carbon sink – a synthesis. *Global Change Biology* **5**: 703–722.

Siegenthaler, U. and Sarmiento, J.L. (1993) Atmospheric carbon dioxide and the ocean. *Nature* **365:** 119–125.

Spiecker, H., Mielikäinen, K., Köhl, M. and Skovsgaard, J.P. (1996). *Growth Trends in European Forests*. EFI Report, Springer, Berlin.

Steffen, W., Noble, I., Canadell, J., Apps, M., Schulze, E.D., Jarvis, P.G. *et al.* (1998) The terrestrial carbon cycle: implications for the Kyoto Protocol. *Science* **280:** 1393–1394.

Swingland, I. (2002). *Capturing Carbon and Conserving Biodiversity: A Market Approach*. Earthscan, London.

Tans, P.P., Fung, Y. and Takahashi, T. (1990) Observational constraints on the global atmospheric CO_2 budget. *Science* **247:** 1431–1438.

Taylor, J.A. and Lloyd, J. (1992) Sources and sinks of atmospheric CO_2. *Australian Journal of Botany* **40:** 407–418.

Titus, B.D. and Malcolm, D.C. (1999) The long-term decomposition of Sitka spruce needles in brash. *Forestry* **72:** 207–221.

Valentini, R., Matteucci, G., Dolman, A.J. *et al.* (2000) Respiration as the main determinant of carbon balance in European forests. *Science* **404:** 861–865.

Waring, R.H., Landsberg, J.J. and Williams, M. (1998) Net primary production of forests: a constant fraction of gross primary production? *Tree Physiology* **18:** 129–134.

Whittaker, R.H. and Likens, G.E. (1975) The biosphere and man. In: Whittaker, R.H. and Likens, G.E. (eds) *Primary Productivity of the Biosphere*, pp. 305–328. Springer, Berlin.

Woodwell, G.M. and Whittaker, R.H. (1968) Primary production in terrestrial ecosystems. *American Zoologist* **8:** 19–30.

Woodwell, G.M., Whittaker, R.H., Reiners, W.A., Likens, G.E., Delwiche, C.C. and Botkin, D.B. (1978) The biota and the world carbon budget. *Science*, **199:** 141–146.

Wullschleger, S.D., Post, W.M. and King, A.W. (1995) On the potential for a CO_2 fertilization effect in forests: estimates of the biotic growth factor based on 58 controlled-exposure studies. In: Woodwell, G.M. and MacKenzie, F.T. (eds) *Biotic Feedbacks in the Global Climate System: Will Warming Feed the Warming*, pp. 85–107. Oxford University Press, New York.

Yin, X.W. (1999) The decay of forest woody debris: numerical modelling and implications based on some 300 data cases from North America. *Oecologia* **121:** 81–98.

3

Carbon sequestration in European croplands

Pete Smith and Pete Falloon

1. Introduction

The Marrakech Accords, resulting from the 7th Conference of Parties (CoP7) to the 1992 United Nations Framework Convention on Climate Change (UNFCCC), allow biospheric carbon sinks and sources to be included in attempts to meet quantified emission limitation or reduction commitments (QELRCs) for the first commitment period (2008–2012) outlined in the Kyoto Protocol (available at www.unfccc.de). Under Article 3.4, the activities forest management, cropland management, grazing land management, and re-vegetation are included. Soil carbon sinks (and sources) can, therefore, be included under these activities. Parties electing to include cropland management, grazing land management and re-vegetation need to account for changes in these soil carbon sinks and sources on a net–net basis; i.e., they must compare the net flux of carbon from a given activity during the commitment period with the equivalent net flux of carbon in the baseline year (usually 1990). Carbon sequestration in cropland soils, or even a reduction in a flux to the atmosphere compared with the baseline year, can therefore be used by a Party to the UNFCCC in helping to meet emission reduction targets. In this paper we review how croplands might be used to help meet the emission reduction targets set at Kyoto.

2. Fluxes of carbon from European croplands

Croplands are estimated to be the largest biospheric source of carbon lost to the atmosphere in Europe each year, but the cropland estimate is the most uncertain among all land-use types (Janssens *et al.*, 2003). It is estimated that European croplands (for Europe to the east as far as the Urals) lose 300 Tg (C) per year (Janssens *et al.*, 2003), with the mean figure for the European Union (EU 15) estimated to be 78 Tg (C) per year (one standard deviation = 37) (Vleeshouwers and Verhagen, 2002). National estimates for some EU countries are of a similar order of magnitude on a per area basis (Sleutel *et al.*, 2003) but other estimates suggest much lower fluxes (Dersch

---

The Carbon Balance of Forest Biomes, edited by H. Griffiths and P.G. Jarvis.
© 2005 Taylor & Francis Group

and Boehm, 1997). These figures suggest that cropland carbon stocks are continuing to lose soil carbon, perhaps as a result of recent (decadal) land-use changes, although figures for net changes in land use over the past 20–30 years do not suggest large-scale conversion to cropland from other land uses. The size of the flux is similar to the flux measured when converting grassland to tilled cropland (as calculated from figures given by Johnston (1973)); this is an extreme land-use change, suggesting that the estimated flux may be too high. The figures for cropland soil carbon loss are highly uncertain (Janssens *et al.*, 2003) and there is clearly scope to reduce the uncertainty surrounding these estimates.

There is significant potential within Europe to decrease the flux of carbon to the atmosphere from cropland, and for cropland management to sequester soil carbon, relative to the amount of carbon stored in cropland soils at present.

3. Management options for carbon sequestration

Management options available to sequester carbon in European cropland considered by Smith *et al.* (2000) and expanded by Freibauer *et al.* (2004) include reduced and zero tillage, set-aside, perennial crops and deep-rooting crops, more efficient use of organic amendments (animal manure, sewage sludge, cereal straw, compost), improved rotations, irrigation, bioenergy crops, extensification, organic farming, and conversion of arable land to grassland or woodland. Other lists of potential soil carbon sequestration options are given by IPCC (2000); Lal *et al.* (1998) and Metting *et al.* (1999). These are listed in *Table 1*.

4. Biological and realistically achievable potential for carbon sequestration for cropland management options in Europe

Freibauer *et al.* (2004) used previous estimates of carbon sequestration potential given by Smith *et al.* (2000) and estimates by Vleeshouwers and Verhagen (2002) to estimate the carbon sequestration potential of cropland management practices in Europe, and estimated other potential options not considered in previous studies. In *Table 1* we give the carbon sequestration potentials limited only by availability of land, biological resources and land suitability, and the potentials estimated to be realistically achievable by 2012, adapted from figures given by Freibauer *et al.* (2004).

On average, the realistically achievable potentials for carbon sequestration estimated by Freibauer *et al.* (2004) are about 10% of the potential estimated when considering only availability of land, biological resources, and land-suitability, and are much less than the biological potential (Freibauer *et al.*, 2004). Cannell (2003) also estimated the biological potential, the realistically achievable potential, and a conservative achievable potential. For Europe, the realistically achievable potential is about 20% of the biological potential, whereas the conservative achievable potential was estimated to be about 10% of the biological potential (Cannell, 2003). Smith *et al.* (2001) attempted to include estimates of the impacts of non-CO_2 greenhouse gases (GHGs) and balance these against the carbon sequestration potential using the equivalent global warming potentials. The impact of including non-CO_2 GHGs was mixed, with some very small (less than 5%) differences for sequestration options such as sewage sludge application, and large potential impacts (greater than 50%) for others such as zero tillage. It is clear

Table 1. *Carbon sequestration potentials limited only by availability of land, biological resources, and land suitability, and the potentials estimated to be realistically achievable by 2012 adapted from figures in Freibauer et al. (2004).*

Practice	Soil carbon sequestration potential (t (C) ha^{-1} per year)	Estimated uncertainty	Total soil carbon sequestration potential for EU 15 (Tg (C) per year)[†]	Realistic soil carbon sequestration potential for EU 15 (Tg (C) per year) by 2012
Zero tillage	0.38 (0.29)*	>50%	24.4	2.4
Reduced tillage	<0.38	≥50%	<24.4	<2.4
Set-aside	<0.38	≥50%	2.4 (maximum)	0
Permanent crops	0.62	≥50%	0?	0?
Deep-rooting crops	0.62	≥50%	0?	0?
Animal manure	0.38 (1.47)*	≥50%	23.7	?
Cereal straw	0.69 (0.21)*	≥50%	5.5	?
Sewage sludge	0.26	≥50%	2.1	?
Composting	0.38	≥50%	3	3?
Improved rotations	>0	Very high	0?	0?
Fertilization	0	Very high	0	0
Irrigation	0	Very high	0	0
Bioenergy crops	0.62	≥50%	4.5	0.9
Extensification	0.54	≥50%	11	?
Organic farming	0–0.54	≥50%	3.9	3.9
Convert cropland to grassland	1.2–1.69 (1.92)*	≥50%	8.7–12.3	0
Convert cropland to woodland	0.62	>>50%	4.5	4.5 (maximum)

[†] Carbon sequestration potentials limited only by availability of land, biological resources, and land-suitability. All estimates based on extrapolation from Smith *et al.* (2000) except those marked *, where the figure in brackets is derived from Vleeshouwers and Verhagen (2002). For full list of assumptions, limitations, and sources, see Freibauer *et al.* (2004).

from these studies that realistically achievable potentials for carbon sequestration need to be distinguished from biological potentials, and that the full GHG budget needs to be considered when assessing the potential of a soil carbon sequestration option, rather than the impact on soil carbon alone.

5. Duration of soil carbon sequestration and permanence of soil carbon sinks

Soil carbon sinks resulting from sequestration activities are not permanent and will continue only for as long as a carbon-sequestering management practice is maintained. If a land-management or land-use change is reversed, the carbon accumulated will be lost, usually more rapidly than it was accumulated (Smith *et al.*, 1996). For the highest potential soil carbon sequestration to be realized, new carbon sinks, once established, need to be preserved in perpetuity. Within the Kyoto Protocol, mechanisms have been suggested to provide disincentives for sink reversal; for example, when land is entered into the Kyoto process it must continue to be accounted for and any sink reversal will result in a loss of carbon credits.

Soil carbon sinks increase most rapidly soon after a carbon-enhancing land-management change has been implemented, but soil carbon amounts may decrease initially if there is significant disturbance, for example as a result of site preparation for afforestation. The rate at which carbon is removed from the atmosphere and transferred into soil (i.e., the sink strength) becomes smaller as time goes on, as the soil carbon stock approaches a new equilibrium, and at equilibrium the sink is said to have become 'saturated'. The carbon stock will have increased, but the potential for further increase (the sink strength) has decreased to zero. This curve is shown schematically in *Figure 1*.

Figure 1. *Graph showing schematic time-course of soil carbon sink development (adapted from Smith 2003a).*

The time taken to reach a new equilibrium (i.e., for sink saturation to occur) is highly variable. The period for soils in a temperate climatic region such as Europe, to reach a new equilibrium after a land-use change is around 100 years (Jenkinson, 1988; Smith *et al.*, 1996), whereas soils in the Boreal region may take centuries. As a global compromise, current IPCC good practice guidelines for GHG inventories use a figure of 20 years for soil carbon to approach a new equilibrium (IPCC, 1997; Paustian *et al.*, 1997). Because net carbon sequestration occurs for a limited duration and is non-permanent, it should not be considered a substitute for reduction of emissions. It might, however, provide part of a raft of implemented short- to medium-term climate mitigation measures, while other, more permanent, emission reduction technologies are developed (such as non-CO_2 emitting renewable energy sources) (Smith, 2004a).

6. Measurement, monitoring, and verification of soil carbon sequestration

The Kyoto Protocol states that sinks and sources of carbon should be accounted for 'taking into account uncertainties, transparency in reporting, verifiability'. Smith

(2004a) examined the issues surrounding the monitoring and verification of soil carbon sequestration. A significant generic problem with the estimation of changes in terrestrial biospheric carbon relates to resolution of the smallest detectable change. Because the rate of change of most biospheric pools is slow, particularly in relation to the size of the pool, resolvable changes in stocks are typically not easily obtained for the larger pools (IPCC, 2000). In a recent paper, the minimum detectable difference in soil organic carbon was calculated as a function of variance and sample size for soil organic carbon changes after five years under a herbaceous bio-energy crop (Garten and Wullschleger, 1999). The authors showed that the smallest difference that could be detected was about 1 Mg (C) ha^{-1} (1 hectare (ha) = 10^4 m^2) (2–3% of the background carbon amount), and adequate statistical power (90% confidence) could only be achieved when using a very large sample size (more than 100 samples). The minimum difference that could be detected with a reasonable sample size (e.g. 16 samples) and a good statistical power (90% confidence) was 5 Mg (C) ha^{-1} (10–15% of the background carbon amount) (see also Chapter 11, this volume). Most agricultural practices will not cause soil carbon accrual rates as high as this over a 5 year commitment period (Smith *et al.*, 1997). Smith (2004b) suggested that in some cases, the costs associated with demonstrating carbon sequestration can outweigh the value of carbon sequestered!

The level of verifiability of soil carbon sequestration depends upon the stringency of the definition of verifiability that is adopted. A stringent definition might require not only that changes are measured but that independent data be provided for verification (IPCC, 2000). A scheme fulfilling these criteria might need to sample each geo-referenced piece of land subject to an Article 3.4 activity at the beginning and end of a commitment period, using a sampling regime that gives adequate statistical power. Soil and vegetation samples and records would need to be archived and the data from each piece of land aggregated to produce a National figure. Separate methods would be required to deliver a second set of independent verification data. Such an undertaking at the national scale would be prohibitively expensive (Smith, 2004b). At its least stringent, verifiability might entail the reporting of areas under a given practice (without geo-referencing) and the use of default values for carbon stock changes for each area (for example from IPCC default values; IPCC, 1997), so as to infer a change for all areas under that practice. An intermediate scheme in the range of verifiability might be one in which areas undergoing a particular practice are geo-referenced (from remote sensing or ground survey), carbon changes are derived from controlled experiments (in representative climatic regions and on representative soils) or are modelled (using a well-evaluated, well-documented, and archived model), and intensively studied benchmark sites are available for verification. Several countries are investing in such sophisticated National carbon accounting systems that could offer this intermediate level of verifiability. Given that the least stringent definition of verifiability is likely to be acceptable to many countries, this may allow low-level verifiability to be achieved by most Parties to the UNFCCC by the beginning of the first commitment period (Smith, 2004b).

7. Carbon sequestration as part of integrated policies promoting sustainability: a 'no regrets' policy

Given that carbon sequestration in soil has a finite potential and is non-permanent (IPCC, 2000) and that it may also be difficult to measure and verify (Smith, 2004b), soil carbon sequestration is a riskier long-term strategy for climate mitigation than direct emission reduction. However, improved agricultural management often has a range of other environmental and economic benefits in addition to climate mitigation, which may make attempts to improve soil carbon storage attractive as part of integrated sustainability policies. Other authors have cited improvements in soil structure, water holding capacity, fertility, and resilience accruing through increased soil organic matter content and have described improvements in soil organic matter management as a 'win–win' strategy (see, for example, Lal *et al.*, 1998). Given that many soil characteristics or indicators benefit from improved management of soil organic matter (to improve soil sustainability, soil quality, soil health, etc.), a 'no regrets' policy for management of soil organic matter is attractive. A 'no regrets' approach entails implementing policies that will be of benefit now, and may possibly also be of increased benefit in the future (Smith and Powlson, 2003). The IPCC (2001) stresses that linkages among local, regional, and global environmental issues, and their interrelationships in meeting human needs, provide an opportunity to address global environmental issues at local, national, and regional scales in an integrated manner that is cost-effective and meets sustainable development objectives. In assessing the role of soil organic matter in long-term soil sustainability, Smith and Powlson (2003) noted that the importance of integrated approaches to deliver on a multitude of environmental issues is becoming ever clearer. Strategies to increase the soil organic carbon content of cropland soils are clearly consistent with 'no regrets' objectives to improve overall agricultural and soil sustainability.

8. Summary

The Marrakech Accords allow biospheric carbon sinks and sources to be included in attempts to meet emission reduction targets for the first commitment period of the Kyoto Protocol. Forest management, cropland management, grazing land management, and re-vegetation are allowable activities under Article 3.4 of the Kyoto Protocol. Soil carbon sinks (and sources) can, therefore, be included under these activities.

Croplands are estimated to be the largest biospheric source of carbon lost to the atmosphere in Europe each year, but the cropland estimate is the most uncertain among all land-use types. It is estimated that European croplands (for Europe as far east as the Urals) lose 300 Tg (C) per year, with the mean figure for the European Union estimated to be 78 Tg (C) per year (with one SD = 37). National estimates for EU countries are of a similar order of magnitude on a per-area basis. There is significant potential within Europe to decrease the flux of carbon to the atmosphere from cropland, and for cropland management to sequester soil carbon, relative to the amount of carbon stored in cropland soils at present.

The biological potential for carbon storage in European (EU 15) cropland is of the order of 90–120 Tg (C) per year, with a range of options available that include reduced and zero tillage, set-aside, perennial crops, deep rooting crops, more efficient use of organic amendments (animal manure, sewage sludge, cereal straw, compost), improved rotations, irrigation, bioenergy crops, extensification, organic farming, and conversion of arable land to grassland or woodland. The sequestration potential, considering only constraints on land use, amounts of raw materials and available land, is up to 45 Tg (C) per year. The realistic potential and the conservative achievable potentials may be considerably lower than the biological potential because of socioeconomic and other constraints, with a realistically achievable potential estimated to be about 20% of the biological potential. As with other carbon sequestration options, potential impacts of non-CO_2 trace gases also need to be factored in.

If carbon sequestration in croplands is to be used in helping to meet emission reduction targets for the first commitment period of the Kyoto Protocol, the changes in soil carbon must be measurable and verifiable. Changes in soil carbon can be difficult to measure over a 5-year commitment period, and this has implications for Kyoto accounting and verification. Currently, most countries can hope to achieve only a low level of verifiability during the first commitment period, whereas those with the best-developed national carbon accounting systems will be able to deliver an intermediate level of verifiability. Very stringent definitions of verifiability would require verification that would be prohibitively expensive for any country.

There is considerable potential in European croplands to reduce carbon fluxes to the atmosphere and to sequester carbon in the soil, but carbon sequestration in soil has a finite potential and is non-permanent. Given that carbon sequestration may also be difficult to measure and verify, soil carbon sequestration is a riskier long-term strategy for climate mitigation than direct reduction of carbon emissions. However, improved agricultural management often has a range of other environmental and economic benefits in addition to climate mitigation potential, and this may make attempts to improve soil carbon storage attractive as part of integrated sustainability policies.

9. Conclusions

There is considerable potential to sequester carbon in European croplands but when assessing the potential, economic, political, and cultural constraints need to be considered and other environmental impacts (such as non-CO_2 GHG emissions) need to be accounted for. Because carbon sequestration acts for a limited duration and is non-permanent, it should not be considered a substitute for emission reduction. It might, however, provide part of a raft of short- to medium-term climate mitigation measures, implemented while other, more permanent, carbon emission reduction technologies (such as non-CO_2 emitting renewable energy sources) are developed.

Acknowledgements

We are grateful to Professor Howard Griffiths and Professor Paul Jarvis, the organizers of the symposium 'Carbon in Forest Biomes' at the Society of Experimental Biology Annual Meeting in Southampton, April 2003, for which this paper was prepared. Rothamsted Research receives grant-aided support from the Biotechnology and Biological Sciences Research Council of the United Kingdom.

References

Cannell, M.G.R. (2003) Carbon sequestration and biomass energy offset: theoretical, potential and achievable capacities globally, in Europe and the UK. *Biomass and Bioenergy* **24:** 97–116.

Dersch, G. and Boehm, K. (1997) Changes in soil carbon in Austrian croplands. In: Blum, W.E.H., Klaghofer, E., Loechl, A. and Ruckenbauer, P. (eds) *Bodenschutz in Österreich*, pp. 411–432. Bundesamt und Forschungszentrum fuer Landwirtschaft, Österreich. [In German.]

Freibauer, A., Rounsevell, M., Smith, P. and Verhagen, A. (2004) Carbon sequestration in European agricultural soils. *Geoderma* **122:** 1–23.

Garten, C.T. and Wullschleger, S.D. (1999) Soil carbon inventories under a bioenergy crop (switchgrass): measurement limitations. *Journal of Environmental Quality* **28:** 1359–1365.

IPCC (1997) *IPCC (Revised 1996) Guidelines for National Greenhouse Gas Inventories. Workbook.* Intergovernmental Panel on Climate Change, Paris.

IPCC (2000) *Land Use, Land-use Change and Forestry.* Watson, R.T., Noble, I.R., Bolin, B., Ravindranath, N.H., Verardo, D.J. and Dokken, D.J. *et al.* (eds). Cambridge University Press, Cambridge, UK. 377 pp.

IPCC (2001) *Climate Change 2001: The Scientific Basis. Contribution of Working Group I to the Third Assessment Report of the Intergovernmental Panel on Climate Change.* Houghton, J.T., Ding, Y., Griggs, D.J., Noguer, M., van der Linden, P.J., Dai, X., Maskell, K. and Johnson, C.A. (eds). Cambridge University Press, Cambridge, UK, and New York, USA. 881 pp.

Janssens, I.A., Freibauer, A., Ciais, P., Smith, P., Nabuurs, G.-J., Folberth, G., *et al.* (2003) Europe's terrestrial biosphere absorbs 7–12% of European anthropogenic CO_2 emissions. *Science* **300** (June 6 2003): 1538–1542.

Jenkinson, D.S. (1988) Soil organic matter and its dynamics. In: Wild, A. (ed) *Russell's Soil Conditions and Plant Growth, 11th Edition*, pp. 564–607. Longman, London.

Johnston, A.E. (1973) The effects of ley and arable cropping systems on the amounts of soil organic matter in the Rothamsted and Woburn ley–arable experiments. In: Rothamsted Report for 1972, Part 2, pp. 131–159.

Lal, R., Kimble, J.M., Follet, R.F. and Cole, C.V. (1998) *The Potential of U.S. Cropland to Sequester Carbon and Mitigate the Greenhouse Effect.* Ann Arbor Press, Chelsea, MI.

Metting, F.B., Smith, J.L. and Amthor, J.S. (1999) Science needs and new technology for soil carbon sequestration. In: Rosenberg, N.J., Izaurralde, R.C. and Malone, E.L. (eds) *Carbon Sequestration in Soils: Science, Monitoring and Beyond*, pp. 1–34. Battelle Press, Columbus, Ohio.

Paustian, K., Andrén, O., Janzen, H.H., Lal, R., Smith, P., Tian, G., Tiessen, H., van Noordwijk, M. and Woomer, P.L. (1997) Agricultural soils as a sink to mitigate CO_2 emissions. *Soil Use & Management* **13:** 229–244.

Sleutel, S., De Neve, S. and Hofman, G. (2003) Estimates of carbon stock changes in Belgian cropland. *Soil Use & Management* **19:** 166–171.

Smith, P. (2004a) Soils as carbon sinks – the global context. *Soil Use & Management* **20:** 212–218.

Smith, P. (2004b) Monitoring and verification of soil carbon changes under Article 3.4 of the Kyoto Protocol. *Soil Use & Management* **20:** 264–270.

Smith, P. and Powlson, D.S. (2003) Sustainability of soil management practices – a global perspective. In: Abbott, L.K. and Murphy, D.V. (eds) *Soil Biological Fertility – A Key To Sustainable Land Use In Agriculture.* Kluwer Academic Publishers, Amsterdam, The Netherlands: 241–254.

Smith, P., Powlson, D.S. and Glendining, M.J. (1996) Establishing a European soil organic matter network (SOMNET). In: Powlson, D.S., Smith, P. and Smith, J.U. (eds) *Evaluation of Soil Organic Matter Models using Existing, Long-Term Datasets,* pp. 81–98. NATO ASI Series I, vol. 38. Springer-Verlag, Berlin.

Smith, P., Powlson, D.S., Glendining, M.J. and Smith, J.U. (1997) Potential for carbon sequestration in European soils: preliminary estimates for five scenarios using results from long-term experiments. *Global Change Biology* 3: 67–79.

Smith, P., Powlson, D.S., Smith, J.U., Falloon, P.D. and Coleman, K. (2000) Meeting Europe's climate change commitments: quantitative estimates of the potential for carbon mitigation by agriculture. *Global Change Biology* 6: 525–539.

Smith, P., Goulding, K.W., Smith, K.A., Powlson, D.S., Smith J.U., Falloon, P.D. and Coleman, K. (2001) Enhancing the carbon sink in European agricultural soils: Including trace gas fluxes in estimates of carbon mitigation potential. *Nutrient Cycling in Agroecosystems* 60: 237–252.

Vleeshouwers, L.M. and Verhagen, A. (2002) Carbon emission and sequestration by agricultural land use: a model study for Europe. *Global Change Biology* 8: 519–530.

4

Estimating forest and other terrestrial carbon fluxes at a national scale: the UK experience

Ronnie Milne and Melvin G.R. Cannell

1. Introduction

The United Nations Framework Convention on Climate Change (UNFCCC) is the basis for individual countries to develop ways of reducing emissions of gases that may contribute to climate change, the so-called 'greenhouse gases' (GHGs). A key element for the UNFCCC is that countries should prepare annual inventories from 1990 onwards of the emission of a range of gases, including CO_2, and should also include the removals of CO_2 by processes in land-use change and forestry. The Intergovernmental Panel on Climate Change (IPCC) has published guidelines (IPCC, 1997a, b, c) on how these inventories can be prepared. From these guidelines it is necessary to report emissions of GHGs in five sectors: (1) Energy; (2) Industrial Processes; (3) Solvents; (4) Agriculture; (5) Land-Use Change and Forestry; and (6) Waste. In sector 5 (Land-Use Change and Forestry), removals of CO_2 from the atmosphere are also to be reported. Within sector 5, the CO_2 data are reported under categories 5A (Changes in Forests and Other Woody Biomass), 5B (Forest and Grassland Conversion), 5C (Abandonment of Managed Lands), 5D (CO_2 Emissions from Soils), and 5E (Other). In the UK, categories 5A and 5D are the most important; 5B (primarily deforestation) has until recently been treated as negligible, and there are no activities relevant to 5C. Here we present a description of the methods that have been used for assessing 5A, changes in forests and other woody biomass, and provide a summary of preliminary work on assessing emissions resulting from the small amount of deforestation that is occurring. We set these carbon flows in the context of emissions and removals of CO_2 in the other categories and sectors (DOE, 1997).

The Carbon Balance of Forest Biomes, edited by H. Griffiths and P.G. Jarvis.
© 2005 Taylor & Francis Group

2. Changes in forests and other woody biomass stocks

2.1 *Methods used for UK national Greenhouse Gas Inventory*

The UK's annual Greenhouse Gas Inventory, including the Land-Use Change and Forestry (LUCF) sector, appears each year as National Inventory Reports (NIRs). The data discussed here covers the period from 1990 to 2001 (Baggott *et al.*, 2003). The methods have been published by Cannell and Milne (1995, 2000), Cannell *et al.* (1999), Milne and Brown (1997), and Milne *et al.* (1998).

The driver for Changes in Forests and Other Woody Biomass has been the expansion in forestry, particularly over the past 50 years. The Forestry Commission (FC) has performed inventories of woodlands in Great Britain at 15–20 year intervals since 1924. The latest forest inventory has been made in two phases: the main section quantifying woodlands of over 2 ha (1 hectare (ha) = 10^4 m^2) completed in 1999 and a survey of woodlands smaller than 2 ha but larger than 0.25 ha completed in 2001 (Forestry Commission, 2002). These forest inventories have not been used directly in estimating the uptake of carbon by UK woodlands or forests but have recently provided information to allow mapping, on a 20 km grid, of removals of atmospheric CO_2 to forests. Future woodland inventories will be performed as part of a 10 year rolling program that started in 2003. Estimates of removals of atmospheric CO_2 to forests in the Greenhouse Gas Inventory are based primarily on annual planting and felling data. The FC also uses these data to update information on the size and age structure of the national forest estate between the periodic inventories. In Northern Ireland equivalent information is reported by the Northern Ireland Department of Agriculture.

The planting information, together with data derived from the growth characteristics of UK forests, is used in a dynamic, carbon-accounting model (C-Flow) (Cannell and Dewar, 1995; Dewar and Cannell, 1992; Milne *et al.*, 1998) to estimate annual uptake and storage of atmospheric carbon by forests. Growth characteristics are kept under review through a national system of mensuration plots and associated yield models. This information will be used to keep the carbon accounting model up to date.

The model C-Flow calculates the net annual change in the mass of carbon in trees, litter, soil and wood products from harvested material in new even-aged plantations that are clearfelled and then replanted at the time of maximum area increment (MAI). Two types of input data and two parameter sets are required for the model (Cannell and Dewar, 1995). The input data are: (i) the areas of new forest planted in each year in the past; and (ii) the stemwood growth rate and harvesting pattern. Parameter values are required to estimate: (i) stemwood, foliage, branch and root masses from the stemwood volume, and (ii) the decomposition rates of litter, soil carbon, and wood products. *Figure 1* shows the structure of the C-Flow model.

For the estimates described here, we used the combined area of new private and state planting from 1921 to 2001 for England, Wales, Scotland, and Northern Ireland, sub-divided into conifers and broadleaves. Restocking was dealt with in the model by including annual changes in carbon in second and subsequent rotations for the afforested areas similarly to the first rotation, but with additional emissions at the time of felling. Hence the areas of forest felled and restocked in any particular year

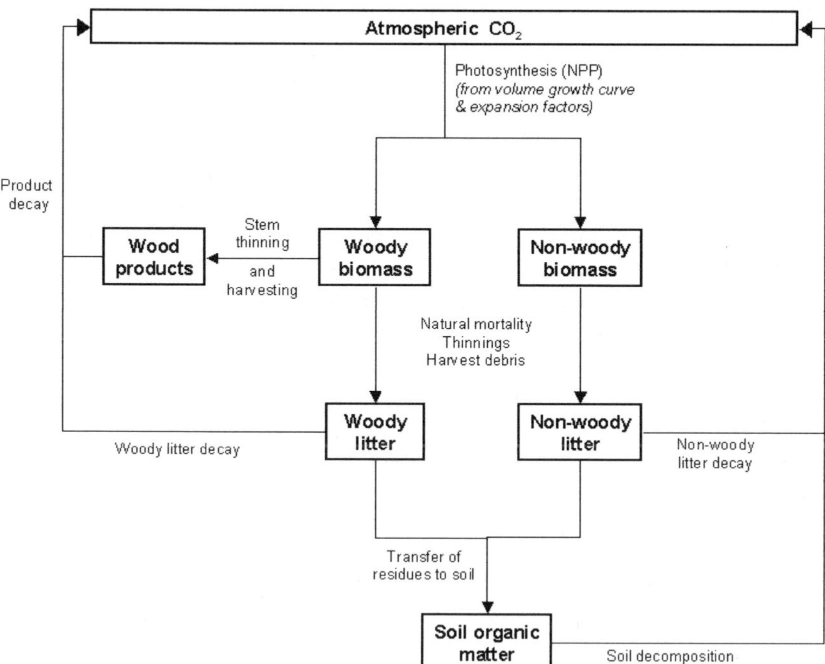

Figure 1. Structure of the C-Flow carbon-accounting model for forests.

were not considered specifically. C-Flow uses FC yield tables (Edwards and Christie, 1981) to describe forest growth in terms of stemwood volume. The volume growth is converted to mass by using basic timber mass/volume relationships from the literature, and the growth of mass in the other tree parts (woody branches, foliage, and roots) are estimated by using expansion factors (that is, multipliers). Non-woody biomass (foliage and roots) is assumed to grow along a logistic curve to achieve an asymptotic value that is provided as a model parameter. The age for felling is set after several years taken from the growth data in the yield tables. The foliage and roots are assumed to have a constant turnover rate and 50% of the decayed material is assumed to become soil organic matter, which in its turn decays at a given rate. Carbon is assumed to be 50% of organic mass in the tree, litter, and soil pools. This is a common assumption for carbon fraction but lies at the upper end of the reported range for plant material (44–50%).

It was assumed that all new conifer plantations have the same growth characteristics as Sitka spruce (*Picea sitchensis* (Bong.) Carr.), with an intermediate thinning management regime. This species accounts for about half the area of coniferous forests in the UK. Intermediate thinning consists of removal of suppressed and subdominant trees at approximately 5-year intervals. Milne *et al.* (1998) have shown that the mean yield class for Sitka spruce varies across Great Britain from 10 to 16 m^3 ha^{-1} per year but that this variation has a less than 10% effect on the estimated removal of CO$_2$ from the atmosphere. The Greenhouse Gas Inventory data have, therefore, been estimated on the assumption that all new coniferous plantations follow the growth pattern of Sitka spruce of yield class 12 m^3 ha^{-1} per year in Great Britain, and yield class 14 m^3 ha^{-1} per year in Northern Ireland.

Data in the most recent inventory of British woodlands (Forestry Commission, 2002) show that the range for broadleaf species in UK forests is larger than the range in conifer species. Beech (*Fagus sylvatica* L.) of yield class 6 m^3 ha^{-1} per year was selected as having characteristics that lie between the long-lived, slow-growing species such as oak, and the fast-growing species such as poplar. Sensitivity analysis of the carbon-accounting model shows that different assumptions on the species of broadleaf planted has little effect on the overall carbon uptake (Milne *et al.*, 1998). Using oak or the sycamore–ash–birch group yield class data instead of those of beech is likely to have an effect of less than 10% on the estimate of removal of carbon from the atmosphere to UK forests.

Irrespective of the assumptions on representative species, variation in atmospheric CO_2 removals from 1990 to the present is determined by the afforestation rate in earlier decades and the effect that this has on age structure in the present forest estate and hence on the average growth rate. Planting rates for conifers and broadleaves in each of the devolved regional areas are presented in *Figure 2*. *Table 1* shows a summary of these data and the present age structure of UK forests. In addition to these planted forests there are about 850 000 ha of woodland that were planted before 1922, or are not of commercial importance. This area is assumed to be at equilibrium and carbon neutral and is therefore not included in the Greenhouse Gas inventory.

Variation from year to year in the reported removals to woody biomass, soils, and harvested products reflect the changing pattern of afforestation over the period of available data. For example, there are increases in removals to harvested products about 50 years after a period of increased planting of conifers, because that is the approximate length of the conifer plantation rotation cycle. It can be shown that if forest expansion continues at the present rate then removals of atmospheric carbon

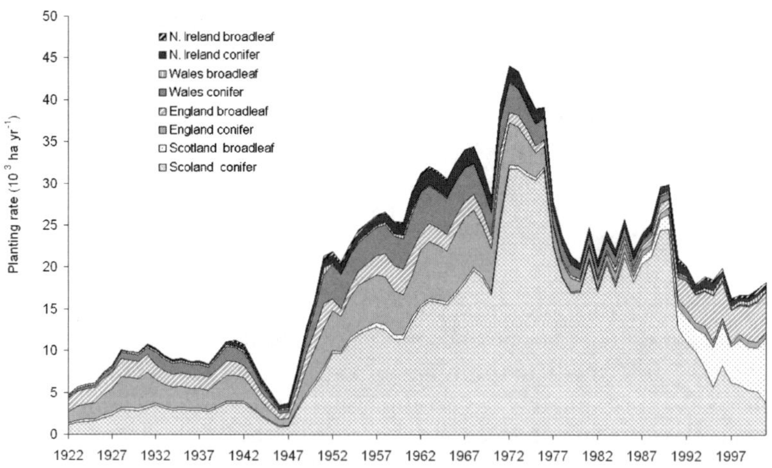

Figure 2. *Annual afforestation rate of conifer and broadleaf forest in the four devolved regions of the UK.*

Table 1. *Afforestation rate and age distribution of conifers and broadleaves in the UK. Data are from Forestry Commission and Northern Ireland Forest Service annual reports as summarized by Milne et al. (2003).*

	Planting rate (10^3 ha per year)		Present age distribution (%)	
	Conifers	Broadleaves	Conifers	Broadleaves
1922–1929	4.9	2.4	2.9	7
1930–1939	7.2	2.2	5.4	9
1940–1949	6.3	1.9	4.7	7
1950–1959	20.0	3.0	14.9	12
1960–1969	28.4	2.9	21.1	12
1970–1979	33.2	1.5	24.8	6
1980–1989	22.5	1.4	16.8	5
1990	26.8	3.1	2.0	1
1991	15.4	5.8	1.1	2
1992	13.4	6.8	1.0	3
1993	11.6	6.5	0.9	3
1994	10.1	8.9	0.8	3
1995	7.4	11.2	0.6	4
1996	9.5	10.5	0.7	4
1997	7.4	8.9	0.6	4
1998	7.0	9.7	0.5	4
1999	6.6	10.1	0.5	4
2000	6.5	10.9	0.5	4
2001	4.9	13.4	0.4	5

will continue to increase until about 2005 and then will begin to decrease, reflecting the reduction in rate of afforestation after the 1970s.

Litter transfer rates from foliage and fine roots to the soil increase to a maximum at canopy closure. Tree species and yield class, but not other factors that vary with location, are assumed to control the decay of litter and soil organic matter. Additional litter is generated at times of thinning and felling. The values of the main parameters in C-Flow for conifers and broadleaves are given in *Table 2.*

In estimating removals of atmospheric carbon to conifer afforestation, Cannell and Dewar (1995) assumed that a net increase in the amount of carbon in soil did not occur, although decaying plant mass may lead to increases in soil organic detritus. The reasoning behind this assumption was that gains in soil organic carbon from the new forests would be balanced by losses resulting from soil disturbance at the time of planting, because most conifer afforestation in the UK is on soils with large organic carbon content. Until recently, little has been known about carbon losses from such soils when ploughed for forest planting. It was generally believed that the losses would be large and continue for very long periods (many decades), especially for forests planted on peat (Cannell *et al.*, 1993). So, although the newly planted forest would add new organic carbon to the existing soil, carbon would also continue to be lost from the existing soil. This approach continues to be used in UK Greenhouse Gas Inventory reporting for most conifer plantations, but with a con-

Table 2. *Main parameters for the forest carbon flow model for species used to estimate carbon uptake by planting of forests of Sitka spruce* (Picea sitchensis) *and beech* (Fagus sylvatica) *in the UK (data from Dewar and Cannell, 1992). YC12 is yield class based on a mean annual increment of 12 m³ ha⁻¹ per year, etc.*

	P. sitchensis	*P. sitchensis*	*F. sylvatica*
	YC12	YC14	YC6
Rotation (years)	59	57	92
Initial spacing (m)	2	2	1.2
Year of first thinning	25	23	30
Stemwood density ($Mg\ m^{-3}$)	0.36	0.35	0.55
Max. carbon in foliage ($Mg\ ha^{-1}$)	5.4	6.3	1.8
Max. carbon in fine roots ($Mg\ ha^{-1}$)	2.7	2.7	2.7
Fraction of wood in branches	0.09	0.09	0.18
Fraction of wood in woody roots	0.19	0.19	0.16
Max. foliage litterfall ($Mg\ ha^{-1}$ per year)	1.1	1.3	2
Max. fine root litter loss ($Mg\ ha^{-1}$ per year)	2.7	2.7	2.7
Dead foliage decay rate ($Mg\ ha^{-1}$ per year)	1	1	3
Dead wood decay rate ($Mg\ ha^{-1}$ per year)	0.06	0.06	0.04
Dead fine root decay rate ($Mg\ ha^{-1}$ per year)	1.5	1.5	1.5
Soil organic carbon decay rate ($Mg\ ha^{-1}$ per year)	0.03	0.03	0.03
Fraction of litter lost to soil organic matter	0.5	0.5	0.5
Lifetime of wood products (year)	57	59	92

tinuing net loss assumed in the case of planting on deep peats. Broadleaved forests are assumed to increase the net amount of carbon in litter and soil and, because normally planted on mineral soils, the emissions from pre-existing soils are assumed to be negligible.

Methane emissions from, or removals to, soils are not required for the Land-Use Change and Forestry sector of the Greenhouse Gas Inventory. The data for removals and emissions of carbon (in CO_2) as estimated by the above procedures and reported in the Changes in Forests and Other Woody Biomass sector of the UK NIR for 2001 (Baggott *et al.*, 2003) are summarized in *Table 3*.

The NIR data show that forest biomass was removing up to 1.563 Tg (C) per year in 1990, with an additional amount of 0.586 Tg (C) per year going to soils and 0.429 Tg (C) per year to wood products, giving total removals from the atmosphere of 2.578 Tg (C) per year. These values have tended to increase since 1990, following the pattern of earlier increases in planting rate. However, in the NIR the afforestation of deep peat is reported as losing continuously 0.400 Tg (C) per year.

The C-Flow model can be used to estimate future removals by making assumptions on the pattern of continued afforestation. If afforestation in the UK continues at 7 000 ha per year for conifers and 10 000 ha per year for broadleaves (approximately the rate in year 2000), it can be shown that removals will continue to rise to about 3.200 Tg (C) per year in 2005, and then begin to fall.

In the context of the Kyoto Protocol, only carbon accumulating in forests planted from 1990 onwards can be used to offset emissions. On the assumption of continued

Table 3. *Removals from the atmosphere resulting from build up of carbon in new trees, litter, soil organic matter, and wood products. Emissions from deep peat are from the peat in existence before forest establishment. Revised columns refer to changes as a result of improved understanding described in section 2.1. Removals from the atmosphere are negative, emissions to atmosphere are positive.*

Tg (C) per year	Inventory removal to forest biomass (incl. roots)	Revised removal to forest biomass (incl. roots)	Inventory removal to forest litter and soils	Revised removal to forest litter and soils	Inventory removal to stock of harvested wood	Inventory emission from drainage of deep peats	Revised emission from drainage of deep peats
1990	−1.563	−1.631	−0.586	−1.829	−0.429	0.400	0.036
1991	−1.587	−1.649	−0.592	−1.849	−0.43	0.400	0.034
1992	−1.724	−1.770	−0.567	−1.834	−0.367	0.400	0.023
1993	−1.872	−1.902	−0.543	−1.819	−0.306	0.400	0.013
1994	−1.935	−1.950	−0.545	−1.829	−0.299	0.400	0.005
1995	−2.075	−2.075	−0.521	−1.808	−0.249	0.400	0.001
1996	−1.984	−1.964	−0.569	−1.859	−0.314	0.400	0.002
1997	−1.965	−1.943	−0.584	−1.875	−0.303	0.400	0.008
1998	−1.904	−1.889	−0.613	−1.906	−0.326	0.400	0.013
1999	−1.862	−1.858	−0.632	−1.925	−0.353	0.400	0.019
2000	−1.951	−1.961	−0.611	−1.898	−0.314	0.400	0.025
2001	−1.867	−1.879	−0.647	−1.925	−0.354	0.400	0.030

afforestation at the present rate, projections from the C-Flow model predict *net* removals of 0.600 Tg (C) per year for the 2008–2012 Kyoto commitment period.

2.2 Fluxes associated with afforestation of deep peats

In the past, UK NIRs have reported CO_2 emissions resulting from drainage of deep peats for afforestation, as referred to in section 2.1. These emissions were based on the areas of forest planted as given by Cannell *et al.* (1993) and emission rates measured in the field one year after planting (Hargreaves and Fowler, 1997; Hargreaves *et al.*, 2003) (*Table 4*). The rate of emission is assumed to continue indefinitely at this early rate (2 Mg (C) ha^{-1} per year), so that the reported emission for the UK has, therefore, been the same for each year of the inventory (i.e., 0.400 Tg (C) per year).

Analysis of measurements taken at an undisturbed deep peat moorland, at similar locations afforested from 1 to 9 years previously and at a 26-year-old conifer forest have recently been completed (Hargreaves *et al.*, 2003). These suggest that long-term losses from afforested deep peat are not as large as had been previously thought, settling to about 0.3 Mg (C) ha^{-1} per year at 30 years after planting. In addition, a short burst of regrowth of moorland plant species occurs before forest canopy closure. The pattern of carbon loss and gain from afforested deep peat moorland is summarised in *Figure 3* and *Table 5*. These new data on fluxes of CO_2 from deep peats after afforestation can be used

Table 4. *Activity and emission factors for deep peat drainage used in published National Inventory reports.*

	Afforested deep peat area (10^3 ha)	Emission rate (Mg (C) ha^{-1} per year)	Annual loss (Gg (C))
England	20	2	40
Wales	10	2	20
Scotland	160	2	320
Northern Ireland	10	2	20
UK	200	2	400

to revise the previously reported Greenhouse Gas Inventory data. Additionally, better estimates of the areas of peat afforested in the decades since 1920 are now available.

The Woodlands Surveys Branch of the FC has provided forest age data for each 20 km × 20 km square in Great Britain (S. Smith, personal communication). These data have been prepared specifically for Greenhouse Gas Inventory purposes from the National Inventory of Woodlands and Trees (NIWT) (Forestry Commission, 2002) using information from 1 ha sample sites. These sample sites fall within 20 km squares in Great Britain and have been summarized to produce tables of areas per species by planting-year class (S. Smith, personal communication). The sample sites are allocated to the 20 km squares by their British National Grid reference. The species reported are those present within each 20 km square using the species

Table 5. *Emissions of carbon from deep peat as a result of ploughing for afforestation based on field measurements (Hargreaves et al., 2003). Negative values indicate uptake of carbon from the atmosphere. Here this results from temporary regrowth of moorland plants between ploughing and forest canopy closure.*

Years after afforestation	Carbon loss (Mg (C) ha^{-1} per year)
1	2.2
2	3.8
3	2.5
4	1.1
5	−0.3
6	−1.2
7	−1.6
8	−1.6
9	−1.3
10	−1.1
15	−0.2
20	0.1
25	0.2
30	0.3

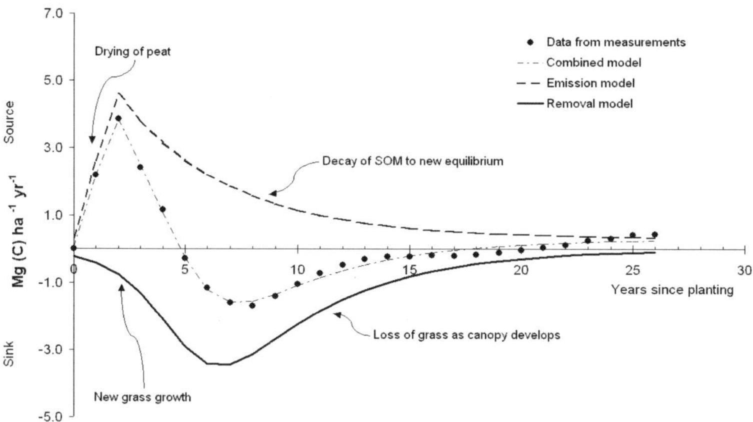

Figure 3. Pattern of carbon loss and gain after conifer afforestation of deep peat excluding the accumulation of carbon in the biomass of the forest and the soil organic matter that is generated. The 'data from measurements' (Hargreaves et al., 2003) are the estimates from combining field measurements of net CO_2 exchange over forests of different ages with estimates from the C-Flow model of the amount of carbon accumulating in the forest. The 'emission model' describes losses of carbon from the pre-existing peat. The 'removal model' describes the uptake of carbon by regrowth and then death of grasses between ploughing and canopy closure. The 'combined model' is the sum of the 'emission' and 'removal' models and has been fitted to the 'data from measurements'.

groupings in the National Inventory reports. Some 20 km squares may contain woodland but may not contain any sample sites. In addition, the total mapped forest area in the square and the area not stocked with trees (e.g., felled) are given. The data take the format shown in *Table 6*, where each entry is the area in hectares for that planting date and species in the square. These areas are estimated from data from all 1 ha sample sites in the 20 km square scaled to the known forest area in that square.

A soil map of Great Britain (1 km grid) was overlaid onto the set of 20 km squares and the area of organic soils planted with conifers in each time period (usually a decade) in each square was extracted. The fractions of all conifer planting on organic soils for each decade since 1920 for England, Wales, and Scotland were then calculated. The locations of the forests within each 20 km square were not available, because the areas were derived from the 1 ha sample sites, so the simplifying assumption was made that woodland was uniformly spread across the square. Patterns of afforestation of organic soils in England, Wales, and Scotland were found to be different (*Table 7*) but reflect known changes in afforestation policy. In each country the fraction of planting on organic soils has increased over the decades, reflecting the expansion of forests to higher elevations. In Scotland the fraction of conifer planting on organic soils has begun to come down again because of the new policies favouring afforestation of land previously in agriculture and reduction in planting on land identified as being of conservation interest.

Table 6. *Example of data on forest age for one 20 km square of the National Inventory of Woodland and Trees. The last class, 1991, extends to the planting year of the most recently planted trees in the 20 km square at the time of survey.*

| Area (ha) | Assumed age in 1995 | 195 | 115 | 90 | 80 | 70 | 60 | 50 | 40 | 30 | 20 | 10 | 2 |
Species	Not stocked	Pre-1860	1860–1900	1901–1910	1911–1920	1921–1930	1931–1940	1941–1950	1951–1960	1961–1970	1971–1980	1981–1990	1991–
Scots pine	0	0	0	0	0	88	216	4	0	44	0	0	
Corsican pine	0	0	0	0	0	62	0	0	0	0	0	0	
Norway spruce	0	0	0	0	0	0	0	0	0	6	0	0	
European larch	0	0	0	0	5	108	13	4	0	0	0	0	
Oaks	0	93	0	32	6	6	18	0	0	0	0	0	
Beech	0	24	0	6	0	0	0	0	0	0	0	0	
Sycamore	0	0	0	9	67	0	186	0	23	10	0	0	
Ash	0	23	0	6	32	0	0	0	5	0	0	0	
Birch	0	0	0	5	4	291	42	5	126	0	0	0	
Poplars	0	0	0	0	0	0	0	0	0	5	0	0	
Sweet chestnut	0	0	0	0	0	0	25	0	70	44	0	0	
Elm	0	0	0	0	0	0	0	0	69	18	0	0	
Other broadleaves	0	32	0	77	33	6	55	0	51	66	0	0	
Mixed broadleaves	0	0	0	0	5	0	6	0	0	5	5	0	
Total mapped woodland area	316	0	171	0	136	152	561	561	13	343	199	5	0

Table 7. *Conifer afforestation of organic soils in Great Britain derived from National Inventory of Woodland and Trees and soil map data.*

	Percentage of conifer planting on organic soils		
Period	England	Scotland	Wales
1921–1930	2	21	5
1931–1940	2	20	4
1941–1950	3	19	4
1951–1960	3	22	4
1961–1970	4	26	5
1971–1980	5	28	6
1981–1990	5	31	7
1991–survey	5	27	4

The C-Flow model was run separately for (a) broadleaved planting, (b) conifer planting on organic soils based on the percentages in *Table 7*, and (c) all other conifer planting. In each case, soil carbon accumulated as the forest grew. Plantations on non-organic soil were assumed to have insignificant loss of carbon from the pre-existing soil but for plantations on organic soil the models in *Figure 3* described losses from the pre-existing peat. The volume growth curves used have also been revised. A curve following an initial exponential growth phase before taking up that given by the yield table has been used (Hargreaves *et al.*, 2003). The effect of the combination of these two revisions on national removals and emissions totals is shown in *Table 3*. The data show a significant increase in the estimated values of removals of atmospheric CO_2 to forest and woody biomass. For example, the estimate of net removal in 1990 increases from 2 178 Mg (C) ha^{-1} per year to 3 853 Mg (C) ha^{-1} per year. It is planned that these revisions will be introduced into the National Inventory Report that will be published in 2005, to cover the period from 1990 to 2003.

2.3 Wood products

It is assumed in the C-Flow carbon-accounting model that harvested material from thinning and felling is made into wood products. These products are then assumed to decay over a period equal to the rotation length of the forest, conifer or broadleaf as appropriate, because products from broadleaves (e.g., furniture) will decay more slowly than those from conifers (e.g., paper, softwood timber). The net change in the carbon content of this pool of wood products is reported as part of the Changes in Forestry and Other Woody Biomass in the UK National Inventory reports. Calculated in this way, that part of the total pool of wood products from UK forests is currently increasing as a result of the continuing expansion in forest area.

2.4 Mapping of UK forest carbon uptake

The national UK data for emissions and removals resulting from land-use change and forestry are based on calculations made for each of the four countries of the UK

(England, Wales, Scotland, and Northern Ireland). The data for the individual countries are not reported to the UNFCCC but are published in other reports (Baggott *et al.*, 2004). However, there is broad interest in preparing estimates of removals and emissions at a more detailed scale. To address this interest, the coding of C-Flow has now been altered so that it can be run for any chosen location in the UK where a time series of forest planting data is available.

The data from the NIWT described above have been used to generate time series of conifer and broadleaf planting for each 20 km grid square. Each age class was assigned specific planting years on the assumption that the data were recorded in 1995. Then the recorded planting area was assumed to have occurred equally across the time period (usually a decade) and 1995 was assumed to be the year of most recent planting (*Table 6*). A time series for each grid square stretching from 1920 to 2000 was estimated, as this period is similar to that used in the regional and national applications of C-Flow. The model was run separately for conifers and broadleaves for each square using these time series. The uptake of CO_2 by the forest in each square, and nationally, was then extracted for 1990. In this study the revisions to the model for afforestation of organic soils were not included.

Table 8 presents the 'national run' results for 1990 for Great Britain, and *Table 9* shows the sum for Great Britain of the outputs from 'the C-Flow run for each 20 km square'. It can be seen that there is general agreement between the two methods. The forest areas appropriate to each calculation are also shown and go some way to explaining the differences between the results of running C-Flow by the two methods. The two agree well for conifer woodlands but the national scale run of C-Flow (*Table 8*) included much less broadleaf forest, even though the 20 km scale woodland data were limited to planting over a similar time span starting in 1920. It is likely, therefore, that the NIWT includes many broadleaved woodlands that fall outside those reported in FC planting statistics, on which the national C-Flow runs are based. These woodlands may be smaller in size or have been planted without grant aid. Amenity and other small private woodlands would appear to fit this description but further investigation is required. This area of woodland, together with woodlands planted before 1920, have, in principle, been taken to be 'in equilibrium' in compilation of data for the UK National Greenhouse Gas Inventory Reports, i.e., with no net removal or loss of carbon. The main purpose of the study reported here is to map these national data but although the use of C-Flow at the 20 km scale appears promising, it is not yet possible to make calculations for just those broadleaf woodlands in the national-scale planting data. The total difference in area can be calculated but the geographical distributions for different-aged woodlands are not known. *Figure 4a* maps the 20 km square data of the CO_2 fluxes from C-Flow for conifer afforestation, and *Figure 4b* shows the equivalent distribution of broadleaf data.

The distribution of forest CO_2 removals shown in *Figure 4* is controlled by two factors: the location of forests and when they were planted. Another factor that may be of importance is the variation in growth rate across the country, i.e., the variation of yield class. Such data on yield class are not readily available, but work is in progress based on the use of geo-climatic land classes overlaid onto the Forest Enterprise Sub-Compartment database. This database holds information on yield class, species, etc. of each patch of forest in the state sector. From this the average yield class of Sitka spruce and beech for each land class has been estimated (S. Smith, personal

Table 8. *Results from national scale run of C-Flow showing total carbon uptake for forests in Great Britain (i.e., does not include Northern Ireland).*

Great Britain 1990

(Tg (C) per year)	Trees	Products	Litter	Soil	Total uptake	Area (10^3 ha)
Conifer	1.24	0.41	0.38	—	2.03	1 185
Broadleaf	0.21	0.01	0.04	0.14	0.40	144
All	1.45	0.41	0.41	0.14	2.42	1 328

Table 9. *Results from 20 km scale run of C-Flow showing total carbon uptake for forests in Great Britain (i.e., does not include Northern Ireland).*

Sum of 20 km squares in Great Britain 1990

(Tg (C) per year)	Trees	Products	Litter	Soil	Total uptake	Area (10^3 ha)
Conifer	1.98	0.19	0.32	—	2.49	1 268
Broadleaf	0.90	0.00	0.14	0.57	1.62	578
All	2.88	0.19	0.46	0.57	4.10	1 846

communication). These yield classes can be assumed to apply also to privately owned forests. Using the land classes present in each 20 km square across Great Britain, the local mean yield class for Sitka spruce and beech can be estimated. Sitka spruce yield class varied downwards from 16 m³ ha⁻¹ per year to very low values in locations with poor growing conditions. Beech yield class was 6 m³ ha⁻¹ per year in the south of England and Wales, the value assumed for national calculations, dropping to 4 m³ ha⁻¹ per year in the northern areas of Wales and England and in Scotland. C-Flow was run for each 20 km square with these estimates of different yield class in addition to the variations in age and coverage. The resulting national and grid data showed only small differences from the estimates obtained when yield class was assumed to be constant across the country. Milne *et al.* (1998) found a similar result by running C-Flow at the national scale and making different choices for assumed, fixed yield class.

3. Forest and grassland conversion (deforestation)

In National Inventory reports, it has been assumed that permanent conversion of forest to non-forest in the UK has been negligible. This assumption was based on stringent government guidelines against deforestation, including the need for approval for any permanent forest felling from the FC, or its equivalent in Northern Ireland. A review of this assumption suggests that some deforestation is happening where urban development is encroaching on old woodlands (Levy and Milne, 2004). This situation is covered by a different set of guidelines and, because of the need for new housing,

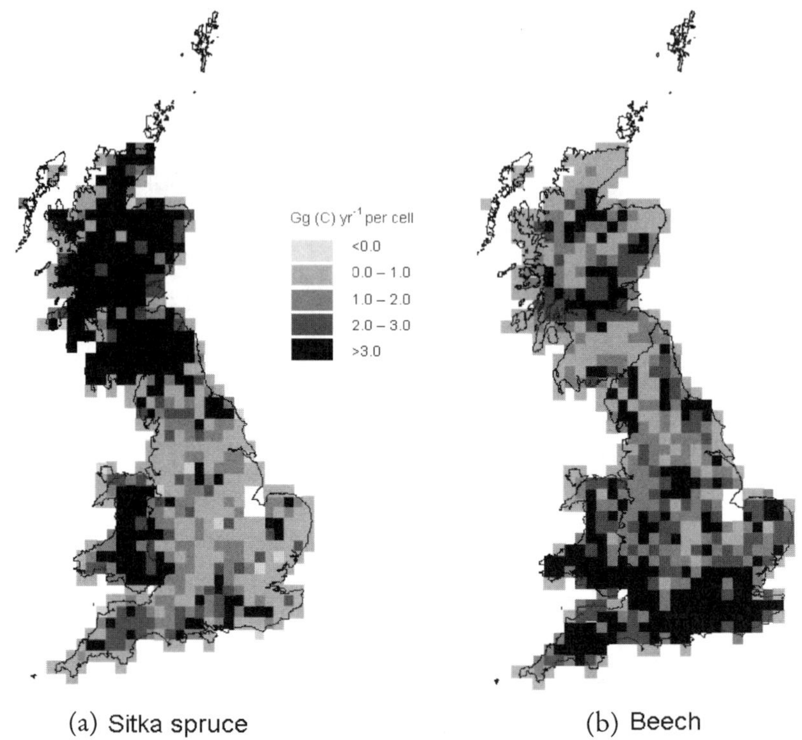

Figure 4. Net uptake of carbon by forest in 20 km square cells in Great Britain (i.e., excluding Northern Ireland) as estimated by the C-Flow model.

permission for felling is more readily obtained. Only local planning authorities hold documentation for agreed felling and this makes estimation of the national total difficult. However, in England the Ordnance Survey (national mapping agency) makes an annual assessment of land-use change from data it collects for map updating (Office of The Deputy Prime Minister, 2003). These data suggest that over the period from 1985 to 1999 about 500 ha per year of woodland were converted to urban use, mostly for housing and outdoor recreation. Scaling by urban areas in Scotland, Wales, and Northern Ireland suggests that an additional 100 ha per year were converted in those regions. *Table 10* shows the range of GHG emissions that would result from deforestation in the range of 500–1000 ha per year in the UK. Appropriate data will therefore be included in future NIRs, on completion of this data analysis.

4. Other removals and emissions of CO_2 within the land-use change and forestry sector

CO_2 removals and emissions also occur as a result of other activities. In this sector of the Greenhouse Gas Inventory, emissions and removals of CO_2 resulting from activities *other than* change in forest biomass have been estimated.

Table 10. *Range of possible emissions resulting from deforestation in the UK. Method based on IPCC 1996 Guidelines (IPCC 1997a, b, c).*

		Low	Mid	High	Units
Carbon in standing forest		120[1]	120[1]	120[1]	Mg (C) ha^{-1}
Fraction in products		0.6[2]	0.6[2]	0.6[2]	
Fraction burned on site		0.4	0.4	0.4	
Area cleared		500	750	1000	ha per year
Fraction oxidized	(1996 Guidelines)	0.9	0.9	0.9	
Immediate emissions of carbon		21.6	32.4	43.2	Gg (C) per year
Immediate emissions of nitrogen		0.2	0.3	0.4	Gg (N) per year
Non-CO$_2$ emissions	Factor (1996 Guidelines)				
CH$_4$	0.012	0.3	0.4	0.5	Gg (C) per year
CO	0.06	1.3	1.9	2.6	Gg (C) per year
N$_2$O	0.007	0.001	0.002	0.003	Gg (N) per year
NO$_x$	0.121	0.026	0.039	0.052	Gg (N) per year

[1]Typical for broadleaf woodland.
[2]Based on data used in estimating CO$_2$ removals to forest biomass in C-Flow.

4.1 *Changes in soil carbon resulting from non-forest land-use change*

The method for assessing changes in soil carbon resulting from land-use change is to use a matrix of change of area from surveys of land linked to a dynamic model of gain or loss of carbon. Matrices from the DETR/ITE Countryside Surveys of 1984 and 1990 are used. Land is placed into four broad groups – natural, farming, woodland and urban – and the matrices detail the areas of change between pairs of these groups over the observation period. A database of soil carbon density for the UK has been constructed (Cruickshank *et al.*, 1998; Milne and Brown, 1997) from information provided by the Soil Survey and Land Research Centre, the Macaulay Land Use Research Institute, and Queen's University Belfast on soil type, land cover, and carbon content of soil cores. The effects of the changes in land use on soil carbon are estimated for each possible transition from the following relationships:

$$C_t = C_f - (C_f - C_0)e^{-kt}$$

where C_t is carbon density at time t, C_0 is carbon density initially, C_f is carbon density after change to new land use, and k is the time constant of change. For example, if the inventory year is 1990 and A_T is the area in a particular land-use transition in any year, T, then the total carbon lost or gained from 1930 to 1990 (X_{1990}) and from 1930 to 1989 (X_{1989}) is given by

$$X_{1990} = \sum_{T=1930}^{T=1990} A_T (C_0 - C_f)(1 - e^{-k(1990-T)}), \text{ and}$$

$$X_{1989} = \sum_{T=1930}^{T=1989} A_T\,(C_0 - C_f)(1 - e^{-k(1989-T)}).$$

Hence the flux of carbon in 1990 is given by the difference:

$$F_{1990} = X_{1990} - X_{1989}.$$

The 1 km scale soil carbon database (Milne and Brown, 1997) was used for the calculations. This showed that carbon densities were much larger on average in Scotland than in England and Wales because of the preponderance of organic soils there. Thus the calculations for change in soil carbon were applied separately to England, Wales, and Scotland. There are few published data on rates of change of soil carbon resulting from land-use change, so a Monte-Carlo method was adopted to explore the range of possible overall changes in soil carbon for a specified range of time constants. Different time constants for the rates of change in carbon were used for losses and gains in soil carbon. For each region losses were assumed to occur at a rate causing 99% of initial soil carbon to be lost after between 50 and 150 years. For England and Wales, soil was assumed to increase, after an appropriate land-use change, at a rate such that 99% of the final value would be reached between 100 and 300 years. To reflect the longer time required to build up the larger amount of soil carbon that occurs in Scotland, the range for Scotland was chosen to lie between 300 and 750 years. A simpler method, described in the IPCC 1996 Guidelines (IPCC, 1997a, b, c), using 20 years for all soil carbon changes to be completed, was used for Northern Ireland.

Calculations were also adjusted to take into account areas specifically being set aside from crop production. This was necessary because this activity did not start until after the land-use surveys providing the matrices. The mean net change in soil carbon for the UK in 1990 across the range of rates of change was estimated to be a loss of 3.821 Tg (C) per year with a range of about ± 4.000 Tg (C) per year. Thus the loss of carbon from soils may be large but it is very uncertain, and may even be a small gain.

4.2 *Emissions of CO_2 resulting from liming agricultural soils*

Emissions of CO_2 as a result of the application of limestone, chalk, and dolomite to agricultural soils are estimated. Data on the application of limestone, chalk and dolomite for agricultural purposes are reported by the British Geological Survey (BGS, 2002). It is assumed that all the carbon contained in the lime is released in the year of use. Emission factors relating emitted CO_2 to applied lime are taken from the IPCC Guidelines (IPCC, 1997a, b, c). For 1990 the loss was estimated to be 0.390 Tg (C) per year, but less lime is now being applied, probably because of the generally poor economic situation in the Agriculture sector.

4.3 *Emissions of CO_2 resulting from drainage of fenlands*

The trend in emissions as a result of changing areas of drainage is based on the work of Bradley (1997). The area of drained lowland wetlands for the UK was taken to be 150 000 ha in 1990. This represents all of the East Anglian fen and Skirtland, and limited areas in the rest of England. This total consists of 24 000 ha of land with thick peat (more than 1 m deep) and the rest with thinner peats. Different loss rates following 1990 were assumed for these two areas. The trend in emissions after 1990 was estimated on the assumptions that no more area has been drained since then but that the existing areas have continued to lose carbon. The annual loss of carbon decreases for a particular location in proportion to the amount of carbon remaining. But, in addition to this, as the peat loses carbon it becomes more mineral in structure. The estimated value for this process was 0.450 Tg (C) per year in 1990 and the rate has been falling slowly since then.

4.4 *Emissions of CO_2 from peat extracted for use as fuel and in horticulture*

Trends in peat extraction in Scotland and England over the period 1990–2000 are estimated from activity data taken from the UK Minerals Handbook (BGS, 2002). In Northern Ireland no recent data on use of peat for horticultural use are available, and a recent survey of extraction of peat for fuel use suggested that there is no significant trend. The contribution to emissions arising from peat extraction in this region is therefore incorporated as constant from 1990 to 2001. Peat extraction is negligible in Wales. Emissions factors have been taken from Cruickshank and Tomlinson (1997). For the UK the emission in 1990 from this activity was 0.216 Tg (C) per year. Variation since 1990 has been about ±0.040 Tg (C) per year.

4.5 *Changes in crop biomass resulting from agricultural management*

Adger and Subak (1996) estimated recent changes in carbon storage in biomass on non-forest lands in the UK, including land used for agriculture, horticulture and urbanization. The land area converted to forest was specifically excluded to avoid overlap with estimates for the change in forest biomass. They used agricultural census statistics for the period 1988–1992 published by the Ministry of Agriculture, Fisheries and Food. These statistics are strongly correlated with agricultural land cover data in 1984 and 1990 UK Countryside Surveys, which were used to calculate changes in soil carbon on non-forest lands, so the two estimates are considered to be compatible. The calculations suggest that about 0.300 Tg (C) per year are removed from the atmosphere each year as a result of agriculture. No variation on this value is reported.

5. Emissions of CO_2 in other sectors and comparison with the Land-Use Change and Forestry sector

CO_2 emissions also occur as a result of other activities, particularly the major emissions resulting from energy use, industry and transport, etc. In this section, emissions in other Sectors taken from the National Inventory Report are summarized in *Table 11* and compared with the removals and emissions from the Land-Use Change and Forestry sector.

Table 11 shows the emissions of CO_2 for key activities within sectors other than Land-Use Change and Forestry. The largest emissions are from fuel combustion (~150 Tg (C) per year in total), with fugitive emissions from fuels (~3 Tg (C) per year) and industrial processes (~4 Tg (C) per year) next in importance; additionally, there is a small emission from incineration of waste (~0.2 Tg (C) per year). The removals and emissions of CO_2 in the Land-Use Change and Forestry sector are therefore small compared with the overall emissions. Their importance is that the portion coming from afforestation after 1990 can help to achieve Kyoto Protocol emission reductions. With continuing afforestation and less intensive agricultural methods, the net emissions in this sector of the National Inventory will tend to become smaller, helping to meet domestic targets for reductions.

Table 11. *Emissions (positive) and removals (negative) of atmospheric carbon in the UK in 1990 (from the National Inventory report, 2002).*

CO_2 fluxes (expressed as mass of carbon)	Tg (C) per year
Energy	
Fuel combustion	
Energy Industries	62.206
Manufacturing industries	25.673
Transport	31.795
Other fuels	32.128
Fugitive emission from fuels	3.310
Industrial processes	3.858
Land-use Change and Forestry	
Changes in total forest biomass[1]	−2.579
Soils	
Land-use change[2]	3.821
Liming of agricultural soils	0.390
Upland drainage	0.400
Lowland drainage	0.450
Other	
Peat extraction	0.216
Changes in crop biomass	−0.300
Waste	0.221
	UK total:
	emission = 164.469
	removal = −2.879

[1]Forest biomass, litter, soils, and products.
[2]Non-forest land-use change and set-aside.

6. Summary

For the UK the best understood terrestrial carbon flux is that associated with the expansion of forest area over the past 80 years. Modelling methods are described that have been used to estimate this uptake of carbon for the Land-Use Change and Forestry sector in the Greenhouse Gas Inventory of the UNFCCC. Measurements of losses of carbon from afforested organic soils are also presented and the consequent effect on the net national carbon sink for forests is discussed. These calculations show that the expansion of forests is currently causing about 2.8–4.0 Tg (C) per year to be removed from the atmosphere, the uncertainty depending on assessment of the situation for plantations on organic soils. Within these totals about 1.9 Tg (C) per year is being added to living trees, about 0.4 Tg (C) per year to the stock of wood products and the rest to soils. A preliminary estimate for emissions of greenhouse gases resulting from removal of forests for residential development is also presented. Methods to map the variation in the net flux across Great Britain are discussed. The methods used, and the values estimated, for other UK terrestrial carbon fluxes are summarized. A comparison with emissions of CO_2 from energy generation, etc. is made.

Acknowledgements

The Department for Environment, Food and Rural Affairs (DEFRA) provided funding for this work under contract EPG 1/1/160. The Forestry Commission supplied the data from the National Inventory of Woodlands and Trees. Steve Smith of the Forestry Commission's Woodlands and Survey Branch provided the advice required to use these data appropriately. Soil data for England and Wales were supplied by National Soils Resource Institute, and for Scotland by the Macaulay Land Use Research Institute. Deena Mobbs drew the figures.

References

Adger, W.N. and Subak, S. (1996) Estimating above-ground carbon fluxes from U.K. agricultural land. *Geographical Journal* **162:** 191–204.

Baggott, S.L., Davidson, L., Dore, C., Goodwin, J., Milne, R. Murrells, T.P. *et al.* (2003) *UK Greenhouse Gas Inventory, 1990 to 2001 Annual Report for submission under Framework Convention on Climate Change.* National Environmental Technology Center, AEA Technology. AEAT/ENV/R/1396.

Baggott, S.L., Milne, R., Misslebrook, T., Murrells, T.P., Thistlethwaite G. and Watterson, J. (2004) *Greenhouse Gas Inventories for England, Scottland, Wales and Northern Ireland: 1990–2001.* National Environment Technology Center, AEA Technology. AEAT/ENV/R/1481.

BGS (2002) *United Kingdom Minerals Yearbook 1.* British Geological Survey, Natural Environment Research Council, Nottingham.

Bradley, I. (1997) Carbon loss from drained lowland fens. In: Cannell, M.G.R. (ed) *Carbon Sequestration in Vegetation and Soils*, DOE/ITE Contract EPG 1/1/3. Final Report March 1997. Department of Environment, London.

Cannell, M.G.R. and Dewar, R.C. (1995) The carbon sink provided by plantation forests and their products in Britain. *Forestry* **68:** 35–48.

Cannell, M.G.R. and Milne, R. (1995) Carbon pools and sequestration in forest ecosystems. *Forestry* **68:** 361–378.

Cannell, M.G.R. and Milne, R. (2000) Kyoto, carbon and Scottish forestry. *Scottish Forestry* **54:** 11–16.

Cannell, M.G.R., Dewar, R.C. and Pyatt, G. (1993) Conifer plantations on drained peatlands in Britain: A net gain or loss of carbon. *Forestry* **66:** 353–369.

Cannell, M.G.R., Milne, R., Hargreaves, K.J., Brown, T.A.W., Cruickshank, M.M., Bradley, R.I. *et al.* (1999) National inventories of terrestrial carbon sources and sinks: the UK experience. *Climatic Change* **42:** 505–530.

Cruickshank, M.M. and Tomlinson, R.W. (1997) Carbon loss from UK peatlands for fuel and horticulture. In: Cannell, M.G.R. (ed) *Carbon Sequestration in Vegetation and Soils*, DOE, Contract EPG 1/1/3. Final Report March 1997. Department of Environment, London.

Cruickshank, M.M., Tomlinson, R.W., Devine, P.M. and Milne, R. (1998) Carbon in the vegetation and soils of Northern Ireland. *Proceedings of the Royal Irish Academy* **98B:** 9–21.

Dewar, R.C. and Cannell, M.G.R. (1992) Carbon sequestration in the trees, products and soils of forest plantations: an analysis using UK examples. *Tree Physiology* **11:** 49–72.

DOE (1997) *Climate Change: The UK Programme; United Kingdom's Second Report under the Framework Convention on Climate Change.* The Stationery Office, London.

Edwards, P.N. and Christie, J.M. (1981). Yield models for forest management. *Forestry Commission Booklet* No. 48. HMSO, London.

Forestry Commission (2002) *National Inventory of Woodland and Trees.* http://www.forestry.gov.uk/forestry/hcou-54pg4d

Hargreaves, K. and Fowler, D. (1997) Short-term CO_2 fluxes over peatland. In: Cannell, M.G.R. (ed) *Carbon Sequestration in Vegetation and Soils.* DOE Contract EPG 1/1/3. Final Report March 1997. Department of Environment, London.

Hargreaves, K.J., Milne, R. and Cannell, M.G.R. (2003) Carbon balance of afforested peatland in Scotland. *Forestry* **76:** 300–317.

IPCC (1997a) *IPCC Revised 1996 Guidelines for National Greenhouse Gas Inventories, Volume 1, Greenhouse Gas Inventory Reporting Instructions.* IPCC WGI Technical Support Unit, Hadley Centre, Meteorological Office, Bracknell, UK.

IPCC (1997b) *IPCC Revised 1996 Guidelines for National Greenhouse Gas Inventories, Volume 2, Greenhouse Gas Inventory Workbook.* IPCC WGI Technical Support Unit, Hadley Centre, Meteorological Office, Bracknell, UK.

IPCC (1997c) *IPCC Revised 1996 Guidelines for National Greenhouse Gas Inventories, Volume 3, Greenhouse Gas Inventory Reference Manual.* IPCC WGI Technical Support Unit, Hadley Centre, Meteorological Office, Bracknell, UK.

Levy, P. and Milne. R. (2004) Estimation of deforestation rates in Great Britain. *Forestry,* **77:** 9–16.

Milne, R. and Brown, T.A. (1997) Carbon in the vegetation and soils of Great Britain. *Journal of Environmental Management* **49:** 413–433.

Milne, R., Brown, T.A.W. and Murray. T.D. (1998) The effect of geographical variation in planting rate on the uptake of carbon by new forests of Great Britain. *Forestry* **71:** 298–309.

Milne, R., Tomlinson, R. and Murray T.D. (2003) Land Use Change and Forestry: The 2001 UK Greenhouse Gas Inventory and projections to 2020. In: *UK Emissions by Sources and Removals by Sinks due to Land Use, Land Use Change and Forestry Activities.* Annual report for DEFRA Contract EPG1/1/160 (Ed. by R. Milne). Available as: http://www.edinburgh.ceh.ac.uk/ukcarbon/docs/DEFRA_Report_2003_Section02.pdf.

Office of The Deputy Prime Minister (2003) *Land Use Change Statistics Index.* http://www.planning.odpm.gov.uk/luc/lucindex.htm.

5

Regional-scale estimates of forest CO$_2$ and isotope flux based on monthly CO$_2$ budgets of the atmospheric boundary layer

Brent R. Helliker, Joseph A. Berry, Alan K. Betts, Peter S. Bakwin, Kenneth J. Davis, James R. Ehleringer, Martha P. Butler and Daniel M. Ricciuto

1. Introduction

Several lines of evidence point to persistent net annual carbon uptake and storage in the temperate latitudes of the Northern Hemisphere (Battle *et al.*, 2000; Bousqet *et al.*, 1999; Ciais *et al.*, 1995; Prentice *et al.*, 2001; Rayner *et al.*, 1999). This large carbon sink is thought to be largely in the northern forests (Goodale *et al.*, 2002; Chapter 8, this volume), although the tropical forest may also represent a substantial sink (Townsend *et al.*, 2002; Chapter 10, this volume). The uncertainties around these sinks are large (Bousqet *et al.*, 1999) and hypotheses concerning their location and mechanism vary (Bousqet *et al.*, 1999; Fan *et al.*, 1998; Rayner *et al.*, 1999; Chapter 2, this volume). Furthermore, the sustainability of these sinks, or the sustained remediation of anthropogenically derived atmospheric CO$_2$ by forest, is also a point of disagreement (Oren *et al.*, 2001). Considering the global expanse of forests, the importance of forests in the annual global carbon cycle and the uncertainty surrounding the size, duration and location of the northern hemisphere sink, it is imperative that we gain an understanding of forest productivity on large spatial and temporal scales (see Chapter 1, this volume).

The Carbon Balance of Forest Biomes, edited by H. Griffiths and P.G. Jarvis.
© 2005 Taylor & Francis Group

Studies of forest productivity have typically relied on stand-scale eddy covariance (EC) measurements or forest inventories (Baldocchi *et al.*, 2001; Goodale *et al.*, 2002). Together, these methodologies are invaluable in assessing changes in forest productivity in response to climate variability, longer-term climate change, and anthropogenic disturbance. EC measurements are ideal for observations of year-to-year variation in productivity. However, this method offers a relatively limited picture of forest productivity because the surface area of integration, or footprint, is typically less than 1 km². Additionally, heterogeneity at larger spatial scales often precludes the extrapolation of smaller scale measurements to larger scales. Forest inventories provide robust estimates of long-term productivity, but the time-scale of robust forest inventories is typically decadal, not annual (Goodale *et al.*, 2002; Chapter 3, this volume). In this chapter we outline the framework for the measurement of net CO_2 flux over large spatial (10^4–10^6 km²) and temporal (months to years) scales through the ABL. This framework rests on the ABL-equilibrium assumption first developed through model approaches by Betts and Ridgway (1989) for the oceans, and extended to land by Betts (2000), Betts *et al.*, (2004) and for observation-based approaches by Helliker *et al.* (2004; see also Chapter 6, this volume). In this synthesis, we describe the application of the ABL equilibrium concepts to net CO_2 flux over land, and we extend these concepts to the stable isotopes of CO_2 measured in the ABL and the overlying free troposphere. The study area of interest is a temperate mixed-forest region in north-central Wisconsin, USA.

2. Atmospheric boundary layer description, scalar dynamics, and the equilibrium assumption

The ABL extends upwards from the Earth's surface about 1–2 km and is separated from the stable overlying free troposphere (which extends upwards an additional 8–9 km) by an inversion. The ABL is most distinct from the free troposphere when high-pressure systems persist and large-scale subsidence pushes free-tropospheric air down onto the ABL, and ABL air diverges horizontally. During these periods, the ABL is defined as the air mass below the capping inversion and, depending on the time of day, may include the daytime CBL, the nocturnal boundary layer and the residual boundary layer (Stull, 1988). During the day, the dissipation of solar radiation by the surface results in convective turbulence that mixes scalars (CO_2, H_2O, heat, momentum, etc.) from the surface to the top of the ABL. As convection slows towards the day's end, a stable nocturnal boundary layer begins to build at the earth's surface. Overlying the nocturnal boundary layer is the residual boundary layer – a remnant of the previous day's CBL – which is separated from the nocturnal boundary layer below and the free troposphere above by discontinuities in density. As convection resumes after sunrise, the residual boundary layer and nocturnal boundary layer rapidly mix to a homogenous CBL. Viewed in this way, the ABL defines an air mass that may accumulate scalars over several days.

The mixing ratio of scalars in the ABL is influenced by the interaction of several processes, the relative balance of which may change dramatically over the synoptic cycle or with the initiation of deep convection (Freedman *et al.*, 2001). Under high-pressure systems, surface evaporation of water moistens the ABL. This is opposed by subsidence of dry air from the free troposphere, divergent flow of the ABL in the horizontal and by the formation of boundary layer clouds (Fitzjarrold, 2002). CO_2 and other scalars are similarly influenced by their surface fluxes and the mass flux of air from the free troposphere through the ABL (Betts *et al.*, 2004). During low pressure, or upon initiation of strong convection, there may be convergent flow in the horizontal, coupled with deep, precipitating convection (Stull, 1988). Cotton *et al.*, (1995) calculate that the ABL volume is, on average, turned over by deep convection every four days. Down-drafts, associated with evaporation of precipitation into free-tropospheric air, during storms and frontal passages can replace the ABL air mass with cool, moist air and reset the concentration of other scalars to that of the free troposphere (Hurwitz *et al.*, 2004). *Figure 1* illustrates these processes on a variety of temporal scales. Individual, clear weather days show strong gradients in $[CO_2]$ (C, parts per million, p.p.m.) and $[H_2O]$ (q) between the ABL (C_m, q_m) and the free troposphere (C_t, q_t) and also clearly demarcate the inversion at the ABL top (*Figure 1a, b*). The data in *Figure 1a, b* were collected directly by airplane during one afternoon under fair-weather or high-pressure conditions. As storm fronts pass, ABL air can rapidly be replaced by free-tropospheric air. In *Figure 1c*, C_m changed more than 20 p.p.m. in less than a minute owing to strong vertical mixing associated with storm downdrafts. *Figure 1d* shows two frontal sequences where $[CO_2]$ slowly declines in the ABL during high pressure conditions and then approaches free troposphere $[CO_2]$ when a low pressure system passes through the area. On longer time scales, high pressure systems are temporally dominant such that when the averaging period of $[CO_2]$ in the ABL and the free troposphere is increased, we see differences that are suggestive of the seasonally predominant surface flux; photosynthesis in summer and respiration in fall, winter, and spring (*Figure 1e, f*).

The ABL represents a large volume that typically moves across the land surface about 500 km d^{-1} (Raupach *et al.*, 1992). Therefore, studies of the ABL CO_2 balance have the potential to provide information on carbon balance of the land surface on a regional-scale. Estimates of the surface area of integration by the ABL range from 10^4 to 10^6 km^2 (Gloor *et al.*, 2001; Raupach *et al.*, 1992). Several studies have used CO_2 budgets in the ABL to estimate regional, net surface flux of CO_2 during the daytime when the full, mixed ABL is coupled to the surface and the overlying free-troposphere (Denmead *et al.*, 1996; Kuck *et al.*, 2000; Levy *et al.*, 1999; Lloyd *et al.*, 2001; Styles *et al.*, 2002). These studies have demonstrated that budget methods for the period of daytime boundary layer growth can be used to obtain estimates of daytime surface fluxes. However, the difficulty in quantifying fair-weather cloud flux, entrainment, subsidence, and horizontal divergence over a day results in widely variable agreement between ABL budget methods and surface-based methods (Fitzjarrald, 2002). Additionally, the ABL is decoupled from the surface at night, which complicates boundary layer budget closure over complete diurnal cycles. Taken together, these difficulties make the standard boundary layer budget approach impractical for long-term CO_2 balance estimates.

To obtain longer-term estimates of net CO_2 flux at the integrative ABL scale, we make a sharp break from previous CO_2 budget attempts. The ABL is an integral component in large-scale circulation as it connects the ascending and descending branches of the atmospheric circulation. Here we assume that the properties of the ABL

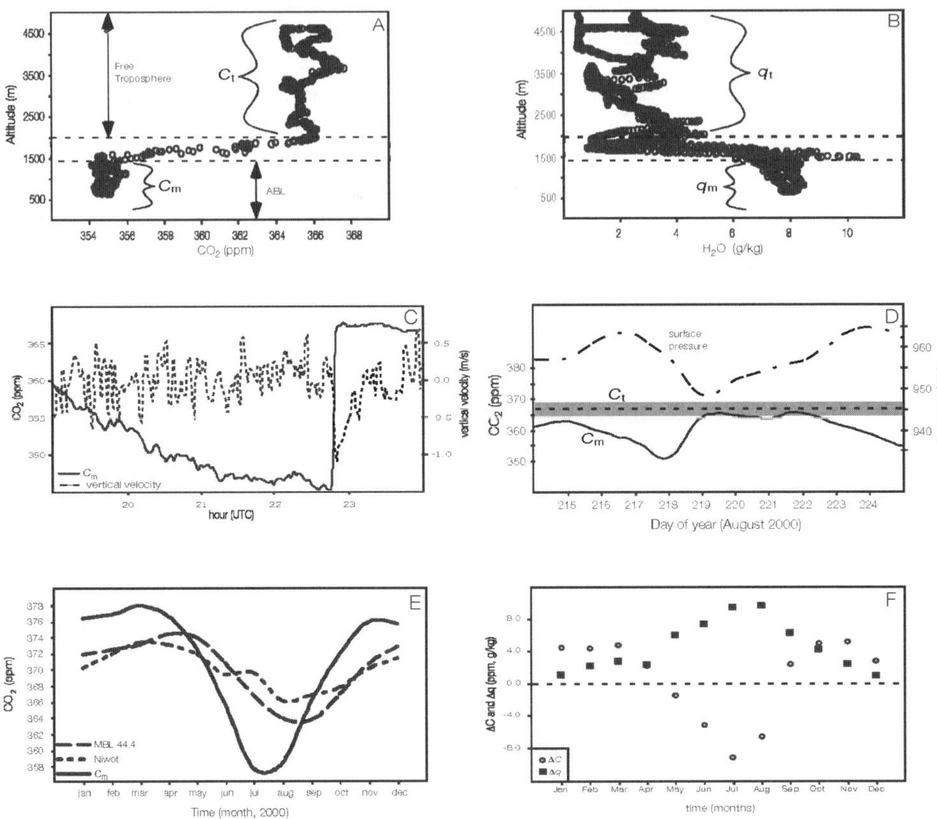

Figure 1. *Representative ABL profiles of (a) [CO_2] and (b) [H_2O] during mid-afternoon. Data were obtained directly by airplane in August 2000 by the COBRA campaign. (c) Five hour period of [CO_2] (solid line) and vertical velocity (dashed line) measured at 396 m from the WLEF tower. (d) [CO_2] (solid line) and surface pressure (dash-dot line) for several days of August 2000 at the WLEF tower site. C_m values are 24 h averages of [CO_2] at 396 m. The dashed line and shaded area represent the mean free-tropospheric CO_2 mixing ratio and standard deviation for all of August (measured directly by COBRA). (e) Monthly averages of [CO_2] in the marine boundary layer (MBL) at 44.4° N, from 3475 m from atop Niwot Ridge, CO, 40.1° N, and from 396 m from the WLEF tower, 45.9° N (C_m). (f) Monthly averages of ΔC ($C_t - C_m$; open circles) and Δq ($q_t - q_m$; closed squares) for the year 2000. C_m and q_m were obtained through continuous measurements at 396 m from the tower. C_t was obtained by monthly averages of [CO_2] collected at Niwot Ridge, CO. RUC weather-forecasting data from the geopotential heights of 3200–7600 m were used to obtain q_t. Data for figure (c) are from Hurwitz et al. (2004). Data from (d) and (e) are from Helliker et al. (2004).*

approach a steady-state or equilibrium between the surface fluxes, cloud effects on radiation and subsidence of the overlying free troposphere over temporal scales longer than a day; an equilibrium which is controlled by larger scale processes that influence air mass ascension (low-pressure systems) and subsidence (high-pressure systems). The equilibrium assumption is a fundamental shift from the focus on the dynamics of the diurnal cycle of the ABL to the slow evolution of the ABL, averaged over the diurnal cycle, and for much of the time nearly in balance with larger-scale subsiding circulations. The non-linear processes of daytime ABL growth and the decoupling of the stable boundary layer at night are superimposed on this slowly evolving mean state.

The concept of an 'equilibrium boundary layer' was first put forth by Betts and Ridgeway (1989) for the marine boundary layer over the open ocean and later applied to land by Betts (2000, 2004) and Betts *et al.* (2004). Betts (2000) observed that the equilibrium boundary layer approach adequately explained the observed water vapor, potential temperature, and boundary layer height of composites representing several synoptic cycles. Betts *et al.* (2004) extended the idealized model of Betts (2000) to show how the mixed layer equilibrium of water vapor, CO_2, and radon was coupled to surface fluxes through mass exchange with the free troposphere during periods of large-scale subsidence. Betts (2004) showed that the 24 hour mean climatic state and fluxes describe the coupling over land using 30 years of the fully time-dependent, European Center for Medium-Range Weather Forecasting (ECMWF) reanalysis model data. Helliker *et al.*, (2004) extended the equilibrium approach to observations of CO_2 and H_2O flux and mixing ratio data. By using observed, long-term averages of H_2O flux and mixing ratio data, they solved for an observation-based estimate of mass exchange of the ABL with the free troposphere. This estimate of mass exchange was then used to estimate the average net surface flux of CO_2 on monthly to annual periods. Further support of the equilibrium assumption was obtained by Bakwin *et al.* (2004) who showed that monthly surface fluxes could be recovered by applying model estimates of vertical exchange to observations of long-term, average differences in CO_2 between the ABL and the free troposphere.

3. Equilibrium atmospheric boundary layer CO_2 flux method

In this section, we outline the underlying equations and discuss some results for equilibrium CO_2 flux in the ABL. The study site was located in sparsely populated, north-central Wisconsin, USA, and the study platform was the 447 m tall WLEF (WLEF are the national public broadcast station call letters) television broadcast tower from which continuous [CO_2], [H_2O], and EC measurements have been made since 1995 (Bakwin *et al.*, 1998; Davis *et al.*, 2003). The tower is located within the Chequamegon-Nicolet National Forest and is a National Ocean and Atmospheric Administration-Climate Monitoring and Diagnostics Lab. (NOAA-CMDL) CO_2 sampling site (Bakwin *et al.*, 1998). The area is largely forested for hundreds of kilometers to the east and west, Lake Superior is approximately 70 km to the north and agriculture begins to dominate about 200 km to the south. The dominant forest types are mixed northern hardwood, aspen, and wetlands.

The tower measurement height of 396 m is well within the convective boundary layer during the day and typically within the residual layer and above the stable nocturnal boundary layer at night. Over a composite diurnal cycle, the CO_2 mixing ratio measured at 396 m typically varies little as compared to measurements at 30 m (Yi *et al.*, 2001). Thus, continuous measurements from this height reflect changes in the daytime convective boundary layer and the night-time residual layer through time

$\partial C_m/\partial t$. Cotton *et al.* (1995) showed that on a global average ABL air is replaced by free-tropospheric air every four days, so on these longer timescales, we assume that horizontal advection becomes less important than vertical advection and mixing across the substantial jump in $[CO_2]$ associated with the capping inversion of the ABL. We therefore use a simplified budget equation for the ABL with depth z and mean $[CO_2]$ C_m in which we neglect horizontal advection. As the ABL deepens in a subsiding mean flow ($\overline{w_+}$) it entrains air from the free troposphere above with properties C_t, and we can write the budget equation for the ABL as (Betts, 1992; Helliker *et al.*, 2004):

$$\frac{\partial}{\partial t}(\rho z C_m) = F_{NEE} + \rho \left(\frac{\partial z}{\partial t}\right) C_t - \rho w_+ (C_t - C_m) \tag{1}$$

where F_{NEE} is the net surface CO_2 flux (including photosynthetic uptake, respiratory and fossil-fuel release) in flux density units (μmol m^{-2} s^{-1}). Rearranging and ignoring the time variation of the molar mean air density, ρ, gives

$$\rho z \frac{\partial C_m}{\partial t} = F_{NEE} - \rho w (C_t - C_m) \tag{2}$$

where $w = \left(\frac{\partial z}{\partial t}\right) - \overline{w_+}$ is the entrainment rate of the ABL, or the rate at which the ABL mixes in air from the free troposphere (i.e., $\overline{w_+}$) is typically negative corresponding to mean subsidence. In strict equilibrium, $\dfrac{\partial C_m}{\partial t} = \dfrac{\partial h}{\partial t} = 0$ and we get

$$F_{NEE} = \rho w (C_t - C_m) = \rho \overline{w_+} (C_t - C_m). \tag{3}$$

A similar equation to (2) can be written for water vapour, and following the same assumptions and derivation we arrive at

$$\rho z \frac{\partial q_m}{\partial t} = F_q - \rho w (q_t - q_m) \tag{4}$$

and

$$F_q = \rho w (C_t - C_m). \tag{5}$$

Note that over longer averaging periods, the time rate of change terms $\left(\dfrac{\partial C_m}{\partial t}, \dfrac{\partial q_m}{\partial t}\right)$ become small compared with the flux terms (Helliker *et al.*, 2004). Following Raupach *et al.* (1992), we assume similarity of transport for C and q. Using long-term averages of observed F_q and $(q_t - q_m)$ and rearranging equation (5) to solve for ρw, we can obtain the observation-based estimate of vertical exchange over the given averaging period. ρw thus obtained can then be used along with observed difference in ABL-to-free-troposphere $[CO_2]$ (for example, from *Figure 1f*) to solve for F_{NEE} through equation (3).

Precipitation and subsequent evaporation processes violate the assumption of similar transport for CO_2 and water vapour, so we filtered data to exclude days when precipitation was more than 1 mm (precipitation (ppt) < 1). However, subsidence-dominant (fair-weather) days are temporally dominant, so we also present data that represent complete monthly averages with no rainy days removed (all days). Monthly estimates of NEE using periods when precipitation days are omitted and periods using all data (thus including periods when low pressure systems pass through the area) are presented in *Figure 2*. F_{NEE} was determined by equation (3) using two independent estimates of ρw: our observation-based, flux-difference estimate ($F_{NEE\text{-}FD}$) and the estimate obtained from the NCEP/NCAR reanalysis model data ($F_{NEE\text{-}\Omega}$). EC estimates of NEE (N_{EC}) measured from the WLEF tower at 122 m are also presented in *Figure 2*. There was remarkable agreement between the independent estimates of NEE. All estimates of NEE showed similar sink-to-source phase shifts and there was good agreement between the estimates on a monthly basis. The overall agreement between the various methods for calculating F_{NEE} and N_{EC} shows that, on longer time scales, the vertical flux of CO_2 from the free troposphere is in near balance with the net CO_2 flux at the surface and the assumptions of the equilibrium ABL appear to be valid.

For all methods of calculating F_{NEE}, $F_{NEE\text{-}FD}$ (all days) had the largest propagated errors (bars shown in *Figure 2*). The errors were relatively small, but they should be considered incomplete as we had to make several assumptions in calculating F_{NEE} and the error associated with these assumptions is not currently quantifiable. First, proxies rather than direct measurements had to be used for C_t and q_t. For C_t we assumed that on a monthly basis, [CO_2] measured in the marine boundary layer, at a similar latitude as the WLEF tower, was representative of mean [CO_2] in the free

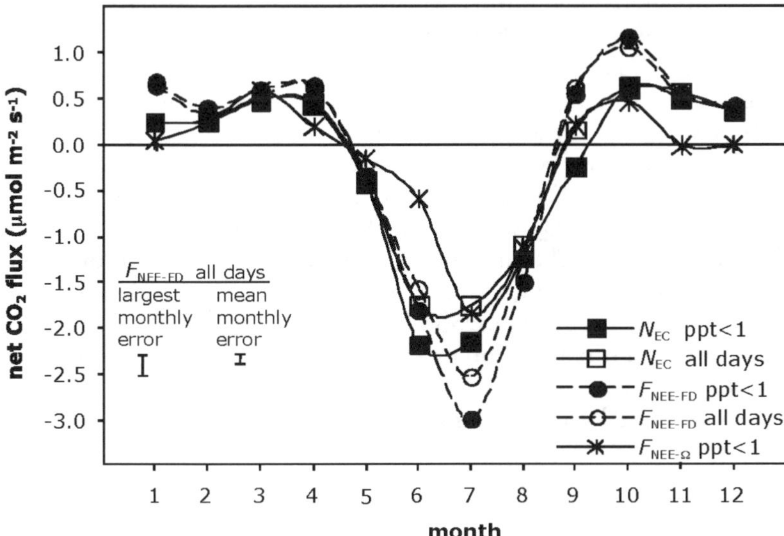

Figure 2. Net ecosystem exchange for CO_2 (NEE) as determined by equation (3) using our flux-difference method with water vapour ($F_{NEE\text{-}FD}$), using equation (3) and NCEP/NCAR reanalysis data for ρw ($F_{NEE\text{-}\Omega}$) and from EC estimates (N_{EC}). For $F_{NEE\text{-}FD}$ and N_{EC} estimates are given for all days in a month and $F_{NEE\text{-}FD}$, N_{EC}, and $F_{NEE\text{-}\Omega}$ estimates are presented with the precipitation filter to remove days with more than 1 mm of precipitation (ppt < 1). $F_{NEE\text{-}FD}$ was calculated using C_t from the MBL (Figure 1e). q_t was obtained from RUC data. F_q, C_m and q_m were obtained by averaging the 24 h observations from the WLEF tower for a given month.

troposphere above Wisconsin (Helliker *et al.*, 2004). For q_t, we had to use monthly averages of model outputs for water vapour from the heights of 2500–3200 m above ground level (rapid update cycle short-range weather forecasting model, http://maps.fsl.noaa.gov/). Second, we assumed that F_q measured by EC at 122 m was fairly representative of the region for calculating ρw through equation (5). For future applications, it would be ideal to have regular flights to build monthly estimates of C_t and q_t from direct measurements. These measurements are currently planned under the implementation of the North American Carbon Plan (Wofsy and Harris, 2002). We obviously need larger estimates of surface water vapour flux than EC measurements provide, and there are more robust land-coverage-based methods for estimating the largely unidirectional flux of water vapour for future tests of the flux-difference method (Anderson *et al.*, 1997, 2000; Mackay *et al.*, 2002). Finally, we do not explicitly include fossil fuel emissions in our regional analysis, but the surface footprint that affects C_m could include a larger fossil fuel flux component than the eddy correlation footprint. It is possible that the differences between F_{NEE-FD} and N_{EC} reflect different fossil fuel emissions within the different footprints. Further study using the direct measurements mentioned above, together with measurements of stable isotopes, are required to quantify these flux differences.

4. Isotopes in the equilibrium atmospheric boundary layer

The analysis of naturally occurring stable isotopes in plant material provides invaluable insight into integrated plant processes (Farquhar *et al.*, 1989). As a corollary, the effect that plants have on the stable isotopes of CO_2 of atmospheric air offers an integration of surface flux processes (Ehleringer *et al.*, 2002; Lloyd *et al.*, 1996; Yakir and Sternberg 2000). Plant and ecosystem responses to short- and long-term climatic changes can be assessed by the analysis of carbon and oxygen isotopes. At ecosystem and larger scales the distinct isotopic signature associated with photosynthesis and respiration allows for the partitioning of NEE into its primary components (Bowling *et al.*, 2001; Bowling *et al.*, 2003; Lloyd *et al.*, 1996; Ogée *et al.*, 2004; Yakir and Wang, 1996). In this section, we use the equilibrium or steady-state assumptions from above to (i) develop the theory behind net isotope effects in an equilibrium boundary layer, (ii) examine some estimates of net carbon isotope composition ($\delta^{13}C_{net}$) for an entire year and, (iii) examine a flux partitioning exercise with measurements of oxygen isotope ratios ($\delta^{18}O$).

As others have defined, the NEE of CO_2 can be represented by

$$F_{NEE} = F_p + F_r \tag{6}$$

where F_p and F_r are the ABL-scale gross fluxes of CO_2 uptake (photosynthesis) and release (respiration and fossil fuel at the ABL scale), respectively. From equations (3) and (6), we can write the following equality:

$$F_{NEE} = \rho w \, (C_t - C_m) = F_p + F_r. \tag{7}$$

For any stable isotope pair, the major isotope (such as ^{12}C or ^{16}O) is in such excess that abundance is directly proportional to concentration as measured from an infra-red gas analyser system. When the stable isotope composition is measured simultaneously, and if we know the magnitude of the fractionations that occur against the heavy isotope (^{13}C or ^{18}O), then we can calculate the relative contribution by isotopic mass balance – for absolute concentrations or fluxes of CO_2 from photosynthesis or respiration. Thus, the measured net isotope composition will reflect the sum of constituent components, and for any heavy isotope we can write,

$$F_{NEE}\delta_{net} = \rho w \, (C_t\delta_t - C_m\delta_m) \tag{8}$$

and

$$F_{NEE}\delta_{net} = F_p\delta_p + F_r\delta_r \tag{9}$$

where δ values are the isotope ratio of an associated CO_2 flux or mixing ratio expressed relative to a standard by,

$$\delta = \left(\frac{R_{sample}}{R_{standard}} - 1 \right) 1000 \tag{10}$$

where R is the molar ratio of heavy to light isotopes ($^{13}C : {}^{12}C$, $^{18}O : {}^{16}O$). For all isotope ratios given here, the standard is Pee Dee belemnite (PDB). The term $F_{NEE}\delta_{net}$ in equations (8) and (9) is often termed the 'isoflux' (Bowling $et\ al.$, 2001).

By solving equation (8) for δ_{net} and recognizing from equation (3) that $F_{NEE}/\rho w = (C_t - C_m)$ we obtain

$$\delta_{net} = \frac{(C_t\delta_t - C_m\delta_m)}{(C_t - C_m)} \tag{11}$$

which explains δ_{net} based solely on the equilibrium differences in CO_2 and isotopes between the ABL and the overlying free troposphere. Equation (11) is the 'delta-notation' form of the equation derived by Farquhar in Evans $et\ al.$ (1986) for measurement of on-line isotope discrimination. Note that equation (11) allows for the estimation of δ_{net} independently of any knowledge of surface flux.

By solving (9) for δ_{net} and inserting (6) we obtain

$$\delta_{net} = \frac{F_p\delta_p + F_r\delta_r}{F_p + F_r} \tag{12}$$

which explains δ_{net} based on the individual surface flux components. δ_{net} can further be expressed as a function of the ratio of photosynthetic uptake to CO_2 release, γ ($\gamma = F_p/F_r$) and inserted into equation (12):

$$\delta_{net} = \frac{\delta_r}{1 + \gamma} + \frac{\delta_p\gamma}{1 + \gamma}. \tag{13}$$

δ_{net} can be obtained through equation (11) with average measurements of the mixing and isotope ratios of ABL and free-tropospheric CO_2. δ_r can be estimated by Keeling plot analysis (Flanagan $et\ al.$, 1996; Keeling, 1958) and, by knowing the environmental inputs, δ_p can be modelled (Farquhar $et\ al.$, 1993; Flanagan $et\ al.$, 1994; Gillon and Yakir, 2000). Hence, by knowing all the isotopic inputs at the ABL-scale and re-arranging equations (12) or (13), we can partition F_{NEE} into the component parts of photosynthesis and respiration in a similar manner to the approaches used at the ecosystem scale (Bowling *et al.*, 2001, 2003; Ogée *et al.*, 2004; Yakir and Wang, 1996).

Figure 3a shows monthly averages of the $\delta^{13}C$ of CO_2 measured at the WLEF tower and from Niwot Ridge ($\delta^{13}C_t$), a mountain top in Colorado, USA. Both sites are part of the NOAA-CMDL flask collection program (Globalview-CO_2, 2003). The $\delta^{13}C$ of ABL CO_2 ($\delta^{13}C_m$) was heavier than the free-tropospheric proxy ($\delta^{13}C_t$) in the summer (i.e., enriched in ^{13}C) and lighter in the winter (i.e., depleted in ^{13}C). Similar to the CO_2 mixing ratio measurements, this pattern is consistent with the predominance of photosynthetic discrimination against ^{13}C in the summer and the respiratory release of CO_2, depleted in ^{13}C, in the winter. Estimations of δ_{net} are shown in *Figure 3b*. For the summer months δ_{net} ranged from –23.3 to –25.2 and was very consistent for July and August. For May, September, and October, the values of δ_{net} varied widely as the differences in CO_2 between the ABL and free troposphere were small. These seemingly odd values of δ_{net} are expected as $\gamma \to 1$ during the seasonal transition from CO_2 sink to source (Miller and Tans, 2002). During the winter, δ_{net} was very negative, possibly reflecting fossil-fuel inputs that would become proportionally larger during the winter months and have isotope ratios that are typically lighter than that of biospheric exchange.

Knowledge of the net flux of CO_2 and δ_{net} for each month can be used to calculate the corresponding flux-weighted isotopic signature associated with the yearly net uptake of CO_2 by this region. This value, which was –9.5‰, may be somewhat surprising because northern Wisconsin is a C_3-dominated region, and carbon fixed by C_3 plants typically has a $\delta^{13}C$ closer to –25‰. However, this apparent discrepancy can be explained if there is an offset between respiration and photosynthesis (i.e., the isotopic composition of carbon fixed is not equal to that respired and $\gamma \approx 1$ for the year (Miller and Tans, 2002)). To illustrate, we offer a sample calculation of the isotopic composition of the gross CO_2 release (δ_r) that would be required to satisfy the mass and isotope balance of equation (13). It was assumed that $\gamma = -1.03$ and that the isotopic composition of carbon taken up by gross photosynthesis, $\delta_P = -25‰$. The calculation indicates that a value of $\delta_r = -25.5‰$ would satisfy the seasonal net storage of –9.5‰. This might occur, for example, if the gross CO_2 released was 90% from biological decomposition with $\delta_R = -25.0‰$ and 10% from fossil fuel combustion with $\delta_{ff} = -29.6‰$. Additionally, the disequilibrium caused by fossil fuel inputs might explain the distinct seasonal pattern in the observed δ_{net} (*Figure 3b*), where the values were considerably more negative in the winter months than in the summer months.

Recent analysis has shown that oxygen isotopes are preferred over carbon isotopes to partition ecosystem fluxes because the differences in isotope ratio between respiration and photosynthesis are considerably larger and result in smaller uncertainties around the partitioned fluxes (Ogée *et al.*, 2004). For a second partitioning exercise we use oxygen isotopes and capitalize on a two-day period in August 2000 when we had direct measurements of ABL and free-tropospheric [CO_2] through the CO_2 Budget and Regional Airborne study (COBRA) campaign (Gerbig *et al.*, 2003). During this period we were able to measure δ_{net}, δ_r, and model δ_p to solve for γ (*Table 1*). We then used γ and estimates of NEE to partition fluxes into photosynthesis and respiration. For this period, we assumed that the fossil-fuel flux was small compared with the size of photosynthetic and respiratory fluxes. Our results for the

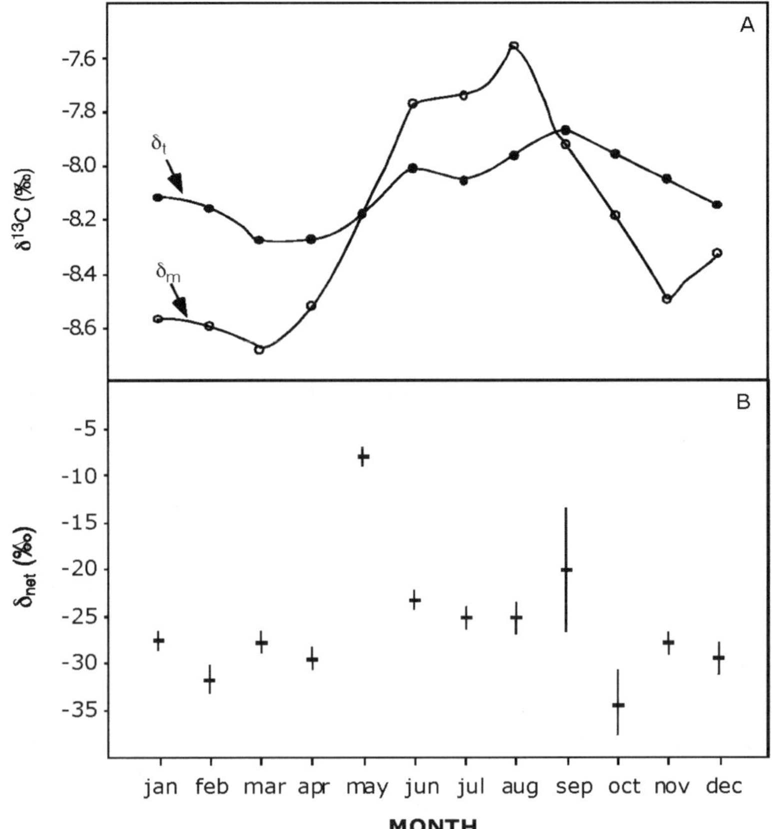

Figure 3. (a) Monthly averages of the $\delta^{13}C$ of CO_2 measured at the WLEF tower ($\delta^{13}C_m$) and from Niwot Ridge ($\delta^{13}C_t$). (b) δ_{net} calculated through equation (11) using isotope ratios in (a) and CO_2 mixing ratios from Figure 1e. The vertical lines represent the propagated error for the calculation of δ_{net}.

partitioned fluxes agreed with independent estimates of flux partitioning through EC-based techniques (*Table 1*). The errors around our estimates are fairly large, but they are comparable to other, accepted methods of flux partitioning. For both of the partitioning exercises detailed here, the input parameters need to be better constrained by measurements before the results of such calculations can be taken seriously. However, this illustrates the additional power that an isotope budget can provide in studies of net carbon exchange.

5. Conclusions

By approaching monthly average CO_2 and water vapour mixing ratios in the ABL as an equilibrium problem, we estimated net CO_2 surface exchange over a forested region for an entire year. These ABL-scale net CO_2 estimates were comparable to measurements made by EC over this same time period. These experiments lend

Table 1. ABL-scale flux partitioning exercise using stable oxygen isotopes.

				ABL-scale estimates			Eddy covariance estimates	
δ_{net} (‰)	δ_r (‰)	δ_p (‰)	γ	NEE (μmol m^{-2} s^{-1})	F_p (μmol m^{-2} s^{-1})	F_r (μmol m^{-2} s^{-1})	F_p (μmol m^{-2} s^{-1})	F_r (μmol m^{-2} s^{-1})
−1.9 ±0.9	−13.5 ±1.4	−11.4 ±1.3	−1.2 ±0.2	−1.2	−6.7 ±1.3	5.5 ±1.3	−7.1 ±1.1	5.9 ±1.1

observational support to the underlying equilibrium hypothesis that over long averaging periods, and during periods of subsidence, the surface flux of CO_2 and water vapour into the ABL are at a near-balance with the mixing-down of air from the free troposphere.

The individual manner in which gross carbon fluxes of photosynthesis, respiration, and fossil fuel use respond to short- and long-term climatic change has large effects on the global carbon cycle. Stable isotopes have been shown to be invaluable tracers of these gross carbon flux processes at leaf, ecosystem, and global scales. However, the uncertainty of these gross processes on the global scale is large because of the lack of knowledge in how leaf-level and ecosystem-level processes play out on regional scales, and how the regional modification of the isotopes in atmospheric CO_2 affect the observations at the global scale. By applying the ABL equilibrium concepts to the measurement of stable isotopes in the ABL and free troposphere, we were able to estimate the regional isotope effects and obtain reasonable estimates of the gross fluxes of CO_2, thus closing a major gap in scales in our current observations.

The ABL flux-difference method presented here can be improved with more systematic mixing ratio and isotope ratio data collection, and more rigorous uncertainty analysis. With these improvements in the method and with more sampling sites, it is not unreasonable to propose that this method could be extended to obtain not only regional, but continental-scale measurements of CO_2 and isotope flux. These improvements have the potential to give us an independent estimate of carbon sources and sinks on an annual basis and help resolve some of the outstanding questions concerning the carbon cycle in forested regions.

References

Anderson, M.C., Norman, J.M., Diak, G.R., Kustas, W.P. and Mecikalski, J.R. (1997) A two-source time-integrated model for estimating surface fluxes using thermal infrared remote sensing. *Remote Sensing of Environment* **60**: 195–216.

Anderson, M.C., Norman, J.M., Meyers, T.P. and Diak, G.R. (2000) An analytical model for estimating canopy transpiration and carbon assimilation fluxes based on canopy light-use efficiency. *Agricultural and Forest Meteorology* **101**: 265–289.

Bakwin, P.S., Tans, P.P., Hurst, D.F. and Zhao, C. (1998) Measurements of carbon dioxide on very tall towers: results of the NOAA/CMDL program. *Tellus Series B – Chemical and Physical Meteorology* **50**: 401–415.

Bakwin, P.S., Davis, K.J., Yi, C., Wofsy, S.C., Munger, J.W., Haszpra, L. and Barcza, Z. (2004) Regional carbon dioxide fluxes from mixing ratio data. *Tellus Series B – Chemical and Physical Meteorology* **56**: 301–311.

Baldocchi, D., Falge, E., Gu, L.H., Olson, R., Hollinger, D., Running, S. *et al.* (2001) FLUXNET: A new tool to study the temporal and spatial variability of ecosystem-scale carbon dioxide, water vapor and energy flux densities. *Bulletin of the American Meteorological Society* **82**: 2415–2434.

Battle, M., Bender, M.L., Tans, P.P., White, J.W.C., Ellis, J.T., Conway, T. and Francey, R.J. (2000) Global carbon sinks and their variability inferred from atmospheric O_2 and $\delta^{13}C$. *Science* **287**: 2467–2470.

Betts, A.K. (1992), FIFE Atmosheric Boundary Layer Budget Methods. *J. Geophys. Res.*, **97**: 18523–189532.

Betts, A.K. (2000) Idealized model for equilibrium boundary layer over land. *Journal of Hydrometeorology* **1**: 507–523.

Betts, A.K. (2004) Understanding hydrometeorology using global models. American Meteorological Society Robert E. Horton Lecture. *Bulletin of the American Meteorological Society* in press.

Betts, A.K. and Ridgway, W.L. (1989) Climatic equilibrium of the atmospheric convective boundary layer over a tropical ocean. *Journal of Atmospheric Sciences* **46**: 2621–2641.

Betts, A.K., Helliker, B.R. and Berry, J.A. (2004) Coupling between CO_2, water vapor, temperature and radon and their fluxes in an idealized equilibrium boundary layer over land. *Journal of Geophysical Research – Atmospheres.* In press.

Bousquet, P., Ciais, P., Peylin, P., Ramonet, M. and Monfray, P. (1999) Inverse modeling of annual atmospheric CO_2 sources and sinks: I. Method and control inversion. *Journal of Geophysical Research*, **104**: 26161–26178.

Bowling, D.R., Tans, P.P. and Monson, R.K. (2001) Partitioning net ecosystem carbon exchanges with isotopic fluxes of CO_2. *Global Change Biology* **7**: 127–145.

Bowling, D.R., Pataki, D.E. and Ehleringer, J.R. (2003) Critical evalutation of micrometeorological methods for measuring ecosystem–atmosphere isotopic exchange of CO_2. *Agricultural and Forest Meteorology* **116**: 159–179.

Ciais, P., Tans, P.P., Trolier, M., White, J.W.C. and Francey, R.J. (1995) A large Northern Hemisphere terrestrial CO_2 sink indicated by the $^{13}C/^{12}C$ ratio of atmospheric CO_2. *Science* **269**: 1098–1102.

Cotton, W.R., Alexander, G.D., Hertenstein, R., Walko, R.L., McAnelly, R.L. and Nicholls, M. (1995) Cloud venting – a review and some new global annual estimates. *Earth-Science Reviews* **39**: 169–206.

Davis, K.J., Bakwin, P.S., Berger, B.W., Yi, C., Zhao, C., Teclaw, R.M. and Isebrands, J.G. (2003) Long-term carbon dioxide fluxes from a very tall tower in a northern forest: Annual cycle of CO_2 exhange. *Global Change Biology* **9**: 1278–1293.

Denmead, O.T., Raupach, M.R., Danin, F.X., Cleugh, H.A. and Leuning, R. (1996) Boundary layer budgets for regional estimates of scalar fluxes. *Global Change Biology* **2**: 275–285.

Ehleringer, J.R., Bowling, D.R., Flanagan, L.B., Fessenden, J., Helliker, B.R., Martinelli, L.A. and Ometto, J.P. (2002), Stable isotopes and carbon cycle processes in forest and grasslands. *Plant Biology* **4**: 181–189.

Evans, J.R., Sharkey, T.D., Berry, J.A. and Farquhar, G.D. (1986) Carbon isotope discrimination measured concurrently with gas exchange to investigate CO$_2$ diffusion in leaves of higher plants. *Australian Journal of Plant Physiology* **13**: 281–292.

Fan, S., Gloor, M., Mahlman, J., Pacala, S., Sarmiento, J., Takahashi, T. and Tans, P. (1998) A large terrestrial carbon sink in North America implied by atmospheric and oceanic carbon dioxide data and models. *Science* **282**: 442–446.

Farquhar, G.D., Ehleringer, J.R. and Hubick, K.T. (1989) Carbon isotope discrimination and photosynthesis. *Annual Review of Plant Physiology and Plant Molecular Biology* **40**: 503–537.

Farquhar, G.D., Lloyd, J., Taylor, J.A., Flanagan, L.B., Syvertsen, J.P., Hubick, K.T., Wong, S.C. and Ehleringer, J.R. (1993) Vegetation effects on the iostope composition of oxygen in atmospheric CO$_2$. *Nature* **363**: 439–443.

Fitzjarrald, D.R. (2002) Boundary layer budgeting. In: Kabat, P., Claussen, M., Dirmeyer, P.A., Gash, J.H., Bravo de Guennin, C., Meybeck, L. *et al.* (eds) *Vegetation, Water, Humans and the Climate: A new Perspective on an Interactive System.* Springer-Verlag, New York, pp. 239–254.

Flanagan, L.B., Brooks, J.R., Varney, G.T., Berry, S.C. and Ehleringer, J.R. (1996) Carbon isotope discrimination during photosynthesis and the isotope ratio of respired CO$_2$ in boreal forest ecosystems. *Global Biogeochemical Cycles* **10**: 629–640.

Flanagan, L.B., Phillips, S.L., Ehleringer, J.R., Lloyd, J. and Farquhar, G.D. (1996) Effect of changes in leaf water oxygen isotopic composition on discrimination against C^{18}O^{16}O during photosynthetic gas exchange. *Australian Journal of Plant Physiology* **21**: 221–234.

Freedman, J.M., Fitzjarrald, D.R., Moore, K.E. and Sakai, R.K. (2001) Boundary layer clouds and vegetation–atmosphere feedbacks. *Journal of Climate* **14**: 180–197.

Gerbig, C., Lin, J.C., Wofsy, S.C., Daube, B.C., Andrews, A.E., Stephens, B.B., Bakwin, P.S. and Grainger, A. (2003) Toward constraining regional-scale fluxes of CO$_2$ with atmospheric observations over a continent: 1. Observed spatial variability from airborne platforms. *Journal of Geophysical Research* **108**: doi:10.1029/2002JDOO3 018.

Gillon, J.S. and Yakir, D. (2000) Internal conductance to CO$_2$ diffusion and C^{18}OO discrimination in C3 leaves. *Plant Physiology* **123**: 201–213.

Globalview-CO$_2$ (2003) Cooperative Atmospheric Data Integration Project – Carbon Dioxide. CD-ROM, NOAA/CMDL, Boulder, Colorado. (Also available on Internet via anonymous FTP at ftp.cmdl.noaa.gov, Path: ccg/co2/GLOB-ALVIEW).

Gloor, M., Bakwin, P., Hurst, D., Lock, L., Draxler, R. and Tans, P. (2001) What is the concentration footprint of a tall tower? *Journal of Geophysical Research* **106**: 17831–17840.

Goodale, C.L., Apps, M.J., Birdsey, R.A., Field, C.B., Heath, J.S., Houghton, J.C. *et al.* (2002) Forest carbon sinks in the Northern Hemisphere. *Ecological Applications* **12**: 891–899.

Helliker, B.R., Berry, J.A., Betts, A.K., Davis, K., Miller, J., Denning, A.S., Bakwin, P., Ehleringer, J., Butler, M.P. and Ricciuto, D. (2004) Estimates of net CO_2 flux by application of equilibrium boundary layer concepts to CO_2 and water vapor measurements from a tall tower. *Journal of Geophysical Research – Atmospheres* In press.

Hurwitz, M.D., Ricciuto, D.M., Davis, K.J., Wang, W., Yi, C., Butler, M.P. and Bakwin, P.S. (2004) Advection of carbon dioxide in the presence of storm systems over a northern Wisconsin forest. *Journal of Atmospheric Sciences* **61**: 607–618.

Keeling, C.D. (1958) The concentrations and isotopic abundances of atmospheric carbon dioxide in rural areas. *Geochimica et Cosmochimica Acta* **13**: 322–334.

Kuck, L.R., Smith, T., Balsley, B.B., Helmig, D., Conway, T.J., Tans, P.P. *et al.* (2000) Measurements of landscape-scale fluxes of carbon dioxide in the Peruvian Amazon by vertical profiling through the atmospheric boundary layer. *Journal of Geophysical Research* **105**: 22137–22146.

Levy, P.E., Grelle, A., Lindroth, A., Molder, M., Jarvis, P.G., Kruijt, B. and Moncrieff, J.B. (1999) Regional-scale CO_2 fluxes over central Sweden by a boundary layer budget method. *Agricultural and Forest Meteorology* **99**: 169–180.

Lloyd, J., Kruijt, B., Hollinger, D.Y., Grace, J., Francey, R.J., Wong, S.-C. *et al.* (1996) Vegetation effects on the isotopic composition of atmospheric CO_2 at local and regional scales: theoretical aspects and a comparison between rain forest in Amazonia and a boreal forest in Siberia. *Australian Journal of Plant Physiology* **23**: 371–399.

Lloyd, J., Francey, R.J., Mollicone, D., Raupach, M.R., Sogachev, A., Arneth, A. *et al.* (2001) Vertical profiles, boundary layer budgets, and regional flux estimates for CO_2 and its $^{13}C/^{12}C$ ratio and for water vapor above a forest/bog mosaic in central Siberia. *Global Biogeochemal Cycles* **15**: 267–284.

Mackay, D.S., Ahl, D.E., Ewers, B.E., Gower, S.T., Burrows, S.N., Samanta, S. and Davis, K.J. (2002) Effects of aggregated classification of forest composition on estimates of evapotranspiration in a northern Wisconsin forest. *Global Change Biology*, **8(12)**: 1253–1265.

Miller, J.B. and Tans, P.P. (2002) Calculating isotopic fractionation from atmospheric measurements at various scales. *Tellus Series B – Chemical and Physical Meteorology* **55**: 207–214.

Ogée, J., Peylin, P., Cuntz, M., Bariac, T., Brunet, Y., Berbigier, P., Richard, P. and Ciais, P. (2004) Partitioning net ecosystem carbon exchange into net assimilation and respiration with canopy-scale isotopic measurements: an error propagation analysis with $^{13}CO_2$ and $CO^{18}O$ data. *Global Biogeochemical Cycles* **18** doi:10.1029/2003 GB002166.

Oren, R., Ellsworth, D.S., Johnsen, K.H., Phillips, N., Ewers, B.E., Maier, C. *et al.* (2001) Soil fertility limits carbon sequestration by forest ecosystems in a CO_2-enriched atmosphere. *Nature* **411**: 469–472.

Prentice, I.C., Farquhar, G.D., Fasham, M.J.R., Goulden, M.L., Heimann, M., Jaramillo, V.J., Kheshgi, H.S., Le Quéré, C., Scholes, R.J. and Wallace, D.W.R. (2001) The carbon cycle and atmospheric CO_2. In *Climate Change 2001: The Scientific Basis. Contribution of Working Group I to the IPCC Third Assessment Report.* Cambridge University Press. Cambride, UK.

Raupach, M.R., Denmead, O.T. and Dunin, F.X. (1992) Challenges in linking atmospheric CO_2 concentrations to fluxes at local and regional scales. *Australian Journal of Botany* 40: 697–716.

Rayner, P.J., Enting, I.G., Francey, R.J. and Langenfelds, R. (1999) Reconstructing the recent carbon cycle from atmospheric CO_2, $d^{13}C$ and O_2/N_2 observations. *Tellus Series B – Chemical and Physical Meteorology* 51: 213–232.

Stull, R.B. (1988) *An Introduction to Boundary Layer Meteorology.* Kluwer, Dordrecht.

Styles, J.M., Lloyd, J., Zolotoukhine, D., Lawton, K.A., Tchebakova, N., Francey, R.J., Arneth, A., Salamakho, D., Kolle, O. and Schulze, E.-D. (2002) Estimates of regional surface carbon dioxide exchange and carbon and oxygen isotope discrimination during photosynthesis from concentration profiles in the atmospheric boundary layer. *Tellus Series B – Chemical and Physical Meteorology* 54: 768–783.

Townsend, A.R., Asner, G.P., White, J.W.C. and Tans, P. (2002) Land use effects on atmospheric ^{13}C imply a sizable terrestrial CO_2 sink in tropical latitudes. *Geophysical Research Letters* 29: 10, doi: 10.1029/2001GL013454.

Wofsy, S.C. and Harris, R.C. (2002) *The North American Carbon Program (NACP). Report of the NACP Committee of the U.S. Interagency Carbon Cycle Science Program.* U.S. Global Change Research Program.

Yakir, D. and Sternberg, L.S.L. (2000) The use of stable isotopes to study ecosystem gas exchange. *Oecologia* **123**: 297–311.

Yakir, D. and Wang, X.F. (1996) Fluxes of CO_2 and water between terrestrial vegetation and the atmosphere estimated from isotope measurements. *Nature* **380**: 515–517.

Yi, C., Davis, K.J. and Berger, B.W. (2001) Long-term observations of the dynamics of the continental planetary boundary layer. *Journal of Atmospheric Sciences* **58**: 1288–1299.

6

Regional measurement and modelling of carbon balances

A.J. (Han) Dolman, Reinder Ronda, Franco Miglietta and Philippe Ciais

1. Introduction

Atmospheric measurements of CO_2 and O_2/N_2 mixing ratios have been important for quantifying the source–sink distribution of CO_2 at global and sub-hemispheric scales (see, for example, Gurney *et al.*, 2002; Roedenbeck *et al.*, 2003), using a globally distributed network of atmospheric observations derived from fewer than 100 stations. The data are used in association with an atmospheric transport model which is run in an inverse mode to infer the 'optimal' source-sink distribution that best matches the observed concentrations, given some *a priori* information (e.g., source–sink initial patterns, correlations, and errors). This global inversion approach yields estimates that cannot be considered to be robust for sub- continental scales. For example, the longitudinal distribution of Northern Hemisphere carbon uptake between the oceans, North America, Europe, and Asia is subject to many investigations and uncertainties (see, for example, Fan *et al.*, 1998; Peylin *et al.*, 2002). In parallel, extensive forest biomass inventories and ecosystem modelling provide estimates of the sources and sinks from a bottom-up perspective (Goodale *et al.*, 2002; McGuire *et al.*, 2001; see also Chapters 4 and 16, this volume) that are highly compatible with inversion results, although there are large errors in both approaches.

The Carbon Balance of Forest Biomes, edited by H. Griffiths and P.G. Jarvis.
© 2005 Taylor & Francis Group

At the local scale (1 km^2) direct flux measurements by the eddy covariance (EC) technique (Baldocchi *et al.*, 2001; Valentini *et al.*, 2000, and Chapter 9, this volume) constrain the net ecosystem exchange (NEE) to within 20%, comparable to the uncertainty estimated from inverse models (see, for example, Janssens *et al.*, 2003). In parallel, intensive field studies can determine the changes in vegetation and soil carbon stocks using biometric techniques, which allow independent quantification of the average carbon balance of ecosystems, also with significant errors (Curtis *et al.*, 2002; Schulze *et al.*, 2000; Wirth *et al.*, 2002; Chapter 15, this volume).

How these two scales interact at the regional level is unknown, and it remains a major challenge, both politically in the context of Kyoto, and scientifically to quantify the carbon balance at this 'missing scale'. It is fair to say that ecosystem measurement sites generally sample growing forests that are not surprisingly sequestering carbon. Infrequent disturbances (e.g., wind break, fire, pest infestations) that open ecosystems to sporadic and rapid loss of CO_2 to the atmosphere, are not captured in most ground-based measurement inventories (Körner, 2003). A major knowledge gap between continental and ecosystem scales therefore requires accounting for disturbances, and more general spatial and temporal heterogeneity, on the biospheric CO_2 fluxes.

Large-scale inversion-based sink/source estimates of CO_2, obtained from a few mostly oceanic stations, suffer from several errors (see, for example, Gerbig *et al.*, 2003). First, solving for continental fluxes is an ill-constrained inverse problem given uneven and sparsely distributed stations with a continental influence. Second, there is a large variability in both atmospheric transport and surface fluxes over vegetated areas, which yields a peak in the CO_2 variance near the ground, so that the signal of mean continental fluxes is difficult to characterize from the noise (variability) when only discrete (flask) sampling is available. Third, vertical mixing and NEE co-vary together on seasonal and diurnal scales to yield rectification gradients in mixing ratios that are especially difficult to capture in large-scale transport models (Denning *et al.*, 1996). Fourth, given the flux heterogeneity, measurements from a single location are not immediately representative of larger regions or grid cells, thus causing representation errors. Fifth, solving for aggregate fluxes that do not evenly influence the overall mixing ratio may cause aggregation errors (Kaminsky *et al.*, 2001). Finally, diurnal and seasonal atmospheric transport processes (e.g., boundary layer mixing and ventilation, orographic effects, frontal uplift of tracers, etc.) are usually poorly represented in large-scale transport models. Similarly, measurements at a single tower site suffer also from representation errors, which require that the information obtained at these sites can only be aggregated up to the scale of a region in a simple fashion.

Gerbig *et al.* (2003) indicated that a significant fraction of the information contained in the signature of boundary layer CO_2 is contained in relatively small spatial and temporal scales. In *Figure 1* the representation error of a series of measurements over the North American continent during the COBRA (CO_2 budget and rectification airborne study in the Northern US) experiment is shown. This error is shown as the variance between samples and the difference in horizontal distance between them, combined with an uncertainty analysis of both, of which details can be found in Gerbig *et al.* (2003). *Figure 1* shows that for this experiment a significant increase in representation errors occurs when horizontal grid size (distance) increases. At very small scales, the uncertainty approaches the measurement error of 0.2 parts per million (p.p.m.). At large grid sizes the representation error rapidly increases: at only

30 km it is around twice the measurement error and the representation error dominates the measurement error; at about 150 km it is already 1 p.p.m. Given the grid size of atmospheric models used in current inversions (see, for example, Gurney *et al.*, 2002), 200–400 km, representation errors of 1–2 p.p.m. can be expected. This suggests that to be able to use the variance of the observed signal during the COBRA study, analysis with grid cells of less than 30 km is required. This requires experimental sampling and model development to resolve the diurnal timescale and to be appropriate at such small spatial scales.

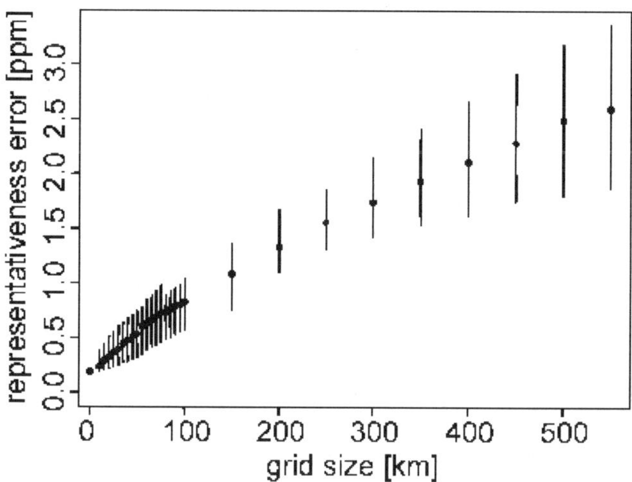

Figure 1. *Total representativeness error of mixed layer averaged CO_2 mixing ratios (combined measurement uncertainty and representativeness error) plotted against the horizontal dimension of the region. Vertical bars indicate the 5–95% range. Gerbig, C. et al, "Constraining regional...", Journal Geophysical Research, in press. Copyright (2003) American Geophysical Union. Reproduced/modified by permission of American Geophysical Union.*

These errors can be substantially reduced when at the regional scale a good link between the measurements obtained at the surface flux stations and those from continental scale inversions can be established. To achieve this, the daytime convective boundary layer (CBL) can be used as a natural integrator of the surface fluxes at the regional scale (Lloyd *et al.*, 1996; Raupach *et al.*, 1992; see Chapter 5, this volume). A problem associated with the CBL budgeting technique is that the spatial footprint of the so-determined fluxes is not explicitly known unless advection is quantified or a Lagrangian approach is chosen (Lloyd *et al.*, 2001; Schmitgen *et al.*, 2003). Further, entrainment with air from the free troposphere by CBL growth and large scale vertical fluxes influence the rate of change of CO_2 as well as surface fluxes, and those terms are also poorly quantified.

Whereas the CBL budget technique gives estimates of fluxes derived from inverting a one-dimensional budget equation, it is also possible to measure regional fluxes directly. The advent of small specialized airplanes in the past decade, measuring fluxes at a resolution of 1–2 km and with comparable accuracy to tower fluxes, has greatly increased the possibilities of providing accurate estimates of spatial heterogeneity. Critical in the development of this type of aircraft is its capability to fly at low altitude and low speed, thereby allowing comparisons between estimates of flux towers and the aircraft and acquiring flux estimates at high spatial resolution.

Atmospheric mesoscale models have become available for studies of regional CO_2 exchange (e.g., Dolman *et al.*, 2003). Non-hydrostatic mesoscale models can model the surface atmosphere exchange of CO_2 at resolutions comparable to that of flux aircraft and single flux towers (Eastman *et al.*, 2001; Nicholls *et al.*, 2003). These models are excellent tools for conducting high-resolution simulations of CO_2 near a detector, and therefore reducing the representation error. In the future, it should be possible to run such models as host platforms for field and remote sensing data assimilation, similar to the use in previous field experiments dealing with land surface hydrology (see, for example, Bougeault *et al.*, 1991). A prime requirement for the use of these models in CO_2 assimilation purposes is the existence of a good spatially and temporally distributed map of fossil fuel sources in the region, and techniques for separating fossil fuel from biospheric contributions (Schmitgen *et al.*, 2003).

With the advance of both regional observations and good high-resolution mesoscale models, inverse modelling of source/sink distribution at regional scale may come within reach (see, for example, Gerbig *et al.*, 2003; Ronda and Dolman, 2003; Uliasz, 2003). This exciting new development has great potential to further elucidate and quantify the role of the 'missing scale' in terrestrial carbon cycle research. This chapter will review new and current developments in the application of CBL budgeting techniques, regional flux aircraft and regional forward and inverse modelling to determine the regional carbon balance of the land surface.

2. Convective boundary layer budgeting

The CBL over the land surface contains a strong signal of both regional terrestrial and anthropogenic sources and sinks of CO_2 (Bakwin *et al.*, 1998; Lloyd *et al.*, 1996; see Chapter 5, this volume). Laubach and Fritsch (2002) give two key properties of the CBL that determine if the CBL can be used as an integrator of those fluxes. Firstly, the CBL must be well mixed, making the exact horizontal position of measurement less important, and secondly, a clear inversion zone that provides an efficient lid on the 'natural chamber' must top it. Under such conditions, CBL budgeting has been used to determine regional ecosystem fluxes during daytime in central Siberia, the Amazon and Europe (Culf *et al.*, 1997; Laubach and Fritsch, 2002; Levy *et al.*, 1999 Lloyd *et al.*, 2001), as well as to determine the regional discrimination of NEE against ^{13}C (Lloyd *et al.* 2001), with varying success. Both Lagrangian approaches, which track a column of air, as well as Eulerian approaches whereby advective terms need to be estimated, have been used by various groups (Gerbig *et al.*, 2003; Laubach and Fritsch, 2002; Schmitgen *et al.*, 2003).

Following Laubach and Fritch (2002) the mass balance of a column of air can be written as:

$$\frac{F_s}{\langle \rho \rangle_i z_i} = \alpha_s \left\{ \frac{\delta \langle s \rangle_i}{\delta t} - \frac{(s_+ - \langle s \rangle_i)}{z_i} \left(\frac{\delta z_i}{\delta t} - w_+ \right) \frac{\rho_+}{\langle \rho \rangle_i} + \frac{\langle v \delta s \rangle}{\delta y} + \frac{\langle u \delta s \rangle}{\delta x} \right\} \tag{1}$$

where F_s is the instantaneous surface flux, α_s is a unit conversion factor for scalar quantity s, ρ the mean molar air density, the subscript i refers to the top of the well-mixed CBL (the inversion height), the subscript + refers to values of quantities just above the inversion at $z = z_i$, and t is the time. The last two terms of the right-hand side represent mean advection in the crosswind and horizontal directions. The first term on the right-hand side of equation (1) represents the momentary change of concentration with time of scalar s, the second term the vertical gradient in s across the top of the CBL, the third term represents the mass flux through the (moving) inversion height $\delta z_i / \delta t$ by 'encroachment' of air above the CBL, and w is a mean vertical velocity. Note that the growth of the CBL by 'encroachment' of air acts to reduce the F_s flux if $(S_+ - S_i) > 0$, as is usually the case, whereas an upward positive vertical velocity acts to imply higher F_s. The last term is a correction allowing the use of a non-constant density with height. Various forms of this equation, or simplifications thereof, are the basis for much of the effort in applying the CBL technique to estimate regional fluxes.

Equation (1) assumes that the daytime atmospheric boundary layer is perfectly mixed vertically, and furthermore, is horizontally homogeneous. To evaluate equation (1) correctly, an estimate of the CBL height and its evolution is required and this appears from simple sensitivity analysis critical to determine the surface source or sink strength. Equally important is an estimate of the uplift or subsidence fluxes at the top of the boundary layer (Levy *et al.*, 1999). In past studies, boundary layer data alone have been identified as insufficient, and large-scale model-estimates of advection, entrainment velocities, subsidence, etc., have been required to close the mass balance of equation (1). In practice, equation (1) is integrated in finite difference form and the time rates of change calculated from profiles of scalars obtained at two different times (Styles *et al.*, 2002).

Figure 2 shows an example of profiles flown above a mixed agricultural and forest landscape in Germany. In the morning profiles the height of the boundary layer is low, about 195 m. The most noticeable feature of the morning profile is the sharp gradient in the concentration of CO_2 indicating considerable build-up of CO_2 during the previous night. The boundary layer has grown to about 1100 m at the time of the second flight at 12.40 h and profiles of all scalar quantities now appear well mixed. At 15.35 h the inversion has moved up to 1365 m. The profiles of temperature suggest advection, since above the inversion height the profiles become warmer during the day. This illustrates one of the key problems in CBL budgeting techniques: that under most natural conditions advection may occur in the boundary layer, certainly in highly heterogeneous regions that are common in Europe.

Laubach and Fritsch (2002) used the profile information to calculate the fluxes of surface fluxes of CO_2 and sensible and latent heat using various assumptions about the state of the air column under consideration. Evaluating the budget for the assumption that the column has a fixed mass, rather than a variable one with the top fixed at the inversion, yields errors of about 10–20% for the CO_2 and sensible heat flux and slightly larger ones for latent heat flux (20–30%). Their attempt to use the information of profiles just above the inversion to quantify advection met with variable success, and no clear physical explanation could be found to explain why in some cases it improved the estimates and in other cases in did not.

To overcome the problems with advection, Schmitgen *et al.* (2003) used a Lagrangian experimental approach whereby the aircraft follows a particular air mass.

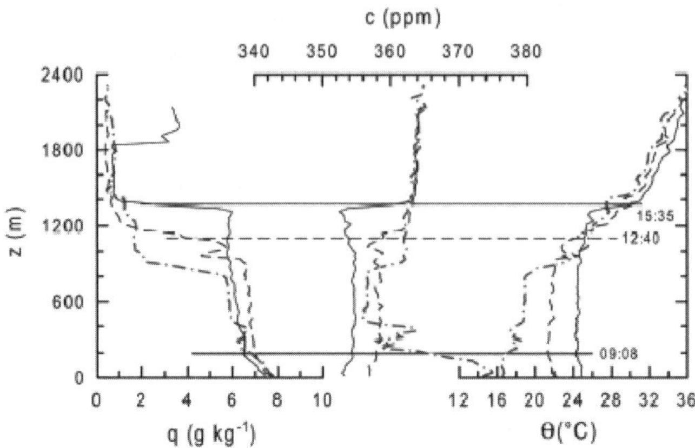

Figure 2. *Profiles of specific humidity (q), CO₂ concentration (c) and potential temperature (θ) on 26 August 2000. Dash-dotted, dashed, and solid lines represent morning, midday, and afternoon flight, respectively. Boundary-layer heights are indicated by horizontal lines. Reprinted from Agricultural and Forest Meteorology, Vol III, Laubach and Fritsch, "Convective boundary layer ...," pp 237–263, Copyright (2002), with permission from Elsevier.*

The Lagrangian experiment took place under favourable conditions over a productive, extended, and spatially homogenous pine forest in southwest France in summer, with flat terrain, negligible anthropogenic sources (checked by concomitant *in situ* measurements of CO), and simple boundary conditions given by homogeneous maritime air moving over the forest. Yet, for only one flight out of four were the proper Lagrangian conditions met, with persistent low wind speeds (6 km h⁻¹) and a stable inversion layer at *ca* 700 m. Their results indicate that during its exposure to surface uptake, the air mass lost CO_2 at a rate of 0.3 p.p.m. km⁻¹, which translates into an estimate for daytime NEE of 16 μmol (C) m⁻² s⁻¹ over the 100 km flight domain. This value compares robustly within its errors with the local-scale measurements of the EC tower located further north of the flight tracks, and the results of a regional remote-sensing model based on 1 km resolution MODIS data.

Having shown that advection was not a significant source of error in the Lagrangian budget flux estimate, Schmitgen *et al.* (2003) quantified instrumental errors from possible drifts in the CO_2 sensor (± 0.5 p.p.m.). Such errors would represent 14% of the signal of fluxes, an aggregation error of 22% (as implied by solving for a mean flux for the whole domain), a bias of 18% (owing to the uncertainties in characterizing the true CBL mean value of $<s_i>$) and a 29% systematic error that takes into account the lack of adequate entrainment data. Adding these error estimates suggests that the order of magnitude of the error is equal to the magnitude of the total flux (16 μmol m⁻² s⁻¹). Schmitgen *et al.* (2003) argued convincingly that these errors can be drastically reduced by more adequate flight plans (both horizontal and vertical stacks) and better sampling and higher resolution measurements. Taking these arguments into account they concluded that an overall error of 22% is reasonable. This may seem comparable to the estimates of Laubach and Fritsch (2002); however, these latter authors neglected to quantify the aggregation and representation errors involved, and referred primarily to the instrumental and method errors.

3. Eddy correlation flux aircraft

Rather than inferring the regional flux from concentration measurements, it is also possible to measure directly the exchange of scalar quantities such as CO_2 and temperature from a moving platform. The technique was pioneered more than 20 years ago by Desjardins *et al.* (1982) and Lenschow *et al.* (1981), and has since been applied in numerous field experiments. The possibility of using small aircraft for measuring surface mass and energy fluxes has been demonstrated in several pilot experiments over the past decade. In particular, Brooks *et al.* (2001) and Crawford *et al.* (1996) have shown that aircraft flux measurements can be very accurate and reliable and still be executed on small, lightweight planes.

However, there are still several unresolved issues that have so far prevented full-scale application. Furthermore, comparison between data from flux tower sites and aircraft has not unequivocally met with success. Two issues stand out. First, the difference in footprint between the fluxes measured from the aircraft and from the tower complicates a direct comparison. Particularly in heterogeneous terrain, this can be a problem, and the precise height of the flight track chosen may influence and almost certainly complicate such a comparison. Secondly, vertical flux divergence in the boundary layer implies that extrapolation to the tower fluxes is needed when aircraft fluxes measured at a certain height well above the surface are compared with measurements from a tower.

The instruments that are now used in state of the art aircraft have reached a high level of perfection and include high-precision pressure spheres on the nose of aircraft, open path infra-red gas analysers for high-frequency measurements of CO_2 and water vapour, differential global positioning systems (GPSs) for estimating the horizontal velocity relative to the ground, and accelerometers to calculate three-dimensional pitch, roll, and heading of the aircraft. An example of such a 'state of the art' plane is the Sky Arrow ERA described in Oechel *et al.* (1998), in essence based on the Long-EZ developed by Crawford and Dobosy (1992). The Sky Arrow also carries on-board photosynthetically active radiation (PAR) and net radiation sensors, and the possibility of a spectral video camera that allows classification of the underlying land surface through indices such as the normalized difference vegetation index (NDVI). This two-seater plane can fly at slow speed (35 m s^{-1}) and at low altitudes (up to 10 m above the surface), so that meaningful comparisons can be made with tower flux measurements. Sampling at 50 Hz, it can detect eddies larger than 1.4 m in windless conditions (Gioli *et al.*, 2003).

The Sky Arrow was recently used in a series of experiments to determine the ability to quantify regional fluxes (Gioli *et al.*, 2004). The aircraft was flown over five regions selected in Europe to account for differences in climate, land cover, and use. A total of 200 flight hours were flown over areas in Spain (rice fields near Valencia), The Netherlands (agriculture and pine forest), central Sweden (spruce forest and agriculture and lakes), central Germany (spruce forest and agriculture) and northwest Italy (mixed Mediterranean forest and agriculture). Flux data on average were calculated as 30 minute averages with a spatial window of 3 km. Large homogeneous patches were used in the comparison. Overall, good agreement (within 10%) was obtained between tower-based estimates and those of the Sky Arrow for several test sites in Europe when the momentum flux was compared.

There was consistent overestimation of the latent heat flux (22%) and consistent underestimation of the CO_2 flux (23%) and sensible heat flux (34%). As an example, *Figure 3* shows the consistent underestimation of daytime CO_2 flux in the European campaigns.

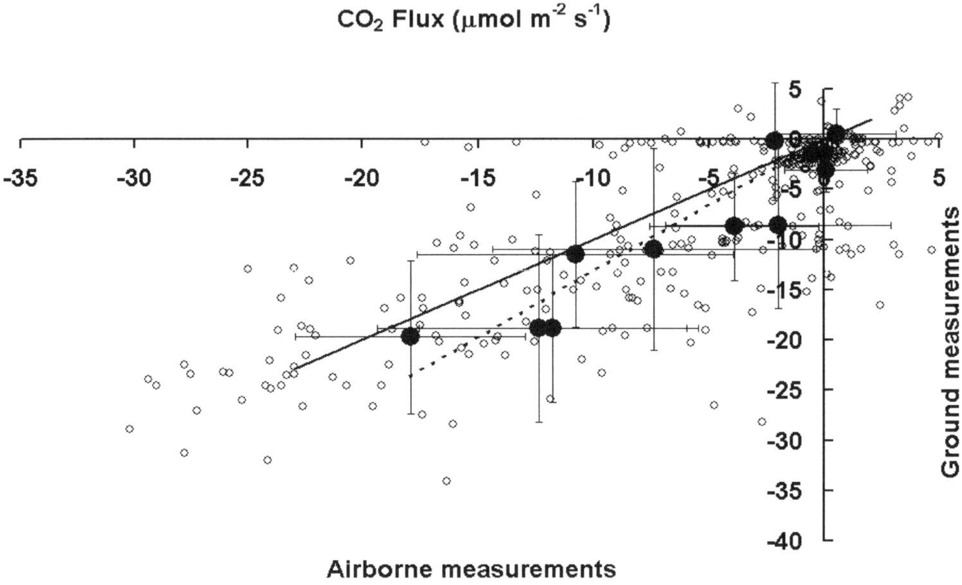

Figure 3. *Comparison of CO_2 fluxes between tower and aircraft measurements for seven campaigns. The large closed circles represent site (daily) averages with corresponding standard deviations for ground and aircraft measurements (from Gioli et al., 2004).*

This difference could be related to flux divergence within the boundary layer that appeared from several flights at different levels within the CBL during their campaign in The Netherlands. A linear decrease in sensible heat flux, as expected from theoretical consideration, was observed during these flights, and by using the measured slope of that decrease, Gioli *et al.* (2004) were able to extrapolate the aircraft flux to the tower flux height and achieve a much better agreement between tower and aircraft measurements. Similar agreement was found between the estimates obtained from flux measurements at 100 m in central Sweden and the aircraft. It is interesting to note that Crawford and Dobosy (2004) claim that the best use for aircraft data is to examine the spatial structure beyond towers and to estimate the non-local terms such as advection.

4. Regional mesoscale modelling

Atmospheric mesoscale models (see, for example, Pielke *et al.*, 1992) or limited-area models are now sufficiently well developed to be coupled to terrestrial ecosystem models that are able to model the surface exchange of carbon. Such forward modelling efforts, focusing on a few selected regions, have the potential to bridge the scale gap between the local and continental scale. Nicholls *et al.* (2003) described such an effort where the Regional Atmospheric Modelling System (RAMS) was used coupled to the Simple Biosphere model (Sib2) in an attempt to model the net ecosystem exchange of an area that is sampled by a high TV tower in Winsconsin, USA, the WLEF tower (see also Chapter 5, this volume). They used multiple grids that allowed resolution of fine-scale features around the tower at 333 m at the highest resolution.

Despite generally good capability of the coupled system to simulate the different synoptic conditions and fine-scale features observed at the Wisconsin tall tower site, they also identified some problems both in the land surface model, which tended to overestimate photosynthetic activity at certain times of the day, and in the turbulence schemes used. In particular, the model at the high-resolution grid appeared sensitive to the closure scheme used, and this influenced the degree of vertical mixing within the model. Similarly, the production of fog and shallow cumulus cloud induced extra variability in CO_2 concentration in the model. Other examples of this type of modelling at somewhat larger scales, also with hydrostatic models, can be found in Chevillard *et al.* (2002) and Eastman *et al.* (2001). A key problem in the performance of these models is the description of the boundary layer. Precise estimation of its height and the strength of the inversion gradient determine to a large extent the resulting concentrations. This requires very high vertical resolution in the boundary layer and near the inversion or very adequate parameterizations of CBL growth and entrainment to produce realistic simulations.

Despite these problems, mesoscale models have currently also been extended to carry the stable isotopes of carbon and other species that may allow determination of the degree of recycling of fossil fuel emissions by the vegetation, and separation of respiration from the net ecosystem exchange (see, for example, Lloyd *et al.*, 1996), thus refining quantification of the biospheric sink and providing a check on the realism of the fossil fuel emission data. This approach requires a precise knowledge of the isotopic composition of the carbon sources (fossil fuel, biospheric, oceanic) that contribute to the atmospheric concentration.

Inverse models at regional scale may provide a further check on the biospheric and anthropogenic emissions. Using mesoscale models combined with measurements from a high tower and using precise data on the structure of the convective boundary layer (Bakwin *et al.*, 1998), the interplay between biospheric sinks and industrial sources at the regional scale can be elucidated. Ronda and Dolman (2003) estimated the regional carbon balance by using Bayesian synthetic inversion on a regional scale. This technique, which has been frequently been used to calculate the global distribution of sources and sinks, calculates the regional distribution of carbon sources and sinks by combining *a priori* estimates of the regional distribution of CO_2 with estimates obtained by inverting a regional transport model. The observations used in the Bayesian synthesis inversion are taken during the intensive field campaigns of the CarboEurope RECAB project. They consist of continuous CO_2 concentration measurements taken at different towers within the domain, atmospheric boundary layer profiles of $[CO_2]$ measured twice a day by a small aircraft at different locations within the domain, and estimates of the CO_2 fluxes taken twice a day along flight legs with an average height above the surface of about 70 m. The *a priori* estimates of the anthropogenic emissions are in this analysis taken from anthropogenic carbon emission inventories, whereas the biospheric fluxes are calculated by using a soil–vegetation–atmosphere transfer scheme included in the mesoscale meteorological RAMS model. By weighting the uncertainty in the *a priori* estimates, and the uncertainty in the modelled and observed concentrations and fluxes, optimal carbon fluxes can be calculated.

In *Figure 4* first results using this technique to optimize anthropogenic emissions are shown for a one day period, 2 February 2002, in central Netherlands. *A priori* estimates are assumed to have an uncertainty of 10% and the standard deviation of the error is calculated accordingly. It can be seen that use of the airborne fluxes reduces the expected error in the optimized anthropogenic emission by about 50%, assuming a standard deviation of the expected error of 3 μmol m^{-2} s^{-1}. Including the concentration measurements taken at two towers in the inversion setup leads to a further decrease of the uncertainty in the optimized anthropogenic emissions. The standard deviation between the expected error between modelled and measured concentration is taken as 1.5 p.p.m. The airborne fluxes and the tower measurements both contain information about the anthropogenic emissions. Unfortunately, the biospheric fluxes are quite small in winter, and this hampers a similar study for biospheric fluxes. However, further study in summer conditions, when biospheric fluxes will be much larger, will reveal whether these techniques can also be used to estimate the regional distribution of biospheric carbon sources and sinks.

At a larger scale, Gerbig *et al.* (2003) used the stochastic time inverted lagrangian transport STILT model (Lin *et al.*, 2003) in a data assimilation framework to determine the regional flux distribution for the northern USA. This model is based on a stochastic simulation of mixed layer turbulence and takes into account cloud

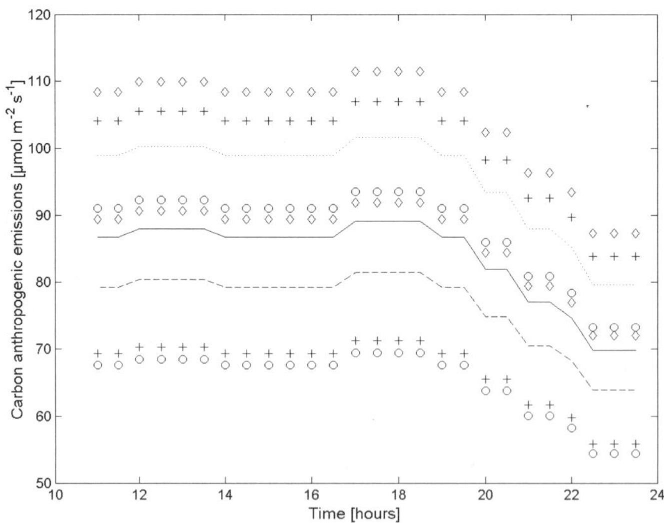

Figure 4. A priori estimates of anthropogenic emissions (solid) + uncertainty (crosses), optimized emissions using airborne fluxes only (dashed line) + uncertainty (circles), and optimized emissions using airborne fluxes and tower concentration measurements (dotted) + uncertainty (diamonds).

venting of CO_2. The model was constrained at its boundaries by concentration data from the global CO_2 observation network. They found overall good agreement between their model and large scale observations of CO_2 from an aircraft. However, differences between the modelled and observed concentrations occur as a result of inadequate representation of atmospheric transport, too simplified a representation of the *a priori* or first guess biospheric fluxes, and uncertainties in the CO_2 observations. Another issue is the use of good fossil fuel inventory data to separate out the contribution of anthropogenic activity to the CO_2 signal.

A crucial problem for inferring regional CO_2 fluxes in populated regions is the superposition of the fossil fuel CO_2 emissions and net CO_2 ecosystem exchange (NEE). To determine the biogenic CO_2 sink, it is essential to separate these contributions. Carbon monoxide, CO, is a promising candidate to separate the anthropogenic contribution from the biogenic surface flux (Bakwin *et al.*, 1998; Gerbig *et al.*, 2003; Meijer *et al.*, 1996; Potosnak *et al.*, 1999).

CO is emitted with CO_2 during combustion processes like fossil fuel and biomass burning. The signal from combustion sources adds to a global background of CO produced from the oxidation of CH_4. Natural CO emissions include a small source from oceans and soils, and the oxidation of biogenically emitted hydrocarbons, such as isoprene and terpenes. However, with a net production of CO averaging 1.2 parts per billion (p.p.b.) h^{-1} according to a summertime experiment in a pine forest in central Greece and chemical box model calculations (Gros *et al.*, 2002), this contribution should be negligible over our experimental time-scale of a few hours. The major loss of CO in the atmosphere is through reaction with OH, although a small amount is also lost by absorption onto soils; but a lifetime of about one month in summer means the photochemical sink of CO during regional campaigns can be ignored. In Europe, the CO surface flux should consequently be dominated by anthropogenic fossil fuel emissions. Multiplying the CO surface flux determined from the measured CO mixing ratio changes with the average $CO_2 : CO$ emission ratio from combustion sources thus gives an estimation of the fossil fuel contribution to the CO_2 surface flux.

However, the anthropogenic $CO : CO_2$ emission ratios may be subject to regional and seasonal variations, which are not yet well characterized. The reason for this is that the emission factors for different combustion sources vary strongly. For instance, the ratio of $CO : CO_2$ emissions from road transport is about a factor 1000 higher than for the energy industry in France (Fontelle *et al.*, 2000). Consequently, the average $CO : CO_2$ emission ratio depends on the relative contribution of different combustion sources reaching the detector, and their regional variability. Preliminary results from an aircraft campaign over the Paris urban area in February 2002 by Schmitgen *et al.* (2003) suggest, however, that near a large city, CO can be used successfully to quantify the anthropogenic component of the CO_2 regional fluxes.

5. Discussion and conclusions

Although some increase in reliability was obtained by Laubach and Fritsch (2002) by considering the mass balance and structure of the full column (also well above the inversion), the few attempts that have tried to use CBL budgeting have not met with unambiguous success. Improvement in reliability was also shown by Schmitgen *et al.* (2003), at least in quantifying the errors involved. Their study suggests that a

Lagrangian approach, provided there is a careful flight path to conserve Lagrangian conditions both horizontally and vertically, with a high-precision CO_2 sensor, can improve the reliability of the regional estimates. Some major problems remain. Predominantly, these include the treatment of advection. Furthermore, serious problems remain with the Lagrangian and Eulerian approaches in that data for the vertical velocity and entrainment are required to reduce the remaining errors. This suggests that CBL budgeting, without additional measurements on the structure and evolution of the boundary layer, is insufficient in itself. These additional measurements should include high-frequency radio sounding or estimation of the boundary layer height with radio acoustic sounders and/or sodars. It is interesting to note that in both the CBL budgeting and the flux aircraft technique, information on the vertical structure of the boundary layer is required to explain fully the results or to improve the reliability of the method. In both cases the gradients across the inversion that determine entrainment are of paramount importance for the success of the technique.

On the basis of the few, scattered studies available, it is hard to conclude definitively in favour of either the Lagrangian or the Eulerian approach. Whereas clearly the Eulerian approach has problems in dealing with advection, Schmitgen *et al.* (2003) showed also severe statistical errors that relate to the variance of the wind speed. They suggested that the main improvement to the technique may come from increased resolution of the airborne sensor (up to 0.15 p.p.m., comparable to that used by Gerbig *et al.* (2003)) and additional vertical flight information. If, however, both the Lagrangian and Eulerian methods have problems in obtaining relevant well-mixed conditions, it is easy to argue that further progress may be obtained by executing carefully planned experiments that use adequate sampling both vertically and horizontally through the domain. When such measurements are executed, assimilation within a regional-scale data assimilation framework based on mesoscale atmospheric models may be possible. First estimates of the flux could then be provided by high-resolution remote sensing data coupled to the land surface schemes. CO and carbon isotopes could then be used to separate further the fossil fuel contribution from the biospheric, and the respiration flux from the assimilation flux. The discussion of Lagrangian versus Eulerian would then become rather obsolete in this case.

Although initial attempts of both forward and inverse modeling of regional CO_2 concentrations are now feasible, several problems still remain. These relate to our current poor ability to represent the boundary layer processes, such as turbulence, growth, and entrainment, adequately in the models, and the amount of high-precision data that are required to check the model's performance. In the case of inverse models, the lack of high-frequency and high-resolution fossil fuel data, and high-precision atmospheric CO_2 data, still provides a serious constraint. However, the possibility of using these models in an inverse and prototype assimilation mode to obtain regional source–sink distributions remains an exciting one. A big advantage of using a Bayesian framework is that the sink–source distributions will be quantified with an associated error. This may help their use in verifying emission reductions in the context of the Kyoto protocol.

Acknowledgements

This paper is partly based on work done under the regional programme of CarboEurope RECAB funded by the European Commission. We thank the organizers of the Society for Experimental Biology meeting in Southampton for inviting us to present our work and contribute to these proceedings.

References

Bakwin P.S., Tans, P.P., Hurst, D.F. and Zhao, C. (1998) Measurements of carbon dioxide on very tall towers: results of the NOAA/CMDL program. *Tellus Series B – Chemical and Physical Meteorology* **50**: 401–415.

Baldocchi, D.D., Falge, E. and Gu, L.H. (2001). Fluxnet: a new tool to study the temporal and spatial variability of ecosystem-scale carbon dioxyde, water, vapor, and energy flux densities. *Bulletin of American Meteorological Society* **82**: 2415–2434.

Bougeault, Ph., Noilhan, J., Laccarrè, P. and Mascart, P. (1991) An experiment with an advanced surface parameterization in a a meso beta model. Part I. Implementation. *Monthly Weather Review* **119**: 2358–2373.

Brooks, S.B., Dumas, E.J. and Verfaillie, J. (2001) Development of the SkyArrow surface/atmosphere flux aircraft for global ecosystem research. *American Institute of Aeronautics and Astronautics Journal and Proceedings,* 39th Aerospace Sciences Meeting and Exhibit, January 8–11, 2001, Reno Nevada. 10 pp.

Chevillard, A., Karstens, U., Ciais, P., Lafont, S. and Heimann, M. (2002) Simulation of atmospheric CO_2 over Europe and Siberia using the regional scale model REMO. *Tellus Series B – Chemical and Physical Meteorology* **54**: 872–894.

Crawford, T.L. and Dobosy, R.J. (1992) A sensitive fast response probe to measure turbulence and heat flux from any airplane. *Boundary Layer Meteorology* **59**: 257–278.

Crawford, T.L. and Dobosy, R.J. (2004) Accuracy and utility of aircraft flux measurements. In: Kabat, O., Claussen, M., Dirmeyer, P.A., Gash, JHC., deGuenni, L.B., Maybeck, M., Pielke, R.A., Vorosmarty, C., Hutjes, R.W.A. and Lutkemeiere, S. (Eds) *Vegetation, water, humans and the climate. A new perspective on an interactive system.* Springer, Berlin. pp 183–197.

Crawford, T.L., Dobosy, R.J., McMillen, R.T., Vogel, C.A. and Hicks, B.B. (1996) Air-surface exchange measurement in heterogeneous regions: extending tower observations with spatial structure observed from small aircraft. *Global Change Biology* **2**: 275–285.

Culf, A.D., Fiosch, G., Mahli, Y. and Nobre, C.A. (1997) The influence of the atmospheric boundary layer on carbon dioxide concentrations over a tropical rain forest. *Agricultural and Forest Meteorology* **85**: 149–158.

Curtis, P.S., Hanson, P.J., Bolstad, P., Barford, C., Randolf, J.C., Scmid, H.P. and Wilson, K.B. (2002) Biomertix and eddy covariance based estimates of annual carbon storage in five eastern North American deciduous forests. *Agricultural and Forest Meteorology* **113**: 3–19.

Denning, A.S., Randall, D.A., Collatz, G.J. and Sellers, P.J. (1996) Simulations of terrestrial carbon metabolism and atmospheric CO_2 in a general circulation model. Part 2: Simulated CO_2 concentrations. *Tellus Series B – Chemical and Physical Meteorology* **48**: 543–567.

Desjardins, R.L., Brach, E.J., Alno, P. and Schuepp, P.H. (1982) Aircraft monitoring of surface carbon dioxide exchange. *Science* **216**: 733–735.

Dolman, A.J., van der Molen, M.L., ter Maat, H.W. and Hutjes, R.A.W. (2003) The effects of forests on mesoscale atmospheric processes. In: Mencucini, M., Grace, J., and McNaughton, K. (eds) *Forest and the Atmosphere*. BIOS publishers, Cambridge. in press.

Eastman, J.L., Coughenour, M.B. and Pielke, R.A. (2001) The regional effects of CO_2 and landscape change using a coupled plant and meteorological model. *Global Change Biology* **7**: 797–815.

Fan, S., Gloor, M., Mahlman, J., Pacala, S., Sarmiento, J., Takahashi, T. and Tans, P. (1998) A large terrestrial carbon sink in north America implied by atmospheric and oceanic carbon dioxide and models. *Science* **282**: 53–74.

Fontelle, J.-P., Chang, J.-P., Allemand, N., Audoux, N., Beguier, S. and Clement, C. (2000) Inventaire des emission de gaz a effet de serre au cours de la periode 1990–1999, Ref. CITEPA 430, Centre Interprofessionnel Technique d'Etudes de la Pollution Atmospherique (CITEPA), http://citepa.org, Paris, 2000.

Gerbig, C., Lin, J.C., Wofsy, S.C., Daube, B.C., Andrews, A.E., Stephens, B.B. *et al.* (2003) Constraining regional to continental scale fluxes of CO_2 with atmospheric observations over a continent; a receptor oriented approach of COBRA data. *Journal of Geophysical Research* **108(D24)**: 4756, doi:10.1029/2002JD003018.

Gioli, B., Miglietta, F., De Martino, B., Hutjes, R.W.A., Dolman, A.J., Lindroth, A. *et al.* (2004) Comparison between tower and aircraft-based eddy covariance fluxes in five European regions. *Agricultural & Forest Meteorology* in press.

Goodale, C. L., Apps, M.J., Birdsey, R.A., Field, C.B., Heath, L.S., Houghton, R.A. *et al.* (2002) Forest carbon sinks in the northern hemisphere. *Ecological Applications* **12**: 891–899.

Gros, V., Tsigaridis, K., Bonsang, B., Kanakidou, M. and Pio, C. (2002) Factors controlling the diurnal variation of CO above a forested area in southeast Europe. *Atmospheric Environment* **36**: 3127–3135.

Gurney, K., Law, R., Denning, S., Rayner, P.J., Baker, D., Bousquet, P. *et al.* (2002) Towards robust regional estimates of CO_2 sources and sinks using atmospheric transport models. *Nature* **415**: 626–630.

Janssens, I.A., Freibauer, A., Ciais, P., Smith, P., Nabuurs, G.J., Folberth, G. *et al.* (2003) Europe's terrestrial biosphere absorbs 7 to 12% of European anthropogenic CO_2 emissions. *Science* **300**: 1538–1542.

Kaminski, T., Rayner, P.J., Heimann, M. and Enting, I.G. (2001) On aggregation errors in atmospheric transport inversions. *Journal of Geophysical Research* **106 (D5):** 4703–4715.

Körner, C. (2003) Slow in, rapid out – carbon flux studies and Kyoto targets. *Science* **300**: 1242–1243.

Laubach, J. and Fritsch, H. (2002) Convective boundary layer budgets derived from aircraft data. *Agricultural and Forest Meteorology* **111**: 237–263.

Lenschow, D.H., Pearson, R. Jr and Stankov, B.B. (1981) Estimating the ozone budget in the boundary layer by use of aircraft measurements of ozone eddy flux and mean concentration. *Journal of Geophysical Research* **86**: 7291–7297.

Levy, P.E., Grelle, A., Lindroth, A., Mölder, M., Jarvis, P.G., Kruijt, B. and Moncrieff, J.B. (1999) Regional scale CO_2 fluxes over central Sweden by a boundary layer budget method. *Agricultural and Forest Meteorology* **98–99**: 169–180.

Lin, J.C., Gerbig, C., Wofsy, S.C., Daube, B.C., Andrews, A.E., Bakwin, P.S., Davis, K.J., Stith, J. and Grainger, A. (2003) A near-field tool for simulating the upstream influence of atmospheric observations: The Stochastic Time-Inverted Lagrangian Transport (STILT) model, *J. Geophys. Res.* **108** (D16): 4493. doi:10.1029/2002JD003161.

Lloyd, J., Kruijt, B., Hollinger, D.Y., Grace, J., Francey, R.J. *et al.* (1996) Vegetation effects on the isotopic composition of atmospheric CO_2 at local and regional scales: theoretical aspects and a comparison between rainforest in Amazonia and a boreal forest in Siberia. *Australian Journal of Plant Physiology* **23**: 371–399.

Lloyd, J., Francey, R., Mollicone, D., Raupach, M.R., Sogachev, A., Arneth, A. *et al.* (2001) Vertical profiles, boundary layer budgets, and regional flux estimates for CO_2 and its $^{13}C/^{12}C$ ratio and for water vapor above a forest/bog mosaic in central Siberia. *Global Biogeochemical Cycles* **15**: 267–284.

McGuire, A.D.C., Sitch, S., Clein, J.S., Dargaville, R., Esser, G., Foley, J. *et al.* (2001) Carbon balance of the terrestrial biosphere in the twentieth century: Analysis of CO_2 climate and land use effects with four process-based ecosystem models. *Global Biogeochemical Cycles* **15**: 183–206.

Meijer, H.A.J., Smid, H.M., Perez, E. and Kreizer, M.G. (1996) Isotopic characterization of anthropogenic CO_2 emissions using isotops and radiocarbon analysis. *Physics and Chemistry of the Earth* **21**: 483–487.

Nicholls, M.E., Denning, A.S., Prihdodko, L., Vidale, P.-L., Baker, I., Davis, K. and Bakwin, P. (2003) A multiple-scale simulation of variations in atmospheric carbon dioxide using a coupled biosphere–atmospheric model. Available at http://biocycle.atmos.colostate.edu/RAMS.3D.1997.pdf.

Oechel, W.C., Vourlitis, G.L., Brooks, S.B., Crawford, T.L. and Dumas, E.J. (1998) Intercomparison between chamber, tower, and aircraft net CO_2 exchange and energy fluxes measured during the Arctic system sciences land—atmosphere–ice interaction (ARCSS-LAII) flux study. *Journal of Geophysical Research* **103**: 28993–29003.

Peylin, P., Baker, D., Sarmiento, J., Cias, P. and Bousquet, P. (2002) Influence of transport uncertainty on annual mean and seasonal inversions of atmospheric CO_2 data. *Journal of Geophysical Research* **107**: doi:10.1029/2001JD000 857.

Pielke R.A., Cotton, W.R., Walko, R.L., Tremback, C.J., Lyons, W.A., Grasso, L.D. *et al.* (1992) A comprehensive meteorological modelling system – RAMS. *Meteorology and Atmospheric Physics* **49**: 69–91.

Potosnak, M.J., Wofsy, S.C., Denning, S., Conway, T.J., Munger, W. and Barnes, D.H. (1999) Influence of biotic exchange and combustion sources on atmospheric CO_2 concentrations at New England from observations at a forest flux tower. *Journal of Geophysical Research* **104**: 9561–9569.

Raupach, M.R., Denmead, O.T. and Dunin, F.X. (1992) Challenges in linking atmospheric CO_2 concentrations to fluxes at local and regional scales. *Australian Journal of Botany* **40**: 697–716.

Roedenbeck, C., Houwelin, S., Gloor, M. and Heimann, M. (2003) CO_2 flux history 1982–2001 inferred from atmospheric data using a global inversion of atmospheric transport. *Atmospheric Chemistry and Physics Discussions* **3**: 2575–2659.

Ronda, R.J. and Dolman, A.J. (2003) Obtaining the regional distribution of sources and sinks of carbon. Poster presented at the CarboEurope Conference: The continental carbon cycle, Lisbon, 19–21 March 2003.

Schmitgen, S., Ciais, P., Geiss, H., Kley, D., Voz-Thomas, A., Neiniger, B., Baeumle, M. and Brunet, Y. (2003) Carbon dioxide uptake of a forested region in southwest France derived from airborne CO_2 and CO observations in a Lagrangian budget approach. *Journal of Geophysical Research* **109**: D14302. doi:10.1029/2003JD004335.

Schulze, E.D., Högberg, P., Van Oene, H., Persson, T., Harrison, A.F., Read, D., Kjoller, A. and Matteucci, G. (2000) Interactions between the carbon and nitrogen cycles and the role of biodiversity: A synopsis of a study along a north–south transect through Europe. In: Schulze (ed) *Carbon and Nitrogen Cycling in European Forest Ecosystems*, pp. 468–491, *Ecological Studies* **142**. Springer Verlag, Heidelberg.

Styles, J.M., Lloyd, J., Zolotoukhine, D., Lawton, K.A., Tchebakova, N., Francey, R.J., Arneth, A., Salamakho, D., Kolle, O. and Schulze, E.-D. (2002) Estimates of regional surface carbon dioxide exchange and carbon and oxygen isotope discrimination during photosynthesis from concentration profiles in the atmospheric boundary layer. *Tellus Series B – Chemical and Physical Meteorology* **54**: 768–783.

Uliasz, M. (2003) A modeling framework to evaluate feasibility of deriving mesoscale surface fluxes of trace gases from concentration data. Available at http://biocycle.atmos.colostate.edu/Marek.mesoinversion7c.pdf.

Valentini, R., Matteucci, G. and Dolman, A.J. (2000) Respiration as the main determinant of carbon balance in European forests. *Nature* **404**: 861–865.

Wirth, C., Czimczik, C.I. and Schulze, E.-D. (2002) Beyond annual budgets : carbon flux at different temporal scales in fire-prone Siberian Scots pine forests. *Tellus Series B – Chemical and Physical Meteorology* **54**: 611–630.

7

The potential for rising CO_2 to account for the observed uptake of carbon by tropical, temperate, and Boreal forest biomes

Philippe Ciais, Ivan Janssens, Anatoly Shvidenko, Christian Wirth, Yadvinder Malhi, John Grace, E.-Detlef Schulze, Martin Heimann, Oliver Phillips and A.J. (Han) Dolman

1. Introduction

Terrestrial ecosystems currently sequester up to one half of annual fossil fuel CO_2 emissions (IPCC, 2001). How and where this uptake takes place is subject to many investigations and uncertainties (Schimel *et al.*, 2001). That an ecosystem sequesters carbon implies that the gain of carbon by net primary productivity (NPP) exceeds the losses to the atmosphere by heterotrophic respiration and episodic disturbances such as fires, pest outbreaks, and harvests. It is as yet uncertain whether the current terrestrial carbon sink is the result of an increase of NPP over respiration, or a decrease in respiration or disturbance regime, with NPP staying more or less constant. On inter-annual time-scales, it is likely that all three one-way fluxes change together in response to climate variation (Duncan *et al.*, 2003; Goulden *et al.*, 1996; Kindermann *et al.*, 1996). On longer time-scales, however, there is convincing evidence that NPP has been increasing over time since the second half of the 20th Century (Mund *et al.*, 2002; Spiecker *et al.*, 1996). Tree ring analysis across Europe has shown that European forest productivity increased by 18% between 1960 and 1990 (Stanners and Bourdeau, 1995). Satellite measurements over the past 20 years have indicated a clear

The Carbon Balance of Forest Biomes, edited by H. Griffiths and P.G. Jarvis.
© 2005 Taylor & Francis Group

greening trend over the mid- and high latitudes of the Northern Hemisphere (Hicke *et al.*, 2002 a and b; Myneni *et al.*, 1997; Nemani *et al.*, 2003). The amplitude of the seasonal cycle of CO_2 in the Northern Hemisphere has increased by almost 25% since 1960 (Keeling *et al.*, 1996). Candidate processes that may cause such a secular NPP increase are the fertilization effect of rising CO_2 concentration, the deposition of nitrogen oxides over industrial regions, changes in climate and radiation (see, for example, Cannell, 1999; Roderick *et al.*, 2001), changes in disturbance regimes (Shvidenko and Nilsson, 2000) and in forestry practices (Mund *et al.*, 2002), and changes in forest area (Goodale *et al.*, 2002). It is of key importance both for improving projections of future atmospheric CO_2 concentrations, and for separating direct from indirect human-induced effects on carbon sinks, as requested in the Kyoto Protocol, to quantify separately the role of each controlling mechanism.

The decadal, spatial-average carbon sink, known as the net biome productivity or NBP (Schulze and Heimann, 1998), relates to NPP and its changes (Lloyd, 1999; Thompson *et al.*, 1996), and to the residence time of carbon within each ecosystem (McGuire *et al.*, 2002). Multiple residence times are associated with the cascade of terrestrial soil carbon pools, ranging from less than a year for the rapidly decomposing detritus up to thousands of years for the refractory soil carbon compounds (Telles *et al.*, 2003; Trumbore, 1993). The novelty of this analysis lies in synthesizing ground-based ecological data with atmospheric composition data and inverting explicitly to obtain a 'beta factor' value, and then seeing if these values are plausible together within a Monte-Carlo error analysis.

Three forest biomes of global significance have been studied in this way: the Amazonian tropical forests; the European mixed evergreen, needle-leaved and deciduous broad-leaved temperate forests; and the largely coniferous Siberian Boreal forests. For each biome, we compiled NPP and NBP at different scales from detailed site studies, extensive forest biomass inventories, and ecosystem-model simulations (Sections 2 to 4). In this 'bottom-up' approach, local information has been extrapolated to the biome scale.

Additionally, we inferred forest NBP based on a 'top-down' inversion approach, in which latitudinal gradients of atmospheric CO_2 concentration are disaggregated into continental-scale CO_2 flux estimates using numerical atmospheric transport models. We have subsequently analysed the relationship between NPP and NBP in terms of the CO_2 fertilization effect (Section 5). Our goal is to determine the extent to which the current carbon sinks can be explained by CO_2 fertilization, as compared with other controlling factors.

2. Observations of net primary productivity

2.1 *Net primary productivity from detailed site studies*

We first discuss NPP information obtained from stand scale biometric studies. Although there are numerous such measurements reported for above-ground NPP (see, for example, Cannell (1982); http://www-eosdis.ornl.gov/NPP/npp_home.html), such estimates are difficult to synthesize because different components of NPP are often reported, and because the exact method used to calculate NPP from primary field data is not always clearly described and may bias the estimates (see Clark *et al.*, 2001a, for tropical stands). In particular, below-ground NPP components are rarely reported in biometric studies. Failure to account for below-ground carbon flows into fine roots may strongly underestimate NPP. For example, fine root turnover and production of root exudates can contribute 18–33% of NPP in Boreal forests (Jackson *et al.*, 1997; Steele *et al.*, 1997; Vogt *et al.*, 1996), 8–29% in temperate European forests (Schulze *et al.*, 2000), and 18–56% in neotropical forests (Clark *et al.*, 2001a; Grace *et al.*, 2001). Thus below-ground NPP often represents a larger contribution to total NPP than annual stem growth. In the *Supplementary data* and *Figure 1* we have selected those studies in which fine-root NPP was either inferred from measurements (Kajimoto *et al.*, 1999; Schulze *et al.*, 2000; Wirth and Schulze, 2002) or indirectly estimated from mass balance calculations, with varying assumptions about root respiration (Grace *et al.*, 2001). Average estimates for each biome are given in *Table 1* (see page 116) and shown in *Figure 2* (see page 120).

For Siberia, we collected biometric NPP data for 606 stands from the database of Bazilevich (1993) and supplemented them with data from the International Institute for Applied Systems Analysis (IIASA). A major part of this database contains results of measurements made from the 1960s to 1990s. The overall average NPP of 2.99 Mg (C) ha^{-1} per year includes data from forests dominated by spruce, fir, and Russian stone pine (*Pinus cembra* var. *sibirica*) (mean 3.01, $n = 223$), larch (2.44, $n = 85$), pine (2.72, $n = 186$), birch and aspen (3.57, $n = 91$), and mixed coniferous–broadleaved forests (largely dominated by Korean pine, *Pinus koraiensis* Sieb. and Zucc.) (5.38, $n = 21$). We did not use these data directly for up-scaling because of lack of co-located NBP estimates and because of the uneven spatial distribution of the sample plots. A disproportionate number of the plots were situated in more productive, southern parts of the species distribution, and there are significant latitudinal differences in NPP between the taiga and Boreal regions (e.g., average NPP for the forests of the northern taiga zone is 2.18 Mg (C) ha^{-1} per year ($n = 107$), whereas the average for the Boreal region is 4.00 Mg (C) ha^{-1} per year ($n = 242$).

For Europe, we collected NPP from the 'elite' data set of European stands in the CANIF project (http://www.bitoek.uni-bayreuth.de/bitoek/en/forschung/proj/; Schulze *et al.*, 1999, 2000). The data show significant site-to-site variation in NPP that can be explained by differences in stand age, nutrient availability, soil fertility, and climate. The average may be somewhat biased towards high values compared with the 'real' average for European temperate forests, both because of the distribution of the sites and because of the selection of healthy 'mature' stands for the CANIF studies. The average NPP value of the European stands (6.2 ± 1.8 Mg (C) ha^{-1} per year, $n = 11$) is about double the value for the Siberian stands (2.99 ± 0.28 Mg (C) ha^{-1} per year, $n = 620$). This is no surprise because trees in Siberia grow in harsher environments with much shorter growing seasons, temperature and nutrient limitations in northern Siberia, and possibly water limitations, caused by summer droughts, in eastern Siberia (Kelliher *et al.*, 1997).

For the neotropics, we collected NPP data from various stands reported in http://www-eosdis.ornl.gov/NPP/npp_home.html (most data being from the 1970s and 1980s) and by Clark *et al.* (2001b). The mean NPP values for these two sources are, respectively, 7.0 ± 0.6 Mg (C) ha^{-1} per year ($n = 6$) and 8.2 ± 1.7 Mg (C) ha^{-1} per year ($n = 13$), i.e., between 15% and 33% higher than the mean value for the European stands. However, these values are probably underestimates of total tropical forest

NPP for several reasons. Firstly, fine root turnover, exudates, harvest by herbivores (Lowman, 1995; McNaughton, 2001), losses of volatile organic compounds (VOCs) to the atmosphere (Kesselmeier *et al.*, 2003) and of dissolved organic carbon (DOC) into the rivers and freshwater systems (Richey *et al.*, 2002) were all neglected. Secondly, the contribution to NPP from very large trees, lianas, and epiphytes are also all generally omitted in currently reported NPP data (Clark *et al.*, 2001a). Taken altogether, inclusion of these omissions may double the current NPP estimates for the tropics, although with very large uncertainties. For example, accounting for fine root turnover and exudation in the tropical forests stands plotted in *Figure 1* (see pages 117–119) would roughly double the NPP of the Amazon stand, if the above-ground NPP were to be scaled by the same ratio of total to above-ground NPP estimated by Grace *et al.* (2001) for a site on infertile soils in central Amazonia.

Despite their inherent uncertainties, stand-scale NPP measurements most likely represent our best knowledge of ecosystem processes. An important point yet to consider is which components of NPP are most relevant to our present purpose. The different components of NPP all have different carbon sink potentials and are connected to carbon pools with significantly different turnover times. In the tropics, fine roots and leaf carbon stores generally have a short residence time, and hence equilibrate rapidly with changing environmental conditions and, therefore, we should not expect them to contribute significantly to NBP. Hence, knowledge of the productivity of coarse wood (above ground and below ground) may be sufficient for calculation of NBP (see Section 4). In temperate and Boreal forests, by contrast, fine roots, twigs and leaves decay much more slowly and consequently have much longer residence times, so that their NPP does contribute strongly to NBP, as illustrated by the significant rates of carbon accumulation observed in soil profiles. On the other hand, for all biomes, root exudates and VOCs have very short residence times (less than one year) and, it could be argued, are almost indistinguishable from autotrophic respiration.

The main source of uncertainty in using NPP from detailed field studies, in the context of their present usage, is the sparse and uneven distribution of stand NPP measurements across the range of disturbance and climate regimes experienced by ecosystems in each biome. Thus bold assumptions are required to up-scale the behaviour of such point measurements to continental scales. To overcome this difficulty, general models of biomass allocation and up-scaling procedures for carbon stocks to stand scale, as developed by Wirth and Schulze (2002) in Europe, need to be developed for other regions.

2.2 *Net primary productivity from spatially extensive forest biomass inventories*

Next, we examine continental-scale NPP estimated from forest biomass inventories (*Supplementary data*; *Figure 1* (pages 117–119)). In biomass inventories, NPP is estimated from the measured volume growth of stem wood and changes in forest area, the modelled fine litter and coarse debris production, corrected for harvest and slash production in managed forests, using book-keeping type of models and geographic information systems for up-scaling. The advantage here is covering and understanding spatial variability, at the expense of detailed knowledge of the component processes. A large uncertainty arises from applying allometric relationships established in a few stands to the thousands of trees counted in an inventory census

(Schelhaas and Nabuurs, 2001; Wirth *et al.*, 2003), although some recent developments do demonstrate increased reliability of multi-dimensional regression models (Shvidenko and Nilsson, 2002; Usoltsev, 2001).

For Siberia, we used the recent NPP estimates of Shvidenko *et al.* (2001) for the period 1988–1992 and Shvidenko and Nilsson (2003) for the period 1961–1998. For Europe, we used the data reported by Nabuurs *et al.* (2003) for the period 1950–1999 for western and central Europe (30 countries). In Amazonia, where there is no systematic basin-wide commercial forest inventory, stem growth measurements have been made by the RAINFOR project (Malhi *et al.*, 2002; http://www.geog.leeds/projects/rainfor/). We present here estimates of above-ground NPP from a new standardized analysis of 100 plots in the New World tropics (Malhi *et al.*, 2003), that builds upon the smaller network of plots analysed by Phillips *et al.* (1998). Here, these values are simply averaged, and not weighted according to geographic coverage.

The inventory approach gives continental-scale NPP values that are by comparison 40% higher for Amazonia, 20% lower for Siberia than stand-scale studies, but comparable to them within their errors (*Figure 2*). Averaged over geographical Europe, forest inventory NPP is 30% lower than the mean of the CANIF stands, which are all in the post harvest 'rebound' phase of growth, or situated at sites with higher-than-average soil fertility, and therefore are not entirely representative of European forests. Even more likely, part of the discrepancy could also be related to systematic underestimation of fine-root NPP in the inventory models. Note, however, that the ratio of NPP between Europe and Siberia for the forest inventory studies (i.e., 2.3) is about the same as the comparable ratio found for stand-scale studies (2.8). This indicates a systematic difference common to both approaches (fine roots?).

3. Estimation of net biome productivity

3.1 *The problems*

Providing defensible estimates of stand-scale NBP is challenging because one must account for soil carbon changes and for the impact of disturbances in controlling the long term average carbon balance (see Figure 8 in Schulze *et al.*, 1999; Körner, 2003). In Boreal forests for example, fluxes caused by disturbances and by consumption of forest products constitute about 20% of NPP and can exceed NBP (Shvidenko and Nilsson, 2003). One cannot simply scale NBP from limited duration (less than 10 years) net ecosystem production (NEP) data obtained by eddy covariance (EC) (Baldocchi *et al.*, 2001), because most eddy-flux towers are located in young or middle-aged stands and do not sample the effects of disturbances, and because the length of such records is too short to average out the large fluctuations in NEP generally observed from one year to the next (Arain *et al.*, 2002; Chen *et al.*, 1999; Wirth *et al.*, 2002). The alternative to attempting to measure stand NBP directly is to make use of chronosequences (Mund *et al.*, 2002).

It is fair to say that there are problems of quantifying NBP at various scales. Firstly, at the spatial scale of stands, NBP can only be defined from the *in situ* carbon balance and thus does not include the fluxes of 'displaced' carbon such as the losses to streams and rivers, or the fate of wood products in harvested forests. A second issue deals with temporal scales in the context of this study. NBP contains 'background' fluxes such as the formation of charcoal in fire-disturbed forests, the inclusion of carbon into mineral soil horizons (see Schulze *et al.*, 1999), or the transport of rock-weathered, dissolved inorganic carbon (DIC) in rivers to the ocean. Such fluxes are part of the 'background' carbon cycle and evolve on long time-scales, whereas we are interested in NBP as the anomalous sink occurring at the spatial scale of the biome on decadal time-scales in response to short-term natural and anthropogenic perturbations of the carbon cycle. It is virtually impossible to separate 'background natural' from 'anthropogenic' NBP in field studies, except maybe when using the bomb-radiocarbon signal to determine the soil component of NBP since the 1960s (Gaudinski *et al.*, 2000; Harrison and Harkness, 1993; Trumbore *et al.*, 1995). On the other hand, inversions of atmospheric gas concentrations do include background NBP components and should, therefore, be 'corrected', so as to be comparable with the field studies.

3.2 *Estimation of net biome productivity using the ecosystem inventory approach*

Very few detailed field studies of NPP have also been used to estimate NBP. As examples, we have selected here three stand-scale estimates of NBP, as given in the *Supplementary data* and *Figure 1*. One is inferred from pine chronosequences in central Siberia (Wirth *et al.*, 2002), the second from radiocarbon measurements in soil in Europe (Schulze *et al.*, 2000), and the third from the net change in above-ground biomass in Amazonian stands (Baker *et al.*, 2004; Phillips *et al.*, 1998). For forest biomass inventory estimates, we assembled NBP data that include corrections for harvest disturbance and the subsequent fate of wood products and other natural and human-induced disturbances. One major source of uncertainty in this method is that the soil carbon balance is rarely estimated using a C/N model validated against field data (Van Oene *et al.*, 2000). Instead, most forest inventory studies use a simple model that has not been validated against field data, or use aggregated estimates of the impacts of changing disturbance regimes on a snapshot of soil inventory data (Shvidenko *et al.*, 2003), or do not account for soil carbon changes at all (Caspersen *et al.*, 2000; Phillips *et al.*, 1998).

3.3 *Estimation of net biome productivity using the atmospheric inversion approach*

In addition, we used the inversion approach to estimate NBP from latitudinal gradients of gas concentrations in the atmosphere (Bousquet, 1999; Gurney *et al.*, 2002; Kaminski and Heimann, 1999; Rayner *et al.*, 1999; Schulze *et al.*, 1999). In this approach, small spatial concentration gradients of CO₂ (and other gases) from not more than 100 globally distributed atmospheric stations are inverted to give estimates of fluxes over large continental regions, using atmospheric transport models of the gases. After subtraction of the source terms from the inverted CO₂ fluxes, the residual term can be called the 'forest NBP'. The source terms, which must be known from independent data, comprise the fossil fuel emissions of CO₂ (with a typical uncertainty of 5%), the net carbon fluxes from agriculture and pasture lands (values for Europe from Vleeshouwers and Verhagen *et al.* (2002); uncertainty of 100%), and the net deforestation fluxes in Brazilian Amazonia (values for the period 1978–1998 from Houghton *et al.* (2000); uncertainty of 50%). There is also a flux correction of 0.3 Mg (C) ha⁻¹ per year for the inversion regions (see below). Note that recent satellite studies for the 1990s suggest that tropical deforestation fluxes may have been overestimated in the past (Achard *et al.*, 2002; DeFries *et al.*, 2002; Chapter 10, this volume). The inversion-based estimate of 'forest NBP' depends on the modelled atmospheric transport, on inversion settings, including concentration data temporal aggregation and error, aggregation of surface fluxes, correlations and prior estimation errors. As a result, inversion-derived estimates of NBP show the largest spread in *Figure 1*.

3.4. *Comparing results from the inventory and inversion approaches*

It is difficult to make a realistic comparison between forest NBP from the atmospheric inversion-derived approach with ecosystem inventory-based estimates. Firstly, inversion-derived fluxes cover very large regions in which additional assumptions are required to separate non-forest and forest NBP. Secondly, atmospheric inversions are notoriously ill-constrained over the *interior* of continents because of sparse and unevenly distributed atmosphere observation sites, making the NBP retrieval very uncertain. Thirdly, regional inversions do not detect non-CO_2 carbon losses from ecosystems (e.g., from biomass burning, VOCs and CH_4 sources) unless the products are oxidized to CO_2 *within* the inversion region. The results of regional inversions will only include any fluxes of DOC and DIC leaching from forest soils into rivers if such laterally displaced carbon is released back to the atmosphere as CO_2 *within* the region covered by the inversion. We have assumed that all DOC is released to CO_2 within each inversion region either in freshwater systems or in the coastal zone and, therefore, that this term is counted in the inversions. For instance, we acknowledge the existence of a large source of CO_2 occurring over flooded areas in the Amazon basin (Richey *et al.*, 2002) that is not accounted for locally as ecosystem respiration but, so long as it is emitted within the inversion region as CO_2, it is accounted for by the inversion. Thus we did not subtract this source from the inverted CO_2 fluxes when estimating the inversion 'forest NBP'. (Note that this flux approximated 0.33 Mg (C) ha^{-1} per year if extrapolated to the inversion region, a source of the same absolute value as the forest inventory sink!) On the other hand, we have assumed that most of the DIC transferred to rivers from the forests gets to the deep ocean where it eventually precipitates as carbonates and releases CO_2. This constitutes a net CO_2 sink that is detected by the inversion approach but is not observed by the field-based NBP estimates (Houghton *et al.*, 2000). It is thus appropriate in this work, where we seek to estimate the anthropogenic NBP, to correct the inversion-derived NBP for this backgound term. To do so, we subtracted from NBP in each inversion region the average consumption of atmospheric CO_2 by rock weathering as given by Amiotte Suchet *et al.* (1994), i.e., 0.024 Mg (C) ha^{-1} per year, 0.023 Mg (C) ha^{-1} per year, and 0.018 Mg (C) ha^{-1} per year for Amazonia, Europe, and Siberia, respectively. Lastly, inversion-derived regional NBP represents a spatial average over a few years of observations (see *Supplementary data*), and thus may not approximate the long-term mean carbon continental balance in the presence of inter-annual variability. For example, terrestrial carbon sinks during the early 1990s were globally stronger, and regionally stronger at least in North America (Bousquet *et al.*, 2000; Hicke *et al.*, 2002 a and b) and in Siberia (Lucht *et al.*, 2002), than during the entire 1980–2000 period.

Table 1. *Mean estimates of net primary productivity (NPP) and net biome productivity (NBP) for Europe, Amazonia and Siberia when estimated by contrasting methods (see Supplementary data for details).*

	Detailed stand-scale studies	Extensive forest biomass inventories	Atmospheric inversions 'forest NBP'	Ecosystem models forested grid cells only
	Europe			
Area	136			
NPP	6.17 (1.82)[a]	4.48 [d]	—	4.77 (1.53)[g]
NBP	1.06 (0.59)[b]	1.07 (0.47) [e]	0.51 (0.55)[f]	0.13 (0.18)[g]
Turnover	8+43/22+125[c]	49 [d]	—	49[d]
	Amazon			
Area	711.6			
NPP	8.2 (1.7)[k] 7.01 (0.55) [l]	9.14 (2.55) [m]	—	9.01 (2.44)[g]
NBP	—	0.62 (0.37)[n]	0.65 (0.87)[f]	0.15 (0.03)[g]
Turnover	15.6 + 16[h] 15.6 + 80[i]	56 (22) + 80[j]	—	56 + 80[j]
	Siberia			
Area	770			
NPP	2.81 (0.94)[o] 2.99 (0.28)[p]	2.28 (0.14) [r]	—	3.04 (1.47)[g]
NBP	1.47[q] 0.17 (0.27)[q]	0.24 (0.67) [r]	0.5 (0.54)[f]	0.10 (0.07)[g]
Turnover	67[s]	93[t]	—	97[u]

a. Average from 11 beech and spruce plots in the European CANIF transect in Schulze (*et al.* 2000).
b. Average from 11 beech and spruce plots in the European CANIF transect; NBP assumed equal to 0.25 times the long-term soil carbon accumulation.
c. For each CANIF stand, a site-specific turnover time was used as reported by Persson *et al.* (2000) for SOM (range 43 to 125 years) and Scarascia-Mugnozza *et al.* (2000) (range 8–33 years). The range between extreme values among stands (biomass + SOM) is reported here.
d. Geographic Europe national inventory data used in the book-keeping model of Nabuurs *et al.* (2003).
e. As reported by Janssens *et al.* (2003).
f. Inversion results from Gurney *et al.* (2002) (15 models), Bousquet *et al.* (2000) (two models), Kaminski and Heimann (1999), and Rayner *et al.* (1999), cited in text. Continental-scale NBP from inversions are corrected into 'forest NBP' by subtracting the land use source in Brazil (Houghton *et al.*, 2000), and crop and pasture net fluxes in geographic Europe (Vleeshouwers and Verhagen, 2002). The error is the squared sum of the spread of the inversion results, Bayesian errors, and bottom-up-based corrections.
g. Four global ecosystem models HRBM, TEM, IBIS, LPJ run under the same protocol including CO$_2$ fertilization, climate variability, and changes in forest area for crop establishment. The model results are averaged over the forested grid cells only. Uncertainty is from the spread of results among the models.
h. Turnover in woody biomass + SOM (Malhi and Grace, 2000).
i. Turnover in woody biomass (Malhi and Grace, 2000) + turnover in SOM active and slow pools (Telles *et al.*, 2003).
j. Turnover in woody biomass averaged from 100 plots (Malhi *et al.*, 2004) + turnover in SOM active and slow pools (Telles *et al.*, 2003).
k. Thirteen plots in the New World tropics without fine-root turnover (Clark *et al.*, 2001b).
l. Six plots in the New World tropics at http://www-eosdis.ornl.gov/NPP/npp_home.html.
m. One hundred plots in the entire Amazon Basin without fine-root turnover (Malhi *et al.*, 2003).
n. Woody NBP only (Phillips *et al.*, 1998); uncertainty = 0.5 (max.–min.).
o. Area weighted NPP from 54% *Larix* stands (Kajimoto *et al.*, 1999) and Wang and Polglase, (1995); 12% pine in central Siberia (Wirth *et al.*, 2002); and 32% spruce stands (Bazilevich, 1993).
p. Average over Siberia (Asian Russia) from 606 sample plots of adapted from Bazilevich, (1993).
q. Landscape average NPP, NBP (Wirth *et al.* 2002), pine chronosequences in central Siberia.
r. Average value over Russia (Shvidenko and Nilsson 2003).
s. Turnover inferred from area-weighted landscape average total ecosystem carbon stocks (98 Mg (C) ha^{-1}) of pine forests (Wirth *et al.*, 2002), calculated from age-class distribution of pine in Kranoyarsk Krai (Alexeyev and Birdsey, 1994).
t. Turnover inferred from Russian forest area carbon stocks (Nilsson *et al.*, 2000).
u. Same as footnote t above, but considering all ecosystems in Russia instead of forest only.

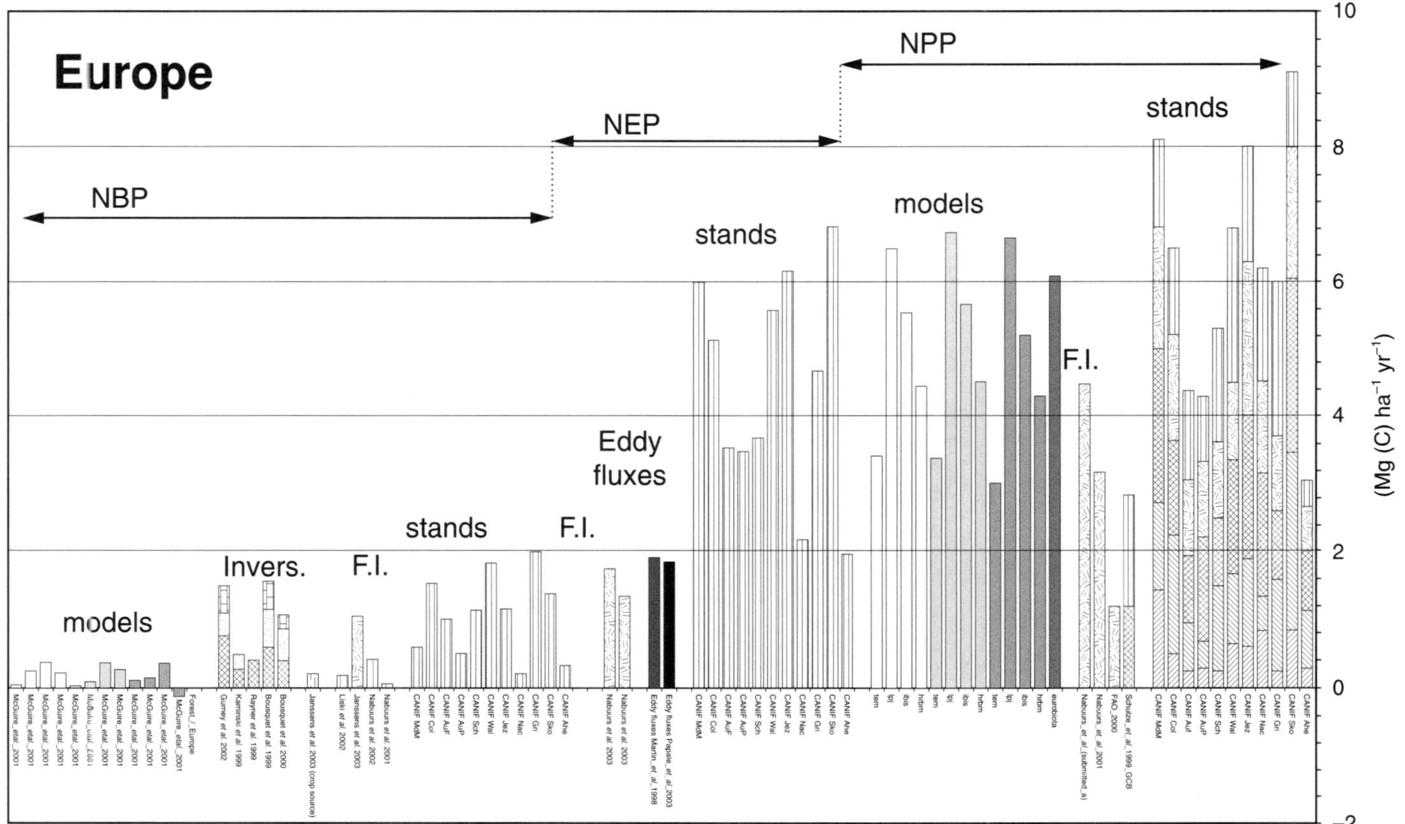

Figure 1 (a). *NPP, NEP, and NBP data over Western Europe obtained from different methods; stands: 11 plots from CANIF transect with NPP components coarse roots (   ), fine roots (   ), stem (   ), twigs and branches (   ), leaves (   ); F.I., forest biomass inventories; models: four global models HRBM, IBIS, LPJ, TEM from left to right prescribed with rising CO_2, climate variability and rising CO_2, climate variability and rising CO_2 and crop establishment (McGuire et al. 2002). Eddy fluxes: up-scaled EUROFLUX towers using satellite and gridded climate fields. Invers., atmospheric inversions are 'forest NBP' obtained from subtraction of cropland and pasture fluxes in red (Janssens et al., 2003).*

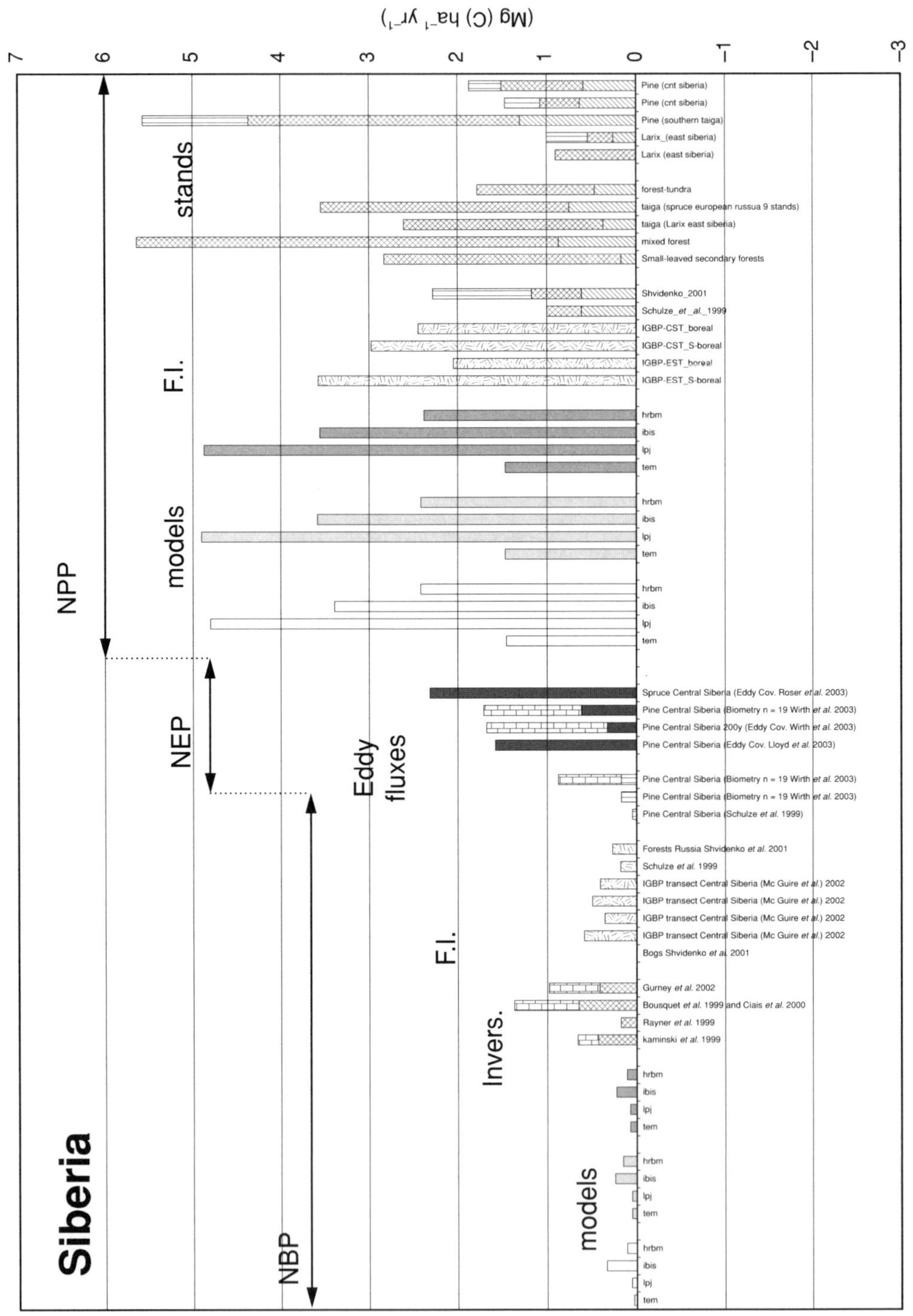

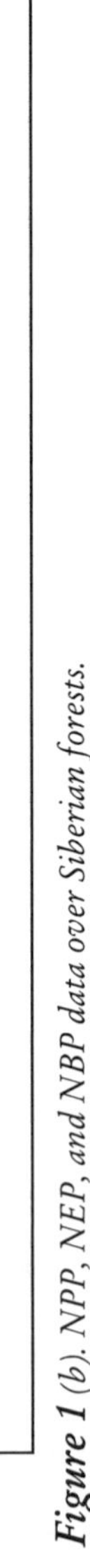

Figure 1 (b). NPP, NEP, and NBP data over Siberian forests.

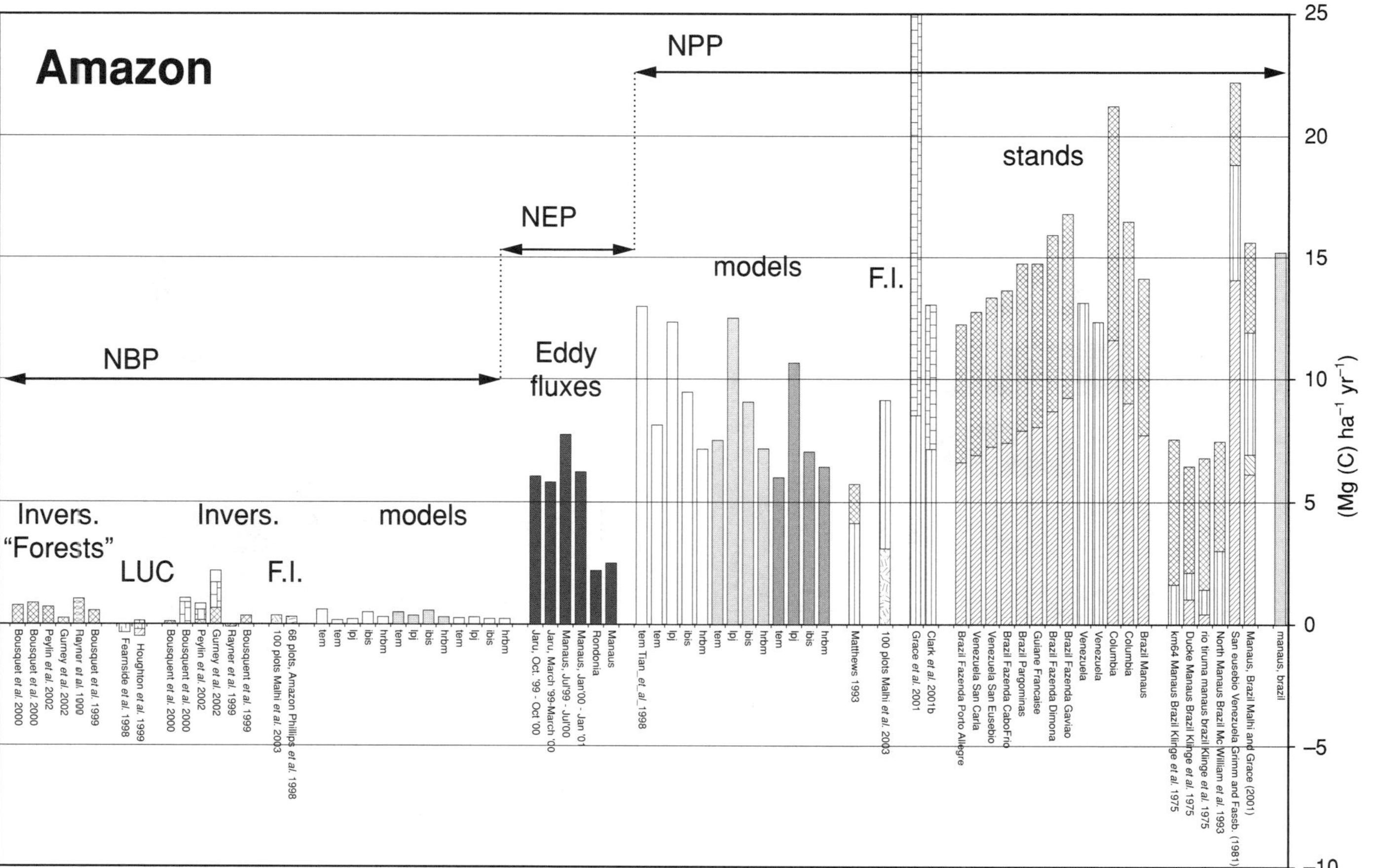

Figure 1 (c). *NPP, NEP, and NBP data over tropical Amazon forests. Invers. 'Forests', inversion fluxes subtracted with deforestation (LUC) source.*

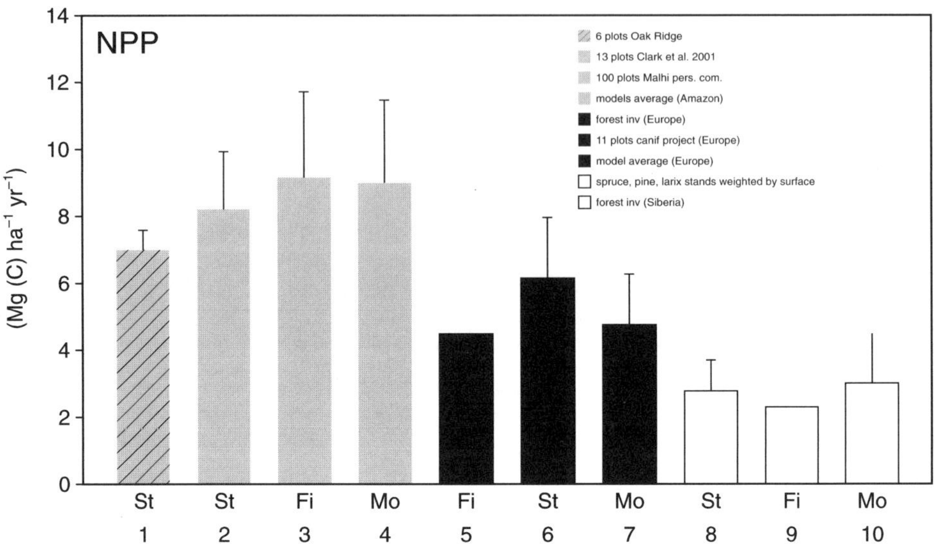

Figure 2a. *Average NPP estimates over the Amazon (bars 1, 2, 3 and 4), Europe (bars 5, 6 and 7) and Siberia (bars 8, 9 and 10). The numbers and errors are those reported in Table 1. Fi, forest biomass inventories; Mo, global dynamic vegetation models; Inv, atmospheric inversions down-scaled for forested area in each continent; St, stand-scale data.*

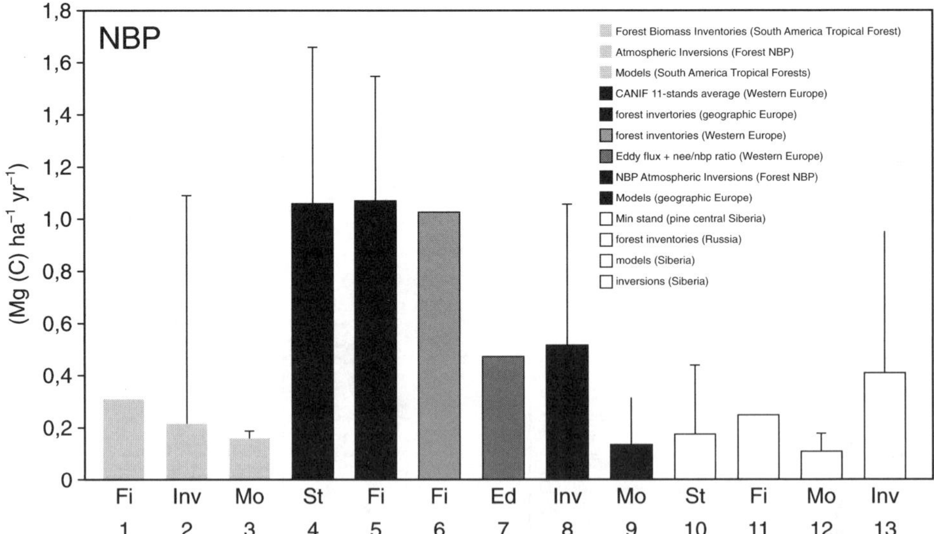

Figure 2b. *Average NBP estimates over the Amazon (bars 1, 2 and 3), Europe (bars 4, 5, 6, 7, 8 and 9) and Siberia (bars 10, 11, 12 and 13). The numbers and errors are those reported in Table 1. Fi, forest biomass inventories; Mo, global dynamic vegetation models; Inv, atmospheric inversions down-scaled for forested area in each continent; St, stand-scale data.*

Although the Boreal, temperate, and tropical forest biomes have very different NPP values, with at least a factor of four between the two extremes of growing conditions in Siberia and Amazon, we have found that NBP does not correlate with NPP (*Figure 2*). It is also remarkable that, despite all the uncertainties in estimation of NBP from biomass inventories and from stand-scale estimations, we have found close agreement between them. For instance, in Europe, average NBP of CANIF stands (1.1 ± 0.6 Mg (C) ha^{-1} per year) is very close to the forest inventories estimate of NBP recently reported by Janssens *et al.* (2003) (1.1 ± 0.5 Mg (C) ha^{-1} per year). The CANIF stand NBP is obtained as a fraction (24%) of soil carbon accumulation to account for the effects of harvest on the soil carbon pools, assuming stems and wood products are not a long-term sink or source through harvesting. Clearly, large uncertainties pertain to this estimation. Because of the limited number of biomass inventories in Amazonia, there is still some uncertainty as to what extent the forest plots fully capture the natural disturbance dynamics. For instance, biomass inventories in Amazonia do not include estimates for change in stocks of coarse woody debris and soil carbon. If the increase in biomass is driven by an external cause (e.g., CO_2 fertilization, climate change, etc.), then the woody debris and soil carbon stocks may also be expected to increase (although the soil may have limited capacity to hold extra carbon) and hence the forest inventories may underestimate NBP. On the other hand, if the observed increase in biomass is the result of recovery from extensive disturbance, then the woody stocks may decrease. Finally, we found that atmospheric inversion-derived values of NBP, although they are the most uncertain of all approaches to estimate NBP, are broadly consistent with the ground-based estimates.

4. Modelled net primary productivity and net biome productivity

Several global models of terrestrial ecosystems have been developed to analyse and predict the response of terrestrial carbon pools and fluxes to changing land use and management practices, climate, and atmospheric composition. We used here two spatially explicit dynamic global vegetation models (DGVMs) of global coverage (IBIS, LPJ) and two spatially explicit terrestrial ecosystem models with prescribed vegetation distribution (HRBM, TEM). All four models have been run under the same protocol including variable climate, rising CO_2 concentration, and crop establishment over the period 1900–1993 (McGuire *et al.*, 2001). Those four models encapsulate biogeochemical processes responsible for biomass production as driven by climate and radiation, and calculate NPP as the difference between photosynthesis and autotrophic respiration components. Modelled NPP estimates at the biome scale that are reported in *Figure 1* are averaged over the forested grid cells within each continent. We found significant model-to-model NPP differences for Siberia, with all models overestimating NPP compared with the measurements except TEM (*Figure 1*), the only model having an explicit N-limitation on plant growth. Yet, the biome scale average NPP using the four models is roughly similar to the estimates from forest inventories and stand-scale studies (*Figure 2*). This is surprising because model NPP calculated from gas exchange processes should, in principle, include all ecosystem components (i.e., fine roots, herbivory, etc.) and thus we would expect that NPP in the models would be *higher* than in the data, but this was not the case. To gain further insights into this apparent discrepancy, one would need to compare separately the root, wood, and foliage components of NPP with field observations.

5. Linking net biome productivity to rising atmospheric CO$_2$ concentration

5.1 *A simple carbon cycle model*

Our next step is to evaluate to what extent NBP is compatible with the effect of rising CO$_2$ on NPP. To do so, we constructed a simple model driven by CO$_2$ fertilization alone, where NPP increases as a function of atmospheric CO$_2$ mixing ratio $C_a(t)$, implying that NBP is a function of the carbon residence times and of the CO$_2$-driven increase in NPP represented by a β-factor. This simple conceptual model comprises a biospheric pool containing living biomass and soil organic matter (SOM) with NPP (N_p) as an input and respiration being proportional to the stock size M and to the inverse average turnover time t_e, i.e.,

$$dM/dt = N_p - M/t_e. \tag{1}$$

As long as we have a series of pools that are connected by linear flows, and if the forcing term (NPP) is given by an exponential, the biosphere behaves exactly as the simple model used here, with a single turnover time for the sum of biomass and SOM turnover. Mathematically, the term t_e is the Laplace transform of the linear system evaluated at $s = t_a$ and could be relabelled as an 'effective' turnover time. The turnover time values were determined from information on carbon stocks and ^{14}C constraints (see *Table 1*). The budget equation (1) corresponds to an ideal case where the only return path of CO$_2$ to the atmosphere is heterotrophic respiration; thus the effect of disturbance is not accounted for. In such an ideal case, we assumed that CO$_2$ fertilization determines the biospheric carbon balance and its evolution, with NPP being stimulated by atmospheric CO$_2$ since pre-industrial times, t_0, as follows:

$$N_p(t) = N_p(t_0)\{1 + \beta[C_a(t) - C_a(t_0)]/C_a(t_0)\}, \tag{2}$$

where C_a is the atmospheric CO$_2$ concentration.

If we approximate the historical increase of atmospheric CO$_2$ since pre-industrial times with an exponential function with a characteristic e-folding time t_a (e.g., 45 years):

$$C_a = a + be^{t/ta}, \tag{3}$$

with $a = 0$ and $b = 280$ parts per million (p.p.m.).

Replacing $C_a(t)$ from (3) into (2) and then solving (1), the NBP carbon sink ($dM/dt = N_b$) in response to CO$_2$ fertilization can be approximated by:

$$N_b = [\beta/(1 + t_a/t_e)][(C_a - C_0)/C_a][N_p/(1 + \beta(C_a - C_0)/C_a)]. \tag{4}$$

Equation (4) gives a relationship between the intensity of the CO_2 fertilization effect (β), and the measured NBP, NPP, and t_e. We have used the values of t_e and NPP, deduced from extensive forest biomass inventories, for inversions with NBP. We inverted the function to obtain the parameter β for pairs of NBP and NPP observations. We have used here a 'linear β-factor' expression denoting a first-order perturbation (Goudriaan and Ketner, 1994); but using a Michaelis–Menten function (Farquhar *et al.*, 1980) or logarithmic expression (Bacastow and Keeling, 1973) would yield qualitatively similar results. Numerous experiments on tree growth in elevated atmospheric CO_2 concentrations all show that β lies between 0.2 and 0.4 (see, for example, Wullschleger *et al.*, 1995). The pairs of NPP and NBP estimates related by equation (4) that fall outside this range can thus be interpreted as reflecting additional sink mechanisms other than CO_2 fertilization alone. *Figure 3* gives a graphic illustration of this where NBP and NPP are plotted over β contours for each biome.

5.2 Results from Siberia

In Siberia, NBP from forest inventories would require $\beta = 0.7$ for that sink to be the result of CO_2 fertilization alone (*Table 2*), whereas β values in the plausible range of 0.2–0.4 would contribute 0.05–0.5 Mg (C) ha^{-1} per year to Siberian NBP, that is 20–50% of the observed sink. This implies that the remainder of the sink is the result of other controlling processes, such as climate changes, which are particularly important in Siberia (Briffa *et al.*, 1995; Pleshikov, 2002; Schulze *et al.*, 1999), or NPP enhancement by nitrogen deposition. Because of the extreme nitrogen limitation to

Table 2. *The magnitude of the CO_2-induced stimulation corresponding to the parameter β. $\beta 1$ is the average β factor obtained by solving equation (4); $\beta 2$ is the average of 10 000 Monte-Carlo simulations in which NPP and NBP were varied within their errors, with uncertainties of 30% on carbon turnover times included.*

$\beta 1$	$\beta 2$	$\sigma_{\beta 2}$	NBP	NBP	NPP	NPP	Method
Siberia							
0.51	0.55	(0.21)	0.15	(0.05)	1.66	(0.50)	Plot studies
0.60	0.65	(0.27)	0.24	(0.07)	2.28	(0.68)	Biomass inventories
1.17	1.29	(3.20)	0.40	(0.54)	2.28	(0.68)	Inverse models
0.17	0.18	(0.06)	0.10	(0.07)	3.05	(1.47)	Ecosystem models
Europe							
1.61	1.94	(2.60)	1.06	(0.60)	6.17	(1.83)	Plot studies
2.78	3.64	(3)	1.07	(0.32)	4.48	(1.34)	Biomass inventories
0.91	1.01	(0.56)	0.51	(0.54)	4.48	(1.34)	Inverse models
0.18	0.19	(0.07)	0.13	(0.18)	4.78	(1.54)	Ecosystem models
Amazonia							
0.33	0.35	(0.12)	0.62	(0.37)	9.15	(2.75)	Biomass inventories
0.06	0.06	(0.02)	0.12	(0.87)	9.15	(2.75)	Inverse models
0.11	0.12	(0.04)	0.13	(0.05)	7.47	(2.14)	Ecosystem models

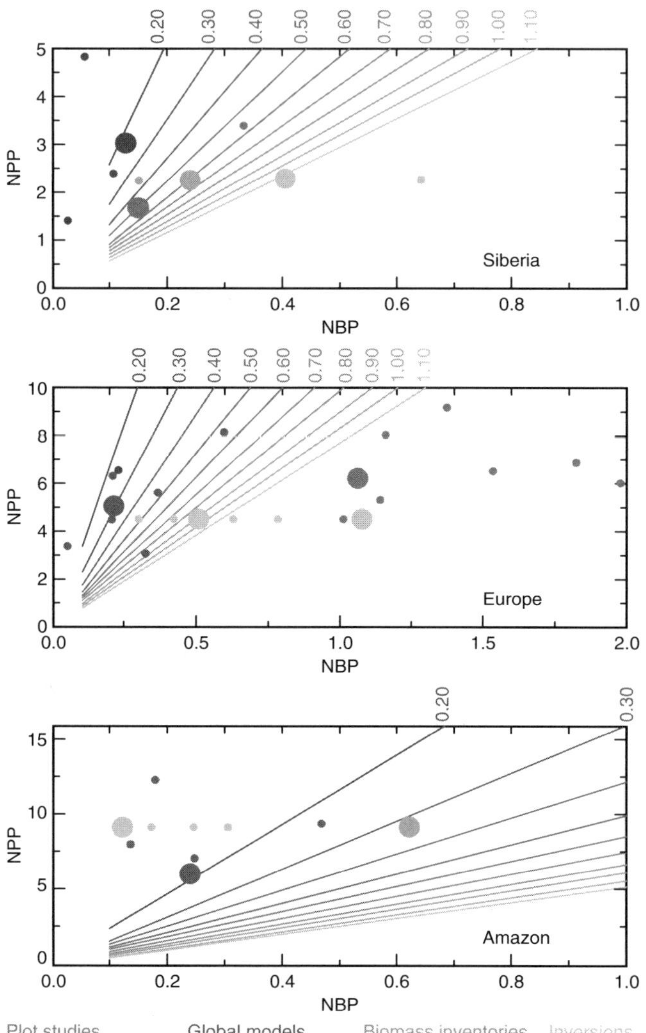

Figure 3. *Graphic representation of β factor within the NBP-NPP space. Small circles: selected individual estimates of NBP and NPP from different methods from Table 1. Units: Mg (C) ha⁻¹ yr⁻¹ Large circles average biome scale NBP and NPP from Table 2. Iso-contours of β indicate which fraction of NBP can be explained by CO₂ fertilization alone.*

growth of the Boreal forests, N deposition is likely to have a significant effect (Chapter 16, this volume), even though deposition rates are low and do not exceed 100–200 kg (N) km⁻² per year, except in a few small areas.

On the other hand, changes in fire regimes are unlikely to have fostered more sequestration in Siberian forests, at least during the early 1990s, because most studies indicate that fire frequency has actually increased over Asian Russia (Ivanova, 1998–1999). Although more fires may possibly result in an increased NPP because of an improved cation supply (Wirth *et al.*, 2002), they also accelerate loss of nitrogen.

Note also that about 50% of Siberian forests are unevenly aged, and this may additionally contribute to the observed disproportion between NPP and NBP compared, for instance, with Western Europe.

We found that a maximum of 50% of stand-scale NBP in Siberia can be explained by CO_2 fertilization. The inversion-derived estimates of NBP are indicative of no large contribution of CO_2 fertilization (less than 10%), but these are the most uncertain estimates of all three methods used, because of the large spread in the inverse model results (*Supplementary data*), and because a derived value of NPP must be assigned to NBP, as atmospheric studies cannot detect NPP separately (for *Figure 3* we used NPP from forest inventories).

By contrast, the outputs of the biosphere DGVM models are compatible with CO_2 fertilization as the single most important driving process (see *Figure 3*). However, the models underestimated NBP (*Figure 2*), and this suggests that they may lack processes stimulating carbon uptake, such as 'favourable' consequences of disturbance and impacts of permafrost, or they may underestimate the sensitivity of NBP to changing climate and radiation in Siberia, perhaps because of the lack of detailed descriptions of decomposition and nitrogen dynamics in the Boreal forests.

5.3 *Results from Europe*

In Europe, where nearly all forests are managed, none of the stand-scale measured NBP estimates is compatible with CO_2 fertilization as the main driver, except for the Nacetin spruce site in the Czech Republic. At Nacetin there has been a very high rate of decomposition, as a result of a large previous accumulation of an organic layer during the times of high SO_2 emissions (1950–1990). These have now decreased and been replaced by high a rate of nitrogen deposition. Our analysis of the CANIF transect stands through equation (4) suggests that forest management and nitrogen deposition are currently the main cause of carbon sequestration in European managed temperate forests, rather than rising CO_2 concentration. Similar conclusions were obtained using forest inventory data in the eastern USA by Caspersen *et al.* (2000), although Joos *et al.* (2002) showed that within the scatter of the forest plot data, one cannot rule out an effect of CO_2 fertilization. However, stand-scale NBP was inferred from the CANIF data as 24% of the soil carbon accumulation rates to account for harvest. Without doubt, uncertainties in the mechanical and ecological effects of harvest on the soil carbon balance are very large in this analysis, as also highlighted in Chapter 16.

Using the stand-scale data from the European forest inventory census, NBP is marginally attributable to CO_2 fertilization according to equation (4). A plausible range for β values of 0.2–0.4 would propagate into a maximal 25% contribution of CO_2 fertilization to the European forest sector carbon uptake. The inversion-derived 'forest NBP' estimates show a large spread across Europe, with four out of five inverted mean fluxes in the *Supplementary data* table falling outside the CO_2 fertilization range.

In contrast to the field studies, the ecosystem models all gave results within the range of β values of 0.2–0.4, suggesting that the models lack adequate description of the sink-controlling processes in managed temperate forests, including, among others, forestry practice affecting forest age structure, and agricultural abandonment during the last four decades of the 20th Century.

5.4 *Results from Amazonia*

In Amazonia, estimates of NBP from stand data analysis, forest inventories and atmospheric inversions are all compatible with β values lying between 0.2 and 0.4 (*Figure 3*), suggesting that, unlike the situations in Siberia and Europe, CO_2 fertilization may explain the current Amazonian carbon sink. One possible explanation for this is that vast areas of forests in Amazonia are still in a pristine state, such that CO_2 fertilization could be the dominant process in driving carbon sequestration. However, interaction with low-frequency variability in climate may also play a role, in addition to CO_2, in controlling present day Amazonian NPP and NBP. For example, Botta *et al.* (2002), using the IBIS ecosystem model forced by long-term fluctuations in climate (New *et al.*, 2000; Victoria *et al.*, 1998), simulated an approximate 5% higher Amazonian NPP in the 1970s and 1980s compared with the past decade. Malhi *et al.* (2002) further showed evidence that there has been sustained warming in Amazonia since the mid-1970s, in synchrony with global warming.

The high NPP of Amazonian forests makes them very responsive to sink enhancement if NPP is stimulated by CO_2 (Lloyd *et al.*, 1996). However, the short turnover time of dead carbon in tropical soils makes any enhancement of NPP through carbon sequestration in fine roots or leaves very unlikely over long periods, and only through stimulation in growth of the woody components is NPP likely to enhance NBP. Thus, the large uncertainty in the magnitude of soft-tissue NPP, outlined in Section 2, may have no significant impact on our attribution of NBP to CO_2 fertilization.

Amazonian NBP deduced from inversions is highly uncertain because of the sparse density of observing stations, and because it was necessary to subtract a poorly quantified land-use source (Houghton, 1998) from the inverted continental fluxes to obtain an inversion 'forest NBP', representative of undisturbed forests. Another caveat concerns possible biases in the presumed Amazonian deforestation source. This has recently been re-appraised to be lower, at least during the 1990s, than had been estimated previously (Achard *et al.*, 2002; DeFries *et al.*, 2002; see Chapter 10, this volume).

5.5 *What are the uncertainties?*

We discuss four distinct sources of uncertainties impacting our results. Firstly, our analysis determines the amount of NBP that CO_2 fertilization *could* contribute, but it does not *demonstrate beyond possible doubt* that this process *must* explain that fraction of NBP. Rather, it is conceivable that CO_2 fertilization does not exist on an ecosystem scale and that NBP is almost entirely determined by other controlling mechanisms, in particular land use (Schimel *et al.*, 2000).

Secondly, there are biases in the different methods used to estimate NPP, and those biases are likely to cause underestimation of NPP, compared with the real world, especially in the tropics. The β value that is inferred from equation (4) is first-order proportional to the inverse of NPP, so that the higher the NPP, the smaller is the effect of CO_2 fertilization. In that sense, our analysis is a conservative approach to quantify the maximum contribution of CO_2 fertilizations to NBP. Furthermore, the woody NPP, which is particularly important for NBP in the tropics, is much less uncertain than the soft tissue NPP. Thirdly, it is only fair that we emphasize the spread around the mean estimates of both NPP and NBP that we have compiled. This uncertainty can be propagated into an error in the estimated value of β.

We performed Monte-Carlo analysis of the uncertainty in β according to the following procedure. For the stand studies, we assumed a normal distribution of errors given by the standard deviation of several sites around the mean, (e.g., the CANIF data). For the forest biomass inventories, which already report pre-averaged quantities, we took the error reported in each original paper and assumed it to be normally distributed. For the atmospheric inversions, we computed the error as the vector sum of three independent terms: (1) the mean of Bayesian 'internal errors' of Gaussian nature returned by each inverse procedure; (2) the spread of the inverted continental fluxes; and (3) the error estimate of the non-forest fluxes that were subtracted from the initial inversion results to yield the 'forest NBP'. For the global dynamic vegetation models, the error was simply computed as the standard deviation of the four ecosystem models and assumed to be normal. When not specified, a normal error of 30% around the mean was assumed. An error of 30% was also assigned to the values of turnover times within each biome. We performed 10 000 Monte-Carlo simulations of β using the above-prescribed errors on NPP, NBP, and t_e and report the results in *Table 2*. One can see that the uncertainties in β are fairly large (between 30% for the stand studies and 100% for the inversions). On the other hand, the average values of β do not change significantly when errors in NBP and NPP are propagated into the analysis. Even more encouraging, varying the turnover times by $\pm 30\%$ did not alter the inference of β.

A fourth source of uncertainty comes from the structure of the simple model itself, implicitly based on a linear flow of carbon among a series of well-mixed pools, with a substitute mean 'effective' residence time. This is far from the case in a forest; in other words, our model could be too simple. Yet, the fact that ecosystem models that contain a more sophisticated description of the effect of CO_2 on plant growth and describe the flow of an excess carbon through multiple pools of various residence times, can simulate NBP in close agreement with our simple model gives us good confidence that our analysis is robust enough to quantify the effects of CO_2 fertilization on carbon sequestration. However, for stand studies, we raise a flag of caution on the issue that the tree biomass stock is certainly not well mixed, with the probability that a mole of carbon incorporated into biomass by NPP will return to the atmosphere being maximal when the stand dies or is harvested. In the case that dead wood remains in the forest, it decays on time-scales of many decades, and this may almost double the time for CO_2 to return to the atmosphere (Vygodskaya *et al.*, 2002). In the case that the stems are regularly harvested, carbon may not decay exponentially with time in pools of wood products, violating the structure of our simple model. For forest inventories encompassing many stands at different stages in their life and management cycles, these effects are less drastic but still present, and could be accounted for by using additional data on the age-class structure of the forests. Ignoring non-exploited, old-grown forests in inventories should then underestimate the biome-averaged turnover. For soil carbon decay, ^{14}C measurements show that there are at least three distinct pools of carbon. Fresh leaf and root detritus are the most active components of SOM with turnover times of a decade or less, and more refractory components have turnover times of many decades to millennia. The latter passive pool, formed with carbon compounds stabilized by clay particles, may be really inert. We tested that hypothesis by assuming that the SOM pools with residence times longer than 100 years are not affected by current perturbations, thus using shorter effective turnover times in equation (4). The results of that sensitivity test lead to even higher, more unrealistic β values in all cases, and this makes our inference of a small contribution of CO_2 fertilization over Europe and Siberia robust, and may allow for other sink mechanisms to explain the Amazonian sequestration.

6. Conclusions

For three representative forest biomes, we have compiled recent NPP and NBP observations and modelling results based on: (1) stand-scale studies; (2) forest biomass inventories; (3) atmospheric inversions, corrected when necessary to account for forest only; and (4) global, spatially explicit, ecosystem models. Results from these compilations and simulations were analysed by using a simple carbon cycle model to identify which fraction of NPP can maximally be sequestered as NBP in response to the effect of CO_2 fertilization. Our conclusions are the following:

- Stand-scale and continental-scale forest biomass inventory data for NPP are difficult to compare one with the other. On the one hand, at the stand scale, the different components of NPP can be constrained by observations, whereas biomass inventories model NPP from observed changes in stem volume. On the other hand, forest inventories cover large spatial scales whereas stand scale behaviour must be boldly extrapolated to the biome scale. Yet, we found that both approaches gave surprisingly similar estimates for both NPP and NBP in all three regions.
- There is a fourfold increase in NPP between Siberia, where temperature, nutrition and length of growing season limit plant growth, and the Amazon basin, where tropical NPP estimates are biased to low values by up to 100% because of underestimation of fine-root turnover and other factors. The NPP of managed, temperate, European forests lies between those two extremes.
- NBP is only a small fraction of NPP in Siberia ($\approx 10\%$) and in the Amazon basin ($\approx 2.5\%$) but a significant fraction of NPP in Europe (approximately up to 30%), indicating that carbon sequestration does not simply correlate with NPP.
- Global ecosystem models appear to underestimate NBP in each biome compared with estimates based on observations.
- A simple analysis suggests that CO_2 fertilization accounts for less than 50% of the NBP sink in Siberia on the basis of forest inventories or stand studies, and less than 10% on the basis of atmospheric inversion. For Europe, CO_2 fertilization probably has a negligible impact on NBP based on stand data and forest inventories, but may account for up to 50% of forest NBP when inferred from inversions, although our downscaling of inversion-derived NBP to forest regions is highly uncertain in Europe.
- In contrast, when the same analysis was applied to the Amazonian forests, CO_2 fertilization accounted for the entire NBP, whatever the source of data used for NPP and NBP.

In the future, improved estimates of NPP at biome scale are expected from the combined use of geo-referenced descriptions of actual landscapes and ecosystems in the form of multi-layer GISs, remote sensing data, ecological process studies in which all components of carbon flows are measured, and forest inventories. Our background philosophy consists of a transition to full carbon accounting and its application in a possible, consistent, holistic way. We regard it as a high priority to collect forest biomass inventory data in tropical forests to enhance the coverage of the existing plot networks. Furthermore, large-scale soil inventories specifically designed to capture changes in soil carbon contents need to be initiated. Improved NBP estimates from atmospheric inversions can be delivered by increasing the density of atmospheric measurement stations over the interiors of continents and by increasing the sampling frequency using, for instance, continuous measurements on tall towers or aircraft. Inverse methods can also be improved by increasing the resolution at which fluxes can be resolved. On the other hand, inventory estimates of NBP should attain a higher temporal resolution to improve cross-validation with inversion estimates. Our simple analysis of the contribution of CO_2 fertilization to regional terrestrial carbon sinks can be also refined by using more realistic biosphere models that combine carbon and nitrogen cycles and describe the flow of NPP-delivered carbon into distinct pools following a spectrum of residence times.

7. Summary

We compiled measured and simulated estimates of NPP and NBP for Amazonian tropical, European temperate, and Siberian Boreal forests from intensive stand-scale field studies, extensive forest biomass inventories, regional atmospheric inversions, and global ecosystem models. We analysed the random and systematic sources of uncertainties pertaining to each approach when comparing their results, and showed that estimates of NPP from different data streams are robustly comparable within their errors. Although NPP increases by a factor of four between Siberia and the Amazon, NBP is larger in Europe than elsewhere, demonstrating that carbon sequestration does not correlate with NPP. We analysed the NPP : NBP ratios in terms of the role of CO_2 fertilization. Our results show that the tropical forest NBP carbon sink can be entirely explained by a CO_2-induced enhancement of NPP, whereas such a mechanism can only account for 10% of the European sink and up to 50% of Siberian sink. Europe and Siberia are the two regions where factors other than CO_2 are likely to be dominant in controlling the sequestration of carbon by forest ecosystems, such as management practice, climate, nitrogen deposition, and variation in disturbance regimes.

***Supplementary data** A meta-analysis of forest biome productivity: estimates derived from the literature for net primary productivity (NPP), net ecosystem productivity (NEP) and net biome productivity (NBP) for (a) Europe (b) Siberia and (c) Amazonia and the neotropics.*

Europe

1. Net Primary Productivity

Intensive field studies (stand scale)

								total NPP	Foliage	Stem	Twig/ branch	Coarse root	Fine root	Soil C MRT yrs	biomass MRT yrs	total MRT yrs
Europe CANIF-Transect	NPP	plot	Aheden, S	Ahe	*Picea*	Schulze pers. com.		3.05	0.40	0.85	0.65	0.30	0.85	43.00	31.15	74.15
Europe CANIF-Transect	NPP	plot	Skogaby, S	Sko	*Picea*	Schulze pers. com.		9.10	1.10	2.60	1.95	0.85	2.60	43.00	8.13	51.13
Europe CANIF-Transect	NPP	plot	Gribskov, DK	Gri	*Fagus*	Schulze pers. com.		6.00	2.30	1.00	1.10	0.25	1.35	60.00	32.33	92.33
Europe CANIF-Transect	NPP	plot	Nacetin, Cz	Nac	*Picea*	Schulze pers. com.		6.30	1.70	1.80	1.35	0.85	0.50	68.00	16.51	84.51
Europe CANIF-Transect	NPP	plot	Jezeri, Cz	Jez	*Fagus*	Schulze pers. com.		8.00	1.70	2.10	2.30	0.60	1.30	56.00	15.38	71.38
Europe CANIF-Transect	NPP	plot	Waldstein, D	Wal	*Picea*	Schulze pers. com.		6.80	2.30	1.65	1.15	0.65	1.05	125.00	22.35	147.35
Europe CANIF-Transect	NPP	plot	Schacht, D	Sch	*Fagus*	Schulze pers. com.		5.30	1.70	1.00	1.10	0.25	1.25	79.00	29.81	108.81
Europe CANIF-Transect	NPP	plot	Aubure, F	AuF	*Fagus*	Schulze pers. com.		4.30	1.00	1.50	1.10	0.30	0.40	85.00	34.19	119.19
Europe CANIF-Transect	NPP	plot	Aubure, P	AuP	*Picea*	Schulze pers. com.		4.40	1.35	1.00	1.10	0.25	0.70	85.00	33.86	118.86
Europe CANIF-Transect	NPP	plot	Collelongo, I	Col	*Fagus*	Schulze pers. com.		6.50	1.30	1.40	1.55	0.50	1.75	115.00	19.69	134.69
Europe CANIF-Transect	NPP	plot	Mt di Mezzo, I	MdM	*Picea*	Schulze pers. com.		8.10	1.30	2.25	1.80	1.45	1.30	66.00	13.21	79.21
						11 plots Canif project (Europe)		*mean*	*std dev*							
								6.17	1.83							

Extensive forest biomass inventories

				total NPP
Geographic Europe	NPP	Forest biomass inventory	Schulze et al. 1999	1.17
EU15	NPP	Forest biomass inventory	FAO 2000	1.21
EU15	NPP	Forest biomass inventory	Nabuurs et al. 2003	3.18
EU15	NPP	Forest biomass inventory	Nabuurs et al. 2003	4.17

Ecosystem Global Models

LCC : Land-use change + CO$_2$ + Climate / CC : CO$_2$ + Climate / C : CO$_2$

				total NPP
Geographic Europe	NPP	HRBM model LCC 1980–1989	McGuire et al. 2001	4.28
Geographic Europe	NPP	IBIS model LCC 1980–1989	McGuire et al. 2001	5.21
Geographic Europe	NPP	LPJ model LCC 1980–1989	McGuire et al. 2001	6.64
Geographic Europe	NPP	TEM model LCC 1980–1989	McGuire et al. 2001	2.98
			mean	4.78
			std dev	1.54

Geographic Europe	NPP	HRBM model CC 1980–1989				McGuire *et al.* 2001	4.49	
Geographic Europe	NPP	IBIS model CC 1980–1989				McGuire *et al.* 2001	5.66	
Geographic Europe	NPP	LPJ model CC 1980–1989				McGuire *et al.* 2001	6.73	
Geographic Europe	NPP	TEM model CC 1980–1989				McGuire *et al.* 2001	3.38	
							mean	*std dev*
							5.07	1.45
Geographic Europe	NPP	HRBM model C 1980–1989				McGuire *et al.* 2001	4.44	
Geographic Europe	NPP	IBIS model C 1980–1989				McGuire *et al.* 2001	5.52	
Geographic Europe	NPP	LPJ model C 1980–1989				McGuire *et al.* 2001	6.50	
Geographic Europe	NPP	TEM model C 1980–1989				McGuire *et al.* 2001	3.39	
							mean	*std dev*
							4.96	1.35
UK	NPP	EuroBiota model				Milne *et al.* 2002	6.10	

2. Net Ecosystem Productivity

Intensive field studies (stand scale)							*mean NEP*	
Europe CANIF-Transect	NEP	plot	Aheden, S	Ahe	*Picea*	Schulze pers. com.	1.95	
Europe CANIF-Transect	NEP	plot	Skogaby, S	Sko	*Picea*	Schulze pers. com.	6.81	
Europe CANIF-Transect	NEP	plot	Gribskov, DK	Gri	*Fagus*	Schulze pers. com.	4.66	
Europe CANIF-Transect	NEP	plot	Nacetin, Cz	Nac	*Picea*	Schulze pers. com.	2.16	
Europe CANIF-Transect	NEP	plot	Jezeri, Cz	Jez	*Fagus*	Schulze pers. com.	6.15	
Europe CANIF-Transect	NEP	plot	Waldstein, D	Wal	*Picea*	Schulze pers. com.	5.58	
Europe CANIF-Transect	NEP	plot	Schacht, D	Sch	*Fagus*	Schulze pers. com.	3.67	
Europe CANIF-Transect	NEP	plot	Aubure, F	AuF	*Fagus*	Schulze pers. com.	3.46	
Europe CANIF-Transect	NEP	plot	Aubure, P	AuP	*Picea*	Schulze pers. com.	3.51	
Europe CANIF-Transect	NEP	plot	Collelongo, I	Col	*Fagus*	Schulze pers. com.	5.11	
Europe CANIF-Transect	NEP	plot	Mt di Mezzo, I	MdM	*Picea*	Schulze pers. com.	5.99	
							mean	*std dev*
							4.46	1.63

Eddy Covariance Towers				*mean NEE*
Western European forests	NEP	From eddy towers with Neural Network	Papale and Valentini 2003	1.85
Western European forests	NEP	From eddy towers with remote sensing	Martin *et al.* 1998	1.90

Extensive forest biomass inventories							*mean NEE*	
EU15	NEP	large scale inventory				Nabuurs *et al.* 2003	1.35	
EU15	NEP	large scale inventory				Nabuurs *et al.* 2003	1.74	

3. Net Biome Productivity

Intensive field studies (stand scale)							*mean NBP*	
Waldstein	NBP	plot	Aheden, S	Ahe	*Picea*	Schulze pers. com.	0.32	
Europe CANIF-Transect	NBP	plot	Skogaby, S	Sko	*Picea*	Schulze pers. com.	1.37	
Europe CANIF-Transect	NBP	plot	Gribskov, DK	Gri	*Fagus*	Schulze pers. com.	1.97	
Europe CANIF-Transect	NBP	plot	Nacetin, Cz	Nac	*Picea*	Schulze pers. com.	0.21	
Europe CANIF-Transect	NBP	plot	Jezeri, Cz	Jez	*Fagus*	Schulze pers. com.	1.16	
Europe CANIF-Transect	NBP	plot	Waldstein, D	Wal	*Picea*	Schulze pers. com.	1.82	
Europe CANIF-Transect	NBP	plot	Schacht, D	Sch	*Fagus*	Schulze pers. com.	1.14	
Europe CANIF-Transect	NBP	plot	Aubure, F	AuF	*Fagus*	Schulze pers. com.	0.49	
Europe CANIF-Transect	NBP	plot	Aubure, P	AuP	*Picea*	Schulze pers. com.	1.01	
Europe CANIF-Transect	NBP	plot	Collelongo, I	Col	*Fagus*	Schulze pers. com.	1.53	
Europe CANIF-Transect	NBP	plot	Mt di Mezzo, I	MdM	*Picea*	Schulze pers. com.	0.59	
							mean	*std dev.*
							1.06	0.60

Extensive forest biomass inventories				*mean NBP*	
EU15	NBP	large scale inventory	Nabuurs *et al.* 2003	0.07	
EU15	NBP	large scale inventory	Nabuurs *et al.* 2003	1.02	
EU15	NBP	large scale inventory	Liski *et al.* 2002, 2003	0.19	
				mean	*std dev*
Geographic Europe	NBP	large scale inventory	Janssens *et al.* 2003	1.07	0.47

Ecosystem Global Models		*LCC : Land-use change + CO_2 + Climate / CC : CO_2 + Climate / C : CO_2*		*mean NBP*	
Forest Geographic Europe	NBP	HRBM model LCC 1980–1989	McGuire *et al.* 2001	−0.08	
Forest Geographic Europe	NBP	IBIS model LCC 1980–1989	McGuire *et al.* 2001	0.35	
Forest Geographic Europe	NBP	LPJ model LCC 1980–1989	McGuire *et al.* 2001	0.14	
Forest Geographic Europe	NBP	TEM model LCC 1980–1989	McGuire *et al.* 2001	0.11	
				mean	*std dev.*
				0.13	0.18
Forest Geographic Europe	NBP	HRBM model CC 1980–1989	McGuire *et al.* 2001	0.26	
Forest Geographic Europe	NBP	IBIS model CC 1980–1989	McGuire *et al.* 2001	0.37	
Forest Geographic Europe	NBP	LPJ model CC 1980–1989	McGuire *et al.* 2001	0.09	

				mean	ext error	int error	area ($\times10^6$ ha)	
Forest Geographic Europe	NBP	TEM model CC 1980–1989	McGuire *et al.* 2001	0.04				
				mean 0.19	*std dev.* 0.15			
Forest Geographic Europe	NBP	HRBM model C 1980–1989	McGuire *et al.* 2001	0.21				
Forest Geographic Europe	NBP	IBIS model C 1980–1989	McGuire *et al.* 2001	0.37				
Forest Geographic Europe	NBP	LPJ model C 1980–1989	McGuire *et al.* 2001	0.23				
Forest Geographic Europe	NBP	TEM model C 1980–1989	McGuire *et al.* 2001	0.05				
				mean 0.21	*std dev.* 0.13			
Atmospheric Inversions (total NBP)				mean	ext error	int error	area ($\times10^6$ ha)	
Geographic Europe	NBP	inversion 8 monthly inversions with 2 transport models 1980–1998	Bousquet *et al.* 2000	0.20	0.20	0.47	1000.00	
Geographic Europe	NBP	inversion 7 monthly inversions with 2 models 1985–1995	Bousquet *et al.* 1999	0.41	0.41	0.54		
Geographic Europe	NBP	inversion 1 monthly inversion; 1 model 1980–1995	Rayner *et al.* 1999	0.20				
Geographic Europe	NBP	inversion 1 monthly inversion with 1 transport model many regions 1981–1987	Kaminski *et al.* 1999	0.08		0.20		
Geographic Europe	NBP	inversion Annual inversions with 15 transport models 1992–1996	Gurney *et al.* 2002	0.56	0.40	0.34		
				mean	mean ext error	mean int error	mean spread	spread+int error+ext error
				0.29	0.34	0.39	0.19	0.55
Agriculture and Grassland carbon balance				mean				
Geographic Europe	NBP	CESAR model	Vleeshouwers *et al.* 2002					
NBP Atmospheric Inversions (Forest NBP)				mean	ext error	int error	area ($\times10^6$ ha)	
Geographic Europe	NBP	inversion x Bousquet et al. 2000	Bousquet *et al.* 2000	0.42	0.20	0.47	1000.00	
Geographic Europe	NBP	inversion x Bousquet et al. 1999	Bousquet *et al.* 1999	0.63	0.41	0.54		
Geographic Europe	NBP	inversion x Rayner et al. 1999	Rayner *et al.* 1999	0.42				
Geographic Europe	NBP	inversion x Kaminski et al. 1999	Kaminski *et al.* 1999	0.30		0.2		
Geographic Europe	NBP	inversion x Gurney et al. 2002	Gurney *et al.* 2002	0.78	0.40	0.34		
				mean	mean ext error	mean int error	mean spread	spread+int error+ext error
				0.51	0.34	0.39	0.19	0.55

Siberia and European Russia
1. Net Primary Productivity

Intensive field studies (stand scale)					*total NPP*	*BG-NPP*	*Woody AG*	*Foliage*
Boreal Siberia	NPP	plot	pine (central Siberia)	Wirth *et al.* 2002	1.86	0.59	0.89	0.38
Boreal Siberia	NPP	plot	pine (central Siberia)	Wirth *et al.* 2002	1.47	0.62	0.43	0.42
Southern Boreal Siberia	NPP	plot	pine (southern taiga)		5.55	1.30	3.06	1.19
Eastern Boreal Siberia	NPP	plot	larch (eastern Siberia)	Wang *et al.* 1995	1.01	0.25	0.27	0.49
Eastern Boreal Siberia	NPP	plot	larch (eastern Siberia)	Kajimoto *et al.* 1999			0.90	
					mean larch	*std dev*		
					0.95	0.07		
					mean pine	*std dev*		
					1.67	0.61		
					total NPP	*BG-NPP*	*AG-NPP*	*stdev total NPP*
Taiga (spruce European Russia)	NPP	9 plots		Bazilevich 1993	3.54	0.74	2.80	1.86
Taiga (larch east Siberia)	NPP	many plots		Bazilevich 1993	2.60	0.36	2.24	0.79
Mixed forest (Russia)	NPP	many plots		Bazilevich 1993	5.63	0.86	4.77	0.53
Small-leaved secondary forests	NPP	many plots		Bazilevich 1993	2.81	0.17	2.64	0.86
					total NPP			
Russian forests	NPP	223 plots spruce, fir, Russian cedar		Shvidenko pers. com.	3.01			
Russian forests (mainly Siberia)	NPP	85 plots larch		Shvidenko pers. com.	2.44			
Russian forests	NPP	186 plots pine		Shvidenko pers. com.	1.72			
Russian forests	NPP	91 plots birch and aspen		Shvidenko pers. com.	3.57			
Russian forests	NPP	21 plots coniferous-broadleaved forests dominated by *Pinus korajensis*		Shvidenko pers. com.	5.38			
Russian forests	NPP	mean across all species			*total NPP*	*std dev*		
					2.99	0.28		

Extensive forest biomass inventories					*total NPP*	*BG-NPP*	*Woody NPP*	*Foliage NPP*
Russian forest lands	NPP	Forest biomass inventory 1988–1992 Area = 882 Mha	Shvidenko pers. com. 2003		2.28	0.61	0.53	1.14
Russian forests	NPP	Forest biomass inventory	Schulze *et al.* 1999		1.00	0.60	0.40	
Central Siberian IGBP Transect	NPP	Forest biomass inventory	McGuire *et al.* 2002		2.42			

				total NPP			
Central Siberian IGBP Transect	NPP	Forest biomass inventory	McGuire *et al.* 2002	2.96			
Far East Siberian IGBP Transect	NPP	Forest biomass inventory	McGuire *et al.* 2002	2.04			
Far East Siberian IGBP Transect	NPP	Forest biomass inventory	McGuire *et al.* 2002	3.57			
Bog area Siberia	NPP		Shvidenko *et al.* 2001	2.50	0.85	0.16	1.49

Ecosystem Global Models		*LCC : Land-use change + CO_2 + Climate / CC : CO_2 + Climate / C : CO_2*		*total NPP*	
Siberian Boreal forests	NPP	HRBM model LCC 1980–1989	McGuire *et al.* 2001	2.36	
Siberian Boreal forests	NPP	IBIS model LCC 1980–1989	McGuire *et al.* 2001	3.53	
Siberian Boreal forests	NPP	LPJ model LCC 1980–1989	McGuire *et al.* 2001	4.85	
Siberian Boreal forests	NPP	TEM model LCC 1980–1989	McGuire *et al.* 2001	1.45	
				mean	*std dev*
				3.05	1.47
Siberian Boreal forests	NPP	HRBM model CC 1980–1989	McGuire *et al.* 2001	2.40	
Siberian Boreal forests	NPP	IBIS model CC 1980–1989	McGuire *et al.* 2001	3.59	
Siberian Boreal forests	NPP	LPJ model CC 1980–1989	McGuire *et al.* 2001	4.89	
Siberian Boreal forests	NPP	TEM model CC 1980–1989	McGuire *et al.* 2001	1.46	
				mean	*std dev*
				3.08	1.48
Siberian Boreal forests	NPP	HRBM model C 1980–1989	McGuire *et al.* 2001	2.41	
Siberian Boreal forests	NPP	IBIS model C 1980–1989	McGuire *et al.* 2001	3.39	
Siberian Boreal forests	NPP	LPJ model C 1980–1989	McGuire *et al.* 2001	4.81	
Siberian Boreal forests	NPP	TEM model C 1980–1989	McGuire *et al.* 2001	1.44	
				mean	*std dev*
				3.01	1.44

2. Net Ecosystem Productivity

Eddy Covariance Towers				*mean NEE*	*std dev*
Black Taiga / Boreal	NEP	plot spruce	Röser *et al.* 2002	2.29	
Pine / Boreal	NEP	plot pine	Wirth *et al.* 2002	0.60	1.11
Central Siberia (Zotino tower site)	NEP	plot pine 200 yrs	Wirth *et al.* 2002	0.31	1.37
Pine / Boreal	NEP	plot pine	Lloyd *et al.* 2002	1.56	
Bog / Boreal	NEP	plot bog	Arneth *et al.* 2002	0.58	

3. Net Biome Productivity

Intensive field studies (stand scale)				mean NBP	std dev
Central Siberia	NBP	19 plots pine chronosequence	Wirth *et al.* 2002	0.17	0.70
Central Siberia (Zotino tower site)	NBP	plot pine 200 yrs	Wirth *et al.* 2002	0.15	
Siberia	NBP	plot Black Carbon formation NBP	Schulze *et al.* 1999	0.02	
European Russian bog	NBP	plot bog from NEE	Schulze *et al.* 2002	0.93	

Extensive forest biomass inventories				mean NBP	std dev
Russian forest lands	NPP	Forest biomass inventory 1988–1992 Area = 882 Mha	Shvidenko pers. com.	0.24	
Forests Siberia	NBP	Forest biomass inventory NPP and NBP/NPP ratio of Shvi. *et al.* 2001	Schulze *et al.* 1999	0.16	
Central Siberian IGBP Transect	NBP	Forest biomass inventory NPP and NBP/NPP ratio of Shvi. *et al.* 2001	McGuire *et al.* 2002	0.38	
Central Siberian IGBP Transect	NBP	Forest biomass inventory NPP and NBP/NPP ratio of Shvi. *et al.* 2001	McGuire *et al.* 2002	0.46	
Far East Siberian IGBP Transect	NBP	Forest biomass inventory NPP and NBP/NPP ratio of Shvi. *et al.* 2001	McGuire *et al.* 2002	0.32	
Far East Siberian IGBP Transect	NBP	Forest biomass inventory NPP and NBP/NPP ratio of Shvi. *et al.* 2001	McGuire *et al.* 2002	0.56	0.35

Ecosystem Global Models		LCC : Land-use change + CO_2 + Climate / CC : CO_2 + Climate / C : CO_2		NBP	
Siberian Boreal forests	NBP	HRBM model LCC 1980–1989	McGuire *et al.* 2001	0.09	
Siberian Boreal forests	NBP	IBIS model LCC 1980–1989	McGuire *et al.* 2001	0.21	
Siberian Boreal forests	NBP	LPJ model LCC 1980–1989	McGuire *et al.* 2001	0.06	
Siberian Boreal forests	NBP	TEM model LCC 1980–1989	McGuire *et al.* 2001	0.05	
				mean 0.10	*std dev* 0.07
Siberian Boreal forests	NBP	HRBM model CC 1980–1989	McGuire *et al.* 2001	0.14	
Siberian Boreal forests	NBP	IBIS model CC 1980-1989	McGuire *et al.* 2001	0.24	
Siberian Boreal forests	NBP	LPJ model CC 1980-1989	McGuire *et al.* 2001	0.05	
Siberian Boreal forests	NBP	TEM model CC 1980-1989	McGuire *et al.* 2001	0.05	
				mean 0.12	*std dev* 0.09

				mean NBP	ext error	int error		
Siberian Boreal forests	NBP	HRBM model C 1980-1989	McGuire *et al.* 2001	0.10				
Siberian Boreal forests	NBP	IBIS model C 1980-1989	McGuire *et al.* 2001	0.33				
Siberian Boreal forests	NBP	LPJ model C 1980-1989	McGuire *et al.* 2001	0.06				
Siberian Boreal forests	NBP	TEM model C 1980-1989	McGuire *et al.* 2001	0.02				
				mean 0.13	*std dev* 0.14			
Atmospheric Inversions				*mean NBP*	*ext error*	*int error*		
Boreal Asia	NBP	Annual inversions with 15 transport models 1992–1996	Gurney *et al.* 2002	0.40	0.50	0.50		
Boreal Asia	NBP	7 monthly inversions with 2 models 1985–1995	Bousquet *et al.* 1999	0.64		0.27		
Boreal Asia	NBP	1 monthly inversion; 1 model 1980–1995	Rayner *et al.* 1999	0.15				
USSR	NBP	1 monthly inversion with 1 transport model many regions 1981–1987	Kaminski *et al.* 1999	0.42		0.23		
			mean	*mean ext error*	*mean int error*	*mean spread*	*spread+ int error+ ext error*	
				0.40	0.50	0.33	0.20	0.54

Amazonia and neotropics
1. Net Primary Productivity

Intensive field studies (stand scale)					total NPP	Fine Root	Coarse Root	Woody	Foliage
Amazonia	NPP	plot (Oak Ridge DB)	Manaus, Brazil	Malhi and Grace (2000)	15.60	6.10	0.80	5.00	3.70
Amazonia	NPP	plot (Oak Ridge DB)	San Eusebio, Venezuela	Grimm and Fassbender (1981)	22.18	14.00		4.80	3.38
Amazonia	NPP	plot (Oak Ridge DB)	north Manaus, Brazil	McWilliam et al. (2003)	7.41			3.00	4.41
Amazonia	NPP	plot (Oak Ridge DB)	Rio Tiruma Manaus, Brazil	Klinge et al. 1975	6.70	0.40		1.00	5.30
Amazonia	NPP	plot (Oak Ridge DB)	Ducke Manaus, Brazil	Klinge et al. 1975	6.40	1.00		1.10	4.30
Amazonia	NPP	plot (Oak Ridge DB)	km64 Manaus, Brazil	Klinge et al. 1975	7.54			1.60	5.94
					mean with FR 18.89	*std dev with FR* 4.65	*mean w/o FR* 7.01	*std dev w/o FR* 0.55	

Intensive field studies (stand scale)					NPP min	NPP max	AG-NPP	BG-NPP min	BG-NPP max
Neotropics	NPP	plot	Brazil, Manaus	Clark et al. 2001b	7.70	14.10	6.40	1.30	7.70
Neotropics	NPP	plot	Columbia	Clark et al. 2001b	9.00	16.50	7.50	1.50	9.00
Neotropics	NPP	plot	Columbia	Clark et al. 2001b	11.50	21.20	9.60	1.90	11.60
Neotropics	NPP	plot	Venezuela	Clark et al. 2001b	2.30	12.30			
Neotropics	NPP	plot	Venezuela	Clark et al. 2001b	6.90	13.10			
Neotropics	NPP	plot	Brazil, Fazenda Gaviao	Clark et al. 2001b	9.20	16.80	7.60	1.60	9.20
Neotropics	NPP	plot	Brazil, Fazenda Dimona	Clark et al. 2001b	8.70	15.90	7.20	1.50	8.70
Neotropics	NPP	plot	Guiane Francaise	Clark et al. 2001b	8.00	14.70	6.70	1.30	8.00
Neotropics	NPP	plot	Brazil, Paragominas	Clark et al. 2001b	8.00	14.70	6.80	1.20	7.90
Neotropics	NPP	plot	Brazil, Fazenda CaboFrio	Clark et al. 2001b	7.40	13.60	6.20	1.20	7.40
Neotropics	NPP	plot	Venezuela, San Eusebio	Clark et al. 2001b	7.30	13.30	6.10	1.20	7.20
Neotropics	NPP	plot	Venezuela, San Carla	Clark et al. 2001b	6.90	12.70	5.80	1.10	6.90
Neotropics	NPP	plot	Brazil, Fazenda Porto Allegre	Clark et al. 2001b	6.70	12.20	5.60	1.10	6.60
				mean	8.22	15.06	6.86	1.35	8.20
				std dev	1.74	2.47	1.12	0.25	1.40

Intensive field studies (stand scale)					total NPP	Fine Root	Coarse Root	Woody	Foliage
Tropical forests	NPP	plot	review from 10 studies	Grace et al. 2001	*mean with FR* 18.10	*std dev with FR* 6.79	*mean w/o FR* 11.20	*std dev w/o FR* 4.20	

Extensive forest biomass inventories					mean AG-NPP	std dev AG-NPP	mean Woody	std dev Woody	mean Foliage	std dev Foliage
Neo tropics	NPP	Forest biomass inventory	100 plots	Malhi *et al.* 2003	9.15	2.55	3.05	0.83	6.09	2.42
					mean total NPP	NPPmin	NPPmax			
Neo tropics	NPP	Forest biomass inventory	68 plots + estimated litterfall	Phillips *et al.* 1998	8.44	5.44	11.26			
Neo tropics	NPP	Forest biomass inventory	59 plots, Amazon	Baker *et al.* 2004	8.61	5.61	8.61			

Litter fall					Woody litter	Foliage litter				
Tropical forests	LITTER	plot review from 3 studies		Matthews 1997	4.10	1.62				

Ecosystem Global Models LCC : *Land-use change* + CO_2 + *Climate* / CC : CO_2 + *Climate* / C : CO_2 total NPP

					total NPP	
Tropical S.America	NPP	HRBM model LCC 1980–1989		McGuire *et al.* 2001	6.36	
Tropical S.America	NPP	IBIS model LCC 1980–1989		McGuire *et al.* 2001	7.00	
Tropical S.America	NPP	LPJ model LCC 1980–1989		McGuire *et al.* 2001	10.61	
Tropical S.America	NPP	TEM model LCC 1980–1989		McGuire *et al.* 2001	5.91	
					mean	std dev
					7.47	2.14
Tropical S.America	NPP	HRBM model CC 1980–1989		McGuire *et al.* 2001	7.08	
Tropical S.America	NPP	IBIS model CC 1980–1989		McGuire *et al.* 2001	9.02	
Tropical S.America	NPP	LPJ model CC 1980–1989		McGuire *et al.* 2001	12.46	
Tropical S.American	NPP	TEM model CC 1980–1989		McGuire *et al.* 2001	7.48	
					mean	std dev
					9.01	2.44
Tropical S.America	NPP	HRBM model C 1980–1989		McGuire *et al.* 2001	7.08	
Tropical S.America	NPP	IBIS model C 1980–1989		McGuire *et al.* 2001	9.42	
Tropical S.America	NPP	LPJ model C 1980–1989		McGuire *et al.* 2001	12.30	
Tropical S.America	NPP	TEM model C 1980–1989		McGuire *et al.* 2001	8.09	
					mean	std dev
					9.22	2.27
Amazonia	NPP	TEM model 1978-1998		Tian *et al.* 1998	12.92	

2. Net Ecosystem Productivity

Eddy Covariance Towers				mean NEE		
Amazon	NEP	mature forest, Manaus	Malhi and Grace, 2000	2.50		
Amazon	NEP	mature forest, Rondonia		Fan et al. 1999		2.19
Amazon	NEP	mature forest, Manaus 2000-2001	B. Kruijt pers. com.	6.20		
Amazon	NEP	mature forest, Manaus 1999-2000	B. Kruijt pers. com.	7.70		
Amazon	NEP	mature forest, Jaru 1999-2000	B. Kruijt pers. com.	5.80		
Amazon	NEP	mature forest, Jaru 1999-2000	B. Kruijt pers. com.	6.00		

3. Net Biome Productivity

Ecosystem Global Models LCC : Land-use change + CO_2 + Climate / CC : CO_2 + Climate / C : CO_2				NBP	
Tropical S.America	NBP	HRBM model LCC 1980-1989	McGuire et al. 2001	0.11	
Tropical S.America	NBP	IBIS model LCC 1980–1989	McGuire et al. 2001	0.14	
Tropical S.America	NBP	LPJ model LCC 1980-1989	McGuire et al. 2001	0.19	
Tropical S.America	NBP	TEM model LCC 1980-1989	McGuire et al. 2001	0.16	
				mean	std dev
				0.15	0.03
Tropical S.America	NBP	HRBM model CC 1980–1989	McGuire et al. 2001	0.26	
Tropical S.America	NBP	IBIS model CC 1980–1989	McGuire et al. 2001	0.52	
Tropical S.America	NBP	LPJ model CC 1980–1989	McGuire et al. 2001	0.31	
Tropical S.America	NBP	TEM model CC 1980-1989	McGuire et al. 2001	0.41	
				mean	std dev
				0.38	0.12
Tropical S.America	NBP	HRBM model C 1980–1989	McGuire et al. 2001	0.25	
Tropical S.America	NBP	IBIS model C 1980–1989	McGuire et al. 2001	0.47	
Tropical S.America	NBP	LPJ model C 1980-1989	McGuire et al. 2001	0.18	
Tropical S.America	NBP	TEM model C 1980-1989	McGuire et al. 2001	0.14	
				mean	std dev
				0.26	0.15
Amazon forests	NBP	TEM model 1978-1998	Tian et al. 1998	0.56	0.30

Extensive forest biomass inventories				mean	std dev
Amazonia	NBP	40 plots, Amazon 1975-1996 (area = 7.116 x10^6ha)	Phillips et al. 1998	0.44	0.26
Amazonia	NBP	59 plots, Amazon forest biomass inventory	Baker et al. 2004	0.61	0.22

Atmospheric Inversions (total NBP)				*mean NBP*	*ext error*	*int error*	*area (x10⁶ha)*	
Tropical S.America	NBP	7 monthly inversions with 2 models 1985–1995	Bousquet *et al.* 1999	−0.07	0.20	0.66	1518.48	
Tropical S.America	NBP	1 monthly inversion; 1 model 1980–1995	Rayner *et al.* 1999	0.07			1518.48	
Tropical S.America	NBP	Annual inversions with 15 transport models 1992–1996	Gurney *et al.* 2002	−0.58	0.49	1.08	1014.53	
Tropical S.America	NBP	8 monthly inversions with 2 transport models 1980–1998	Bousquet *et al.* 2000	0.01	0.13	0.57	1518.48	
				mean error	*mean ext error*	*mean int*	*spread*	*spread+int error+ext error*
				−0.14	0.27	0.77	0.30	0.87

Deforestation		*Mean flux calculated over total inversion area: 1000 Mha (note : deforested area is: 20 x10⁶ha)*		*deforestation*	*fires*	*logging*	*error*	*sum*
Brazilian Amazon	LUC	remote sensing + bookkeeping model 1978–1998	Houghton *et al.* 2000	−0.18	−0.18	−0.01	0.10	0.36
Brazilian Amazon	LUC	FAO statistics + bookkeeping model	Fearnside *et al.* 1998	−0.26				
		remote sensing + bookkeeping model 1990–2000	Achard *et al.* 2002	−0.19				

Atmospheric Inversions (Forest NBP)				*mean*	*ext error*	*int error*	*area (x10⁶ha)*	
Tropical S.America	forest NBP	inversion + deforestation	Bousquet *et al.* 1999	0.17	0.20	0.66	1518.48	
Tropical S.America	forest NBP	inversion + deforestation	Rayner *et al.* 1999	0.30			1518.48	
Tropical S.America	forest NBP	inversion + deforestation	Gurney *et al.* 2002	−0.23	0.49	1.08	1014.53	
Tropical S.America	forest NBP	inversion + deforestation	Bousquet *et al.* 2000	0.24	0.13	0.57	1518.48	
				mean	*mean ext error*	*mean int error*	*spread*	*spread+int error+ext error*
				0.12	0.27	0.77	0.24	0.85

Rivers CO₂ outgassing		*For information – Not substracted from inversions (see text)*						
Amazonia seas flooded areas	RESP 1	77 Mha	Richey *et al.* 2002	−0.30				

References

Achard, F., Eva, H.D., Stibig, H.-J., Mayaux, P., Gallego, J., Richards, T. and Malingreau, J.-P. (2002) Determination of deforestation rates of the world's humid tropical forests. *Science* **297**: 999–1002.

Alexeyev, V.A. and Birdsey, R.A. (1994) Carbon in ecosystems of forests and peatlands of Russia. Sukachev Institute for Forest Research, Siberian Division of the Russian Academy of Sciences, Krasnoyarsk/North Eastern Forest Experiment Station, USDA Forest Service.

Amiotte Suchet, P. and Probst, J.L. (1994) A global model for present day atmospheric / soil CO$_2$ consumption by chemical erosion of continental rocks (GEM-CO$_2$). *Tellus Series B – Chemical and Physical Meteorology* **46**: 1–8.

Arain, M.A., Black, T.A., Barr, A.G., Jarvis, P.G., Massheder, J.M., Verseghy, D.L. and Nesic, Z. (2002) Effects of seasonal and interannual climate variability on net ecosystem productivity of boreal deciduous and conifer forests. *Canadian Journal of Forest Research* **32**: 878–891.

Arneth, A., Kurbatova, J., Kolle, O., Shibistova, OB., Lloyd, J., Vygodskaya, N.N. and Schulze, E.D. (2002) Comparative ecosystem-atmosphere exchange of energy and mass in a European Russian and a central Siberian bog II. Interseasonal and interannual variability of CO$_2$ fluxes. *Tellus B-* **54**: 514–530.

Bacastow, R. and Keeling, C.D. (1973) Atmospheric carbon dioxide and radiocarbon in the natural carbon cycle, II, Changes from A.D. 1700 to 2070 as deduced from a geochemical reservoir. In: Woodwell, M. and Pecan, E.V. (eds) *Carbon and the Biosphere*, pp. 86–135. 24th Brookhaven Symposium May 16–18, 1972. AEC Symposium Series 30, 86135. NTIS U.S. Dept. of Commerce. Springfield, Virginia.

Baker, T.R., Phillips, O.L., Malhi, Y., Almeida, S., Arroyo, L., Di Fiore, A. *et al.* (2004) Increasing biomass? Amazonian forest plots. *Phil. Trans. R. Soc. London.* **B359**: 353–365.

Baldocchi, D.D., Falge, E. and Gu, L.H. (2001) Fluxnet: a new tool to study the temporal and spatial variability of ecosystem-scale carbon dioxide, water, vapor, and energy flux densities. *Bulletin of the American Meteorological Society* **82**: 2415–2434.

Bazilevich, N.I. (1993) Biological productivity of ecosystems of Northern Eurasia. Nauka, Moscow, pp. 294. [in Russian].

Botta, A., Ramankutty, N. and Foley, J. (2002) Long-term variations of climate and carbon fluxes over the Amazon basin. *Geophysical Research Letters* **29**: 331–334.

Bousquet, P. Ciais, P., Peylin, P., Ramonet, M. and Monfray, P. (1999) Inverse modeling of annual atmospheric CO$_2$ sources and sinks – Part 1 – Method and control inversion. *Journal of Geophysical Research* **104**: 26161–26178.

Bousquet, P., Peylin, P., Ciais, P., Le Quere, C., Friedlingstein, P. and Tans, P.P. (2000) Interannual changes in regional CO$_2$ fluxes. *Science* **290**: 1342–1346.

Briffa, K., Jones, P., Schweingruber, F., Shiyatov, S. and Cook, E. (1995) Unusual twentieth century summer warmth in a 1000 years temperature record from Siberia. *Nature* **376**: 156–159.

Cannell, M.G.R. (1982) *World Forest Biomass and Primary Production Data.* Academic Press, London. 391 pp.

Cannell, M.G.R. (1999) Relative importance of increasing atmospheric CO_2 N deposition and temperature in promoting european forest growth. In: Karjalainen, T., Spiecker, H. and Laroussine, O. (eds) *Causes and Consequences of Accelerating Tree Growth in Europe.* Proceedings of the International Seminar held in Nancy, France 14–16 May, 1998.

Caspersen, J.P., Pacala, S.W., Jenkins, J.C., Hurtt, G.C., Moorcroft, P.R. and Birdsey, R.A. (2000) Contributions of Land-Use History to Carbon Accumulation in U.S. Forests. *Science* **290**: 1148–1151.

Chen, W. J., Black, T.A. and Yang. P.C. (1999) Effects of climatic variability on the annual carbon sequestration by a boreal aspen forest. *Global Change Biology* **5**: 41–53.

Ciais, P., Peylin, P. and Bousquet, P. (2000) Regional biospheric carbon fluxes as inferred from atmospheric CO_2 measurements. *Ecological Applications* **10**: 1574–1589.

Clark, D.A., Brown, S., Kicklighter, D.W., Chambers, J.Q., Thomlinson, J. R., and Ni, J., (2001a) Measuring net primary production in forests : concepts and field methods. *Ecological Applications* **11**: 356–370.

Clark, D.A., Brown, S., Kicklighter, D.W., Chambers, J.Q., Thomlinson, J.R., Ni, J. and Holland, E.A. (2001b) NPP in tropical forests : An evaluation and synthesis of existing field data. *Ecological Applications* **11**: 371–384.

DeFries, R., Houghton, R.A. and Hansen, M. (2002) Carbon emissions from tropical deforestation and regrowth based on satellite observations for the 1980s and 1990s. *Proceedings of the National Academy of Sciences of the USA* **99**: 14256–14261.

Duncan, B.N., Martin, R.V., Staudt, A.C., Yevich, R. and Logan, J.A. (2003) Interannual and seasonal variability of biomass burning emissions constrained by satellite observations. *Journal of Geophysical Research* (Atmospheres) **108(D2)**: ACH 1-1, 4100 doi:10.1029/2002JD002378.

Fan, S.M., Blaine, T.L. and Sarmiento, J.L. (1999) Terrestrial carbon sink in the Northern Hemisphere estimated from the atmospheric CO^2 difference between Manna Loa and the South Pole since 1959. *Tellus B-* **51**: 863–870.

FAO (2000) Forest Resources Assessment 2000 (FRA 2000). *FAO* Rome.

Farquhar, G.D., Caemmerer, S. and Berry, J.A. (1980) A biochemical model of photosynthetic CO_2 assimilation in leaves of C3 species. *Planta* **149**: 78–90.

Fearnside, P.M. and Barbosa, R.I. (1998) Soil carbon changes from conversion of forest to pasture in Brazilian Amazonia. *Forest Ecology and Management* **108**: 147–166.

Gaudinski, J.B., Trumbore, S.E., Davidson, E.A. and Zheng, S. (2000) Soil carbon cycling in a temperate forest: radiocarbon-based estimates of residence times, sequestration rates and partitioning of fluxes. *Biogeochemistry* **51**: 33–69.

Goodale, C.L., Apps, M.J., Birdsey, R.A., Field, C.B., Heath, L.S., Houghton, R.A. *et al.* (2002) Forest carbon sinks in the northern hemisphere. *Ecological Applications* **12**: 891–899.

Goudriaan, J. and Ketner, P. (1994) A simulation study for the global carbon cycle, including man's impact on the biosphere. *Climate Change* **6**: 167–192.

Goulden, M.L., Munger, J.W., Fan, S.M., Daube B.C. and Wofsy, S.C. (1996) Exchange of carbon dioxide by a deciduous forest: response to interannual climate variability. *Science* **271**: 1576–1578.

Grace, J., Malhi, Y., Higuchi, N. and Meir, P. (2001) Productivity of tropical rain forests. In: Roy, J., Saugier, B. and Mooney, H.A. (eds) *Terrestrial Global Productivity: Past, Present and Future*, pp. 401–426. Academic Press Incorporated, San Diego, CA; Orlando, FL:, Troy, MO.

Grimm, U. and Fassbender, H.W. (1981) Biogeochemical cycles in a forest escosystem of the western Venezuelan Andes 3. Hydrological Cycle and Chemical Elements transfer with water. *Turrialba* **31:** 89–99.

Gurney, K.R., Law, R.M., Denning, A.S., Rayner, P.J., Baker, D., Bousquet, P. *et al.* (2002) Towards robust regional estimates of CO$_2$ sources and sinks using atmospheric transport models. *Nature* **415:** 626–630.

Harrison, A. F. and D. D. Harkness (1993) The potential for estimating carbon fluxes in forest soils using ^{14}C techniques. *New Zealand Journal of Forestry Science* **23:** 367–379.

Hicke, J.A., Asner, G.P., Randerson, J.T., Tucker, C., Los, S., Birdsey, R., Jenkins, J.C. and Field, C. (2002a) Trends in North American net primary productivity derived from satellite observation, 1982–1988. *Global Biogeochemical Cycles* **16:** 10.1029/2001B001550.

Hicke, J.A., Asner, G.P., Randerson, J.T., Tucker, C., Los, S., Birdsey, R., Jenkins, J.C., Field, C. and Holland, E. (2002b) Satellite-derived increases in net primary productivity across North America, 1982–1998. *Geophysical Research Letters* **29:** pp. 69–1, 1427, D0110.1029/2001GL013578.

Houghton, R.A. (1998) Historic role of forests in the global carbon cycle. In: Kohlmaier, G.H., Weber, M. and Houghton, R.A. (eds) *Carbon Dioxide Mitigation in Forestry and Wood Industry*, pp. 1–24. Springer-Verlag, Berlin.

Houghton, R.A., Skole, D.L., Nobre, C.A., Hackler, J.L., Lawrence, K.T. and Chomentowski, W.H. (2000) Annual fluxes of carbon from deforestation and regrowth in the Brazilian Amazon. *Nature* **403:** 301–304.

IPCC (2001) *Climate Change 2001*: The Scientific Basis Contribution of Working Group I to the Third Assessment Report of the Intergovernmental Panel on Climate Change (IPCC). J.T. Houghton, T. Ding, D.J. Griggs, M. Noguer, P.J. van der Linden and D. Xiaosu (eds.) Cambridge University Press, UK. pp 944.

Ivanova, G.A. (1998–1999) The history of forest fire in Russia. *Dendrochronologia* **16–17:** 147–161.

Jackson, R.B., Mooney, H.A. and Schulze, E.D. (1997) A global budget for fine root biomass, surface area, and nutrient contents. *Ecology* **94:** 7362–7366.

Janssens, I., Freibauer, A., Ciais, P., Smith, P., Nabuurs, G., Schlamadinger, B. *et al.* (2003) Carbon uptake by European Ecosystems. *Science*, submitted.

Joos, F., Prentice, I.C. and House, J.I. (2002) Growth enhancement due to global atmospheric change as predicted by terrestrial ecosystem models: consistent with US forest inventory data. *Global Change Biology* **8:** 299–303.

Kajimoto, T., Matsuura, Y., Sofronov, M.A., Volokitina, A.V., Mori, S., Osawa, A. and Abaimov, A.P. (1999) Above- and belowground biomass and net primary productivity of a *Larix gmelinii* stand near Tura, central Siberia. *Tree Physiology* **19:** 815–822.

Kaminski, T. and Heimann, M. (1999) A coarse grid three-dimensional global inverse model of the atmospheric transport. 2- Inversion of the transport of CO$_2$ in the 1980s. *Journal of Geophysical Research* **104(D15):** 18555–18581.

Keeling, C.D., Chin, J.F.S. and Whorf, T.P. (1996) Increased activity of northern vegetation inferred from atmospheric CO_2 measurements. *Nature* **382**: 146–149.

Kelliher, F.M., Hollinger, D.Y. and Schulze, E.-D. (1997) Evaporation from an eastern Siberian larch forest. *Agricultural and Forest Meteorology* **85**: 135–147.

Kesselmeier, J., Ciccioli, P., Kuhn, U., Stefani, P., Biesenthal, T., Rottenberger, S. *et al.* (2003) Volatile organic compound emissions in relation to plant carbon fixation and the terrestrial carbon budget. *Global Biogeochemical Cycles*, **16(4)**: 1126, doi:10.1029/2001GB001813.

Kindermann, J., Wurth, G., Kohlmaier, G.H. and Badeck, F.W. (1996) Interannual variation of carbon exchange fluxes in terrestrial ecosystems. *Global Biogeochemical Cycles* **10**: 737–755.

Klinge, H., Rodrigues, W.A., Brunig, E. and Fittkau, E.J. (1975) Biomass and structure in a Central Amazonian rain forest. pp 115–122 in Golley, F.B. and Medina, E. (eds), *Tropical Ecological Systems Trends in Terrestrial and Aquatic Research.* Springer Verlag, New York.

Körner, C. (2003) Slow in, rapid out-carbon flux studies and Kyoto targets. *Science* **300**: 1242–1243.

Liski, J., Korotkov, A.V., Prins, C.F.L., Karjalainen, T., Victor, D.G. and Kauppi, P.E. (2002) Increased carbon sink in temperate and boreal forests. *Climatic Change* **61**: 89–99.

Liski, J., Nissinen, A., Erhard, M. and Taskinen, O. (2003) Climatic effects on litter decomposition from arctic tundra to tropical rainforest. *Global Change Biology* **9**: 575–584.

Lloyd, J. (1999) The CO_2 dependence of photosynthesis, plant growth responses to elevated CO_2 concentrations and their interactions with soil nutrient status II. Temperate and boreal forest productivity and the combined effects of increasing CO_2 concentrations and increased nitrogen deposition at a global scale. *Functional Ecology* **13**: 439–459.

Lloyd, J., Grace, J., Miranda, A.C., Meir, P., Wong, S.C., Miranda, H.S., Wright, I.R., Gash. J.H.C. and McIntyre, J. (1996) A simple calibrated model of Amazon rain forest productivity based on leaf biochemical properties. *Plant Cell and Environment* **18**: 1129–1145.

Lloyd, J., Shibistova, O., Zolotoukhine, D., Kolle, O., Arneth, A., Wirth, C., Styles, J.M., Tchebakova, N.M. and Schulze, E.D. (2002) Seasonal and annual variations in the photosynthetic productivity and carbon balance of a central Siberian pine forest. *Tellus B-* **54**: 590–610.

Lowman, M.D. (1995) Herbivory as a canopy process in rain forest trees. In: Lowman, M.D. and Nadkarni, N. (eds) *Forest Canopies*, pp. 431–452. Academic Press. San Diego.

Lucht, W., Prentice, I.C., Myneni, R.B., Sitch, S., Friedlingstein, P., Cramer, W., Bousquet, P., Buermann, W. and Smith, B. (2002) Climatic control of the high-latitude vegetation greening trend and Pinatubo effect. *Science* **296**: 1687–1689.

Malhi, Y. and Grace, J. (2000) Tropical forests and atmospheric carbon dioxide. *Ecology and Evolution* **15**: 332–337.

Malhi, Y., Phillips, O.L., Lloyd, J., Baker, T., Wright, J., Almeida, S. *et al.* (2002) An international network to monitor the structure, composition and dynamics of Amazonian forests (RAINFOR). *Journal of Vegetation Science* **13**: 439–450.

Malhi, Y., Baker, T.R., Phillips, O.L., Almeida, S., Alvarez, E., Arroyo, L. *et al.* (2003) The above-ground wood productivity and net primary productivity of 104 Neotropical forests. *Global Change Biology*: 10: 563–591.

Matthews, G. (1993) The carbon content of trees. Forestry commission, Technical Paper 4, HMSO, London, 21pp.

Martin, P.H., Valentini, R., Jacques, M., *et al.* (1998) New estimate of the carbon sink strength of European forests integrating flux measurements field surveys and space observations: 0.17–0.35 Gt C. *Ambio* 27: 582–584.

McGuire, A.D.C., Sitch, S., Clein, J.S., Dargaville, R., Esser, G., Foley, J. *et al.* (2001) Carbon balance of the terrestrial biosphere in the twentieth century : Analysis of CO₂ climate and land use effects with four process-based ecosystem models. *Global Biogeochemical Cycles* 15: 183–206.

McGuire, A.D., Wirth, C., Apps, M., Beringer, J., Clein, J., Epstein, H. *et al.* (2002) Environmental variation, vegetation distribution, carbon dynamics and water / energy exchange at high latitudes. *Journal of Vegetation Science* 13: 301–314.

McNaughton, S.J. (2001) Herbivory and trophic interactions. In: Roy, J., Saugier, B. and Mooney, H.A. (eds) *Terrestrial Global Productivity: Past, Present, Future*, pp. 101–122. Academic Press, San Diego.

McWilliam, A.L.C., Roberts, J.M., Cabral, O.M.R., Leitao, M., Decosta, A., Maitelli, G.T. and Zamparoni, C. (1993) Leaf-area index and aboveground biomass of Terra-Firme rain-forest and adjacent clearings in Amazonia. *Functional Ecology* 7: 310–317.

Milne, R. and Brown, T.A.W. (2002) Mapping of carbon uptake in British woodlands and forests using EuroBiota on C-Flow. In, *UK Emissions by Sources and Removals by Sinks due to Land Use, Land-Use Change and Forestry Activities.* Report, Centre of Ecology and Hydrology, Edinburgh.

Mund, M., Kummetz, E., Hain, M., Bauer, G.A. and Schulze, E.-D. (2002) Growth and carbon stocks of a spruce forest chronosequence in central Europe. *Forest Ecology and Management* 171: 275–296.

Myneni, R.B., Kelling, C.D., Tucker, C.J., Asrar, G. and Nemani, R.R. (1997) Increased plant growth in the northern high latitudes from 1981–1991. *Nature* 386: 698–701.

Nabuurs, G. J., Schelhaas, M.J., Godefridus, F., Mohren, M.J., and Field, C.B. (2003) Temporal evolution of the European forest sector carbon sink from 1950 to 1999. Global Change Biology, 9: 152–160.

Nemani, R. R., Keeling, C.D., Hashimoto, H., Jolly, W.M., Piper, S.C., Tucker, C.J., Myneni, R.B. and Running, S.W. (2003) Climate-driven increases in global terrestrial net primary production from 1982 to 1999. Science, 300: 1560–1563.

New, M., Hulme, M. and Jones, P. (2000) Representing twentieth-century space-time climate variability. Part II: Development of 1901–96 monthly grids of terrestrial surface climate. *Journal of Climate* 13: 2217–2238.

Nilsson, S., Shividenko, A., Stolbovoi, V., Gluck, M., Jonas, M. and Oberteiner, M. (2000) Full carbon account for Russia, IIASA Interim Report IR-00-021, International Institute for Applied Systems Analysis, Laxenburg, Austria. http://www.iiasa.ac.at/Publications/Documents/IR-00-021.pdf *Journal of Climate* 13: 2217–2238.

Papale, D. and Valentini, R. (2003). A new assessment of European forests carbon exchanges by eddy fluxes and artificial neural network spatialization. *Global Change Biology* **9**: 525–535.

Persson, T., Van Oene, H., Harrison, A. F., Karlsson, P.S., Bauer, G.A., Cerny, J. *et al.* (2000) Experimental sites in the NIPHYS/CANIUF project. In: Schulze, E.-D. (ed) *Carbon and Nitrogen Cycling in European Forest Ecosystems*, 142 pp. 14–46. Ecological Studies, Springer-Verlag. Heidelberg.

Peylin, P., Baker, D., Sarmiento, J., Ciais, P. and Bousquet, P. (2002) Influence of transport uncertainty on annual mean and seasonal inversions of atmospheric CO_2 data. *Journal of Geophysical Research* 107(D19), ACH5-1, 4385, Doi10.1029/2001JD000857.

Phillips, O.L., Malhi, Y., Higuchi, N., Laurance, W.F., Nunez, P.V., Vasquez, R.M. *et al.* (1998) Changes in the carbon balance of tropical forests: Evidence from long-term plots. *Science* **282**: 439–442.

Pleshikov, F.I. (2002) *Forest Ecosystems of the Yenisei Meridian.* V.N. Sukachev Institute of Forest. Krasnoyarsk, Institut lesa SO RAN. 357 pp. [In Russian.]

Rayner, P.J., Enting, I.G., Francey, R.J. and Langenfelds, R. (1999) Reconstructing the recent carbon cycle form atmospheric CO_2, $\delta^{13}C$ and O_2/N_2 observations. *Tellus Series B – Chemical and Physical Meteorology* **51**: 213–232.

Richey, J.E., Melack, J.M., Aufdenkampe, A.K., Ballester V.M. and Hess, L.L. (2002) Outgassing from Amazonian rivers and wetlands as a large tropical source of atmospheric CO_2. *Nature* **416**: 617–620.

Roderick, M.L., Farquhar, G.D., Berry, S.L. and Noble, I.R. (2001) On the direct effect of clouds and atmospheric particles on the productivity and structure of vegetation. *Oecologia* **129**: 21–30.

Roser, C., Montagnani, L., Schulze, E.D., Mollicone, D., Kolle, O., Meroni, M., Papale, D., Marchesini, L.B., Federici, S. and Valentini, R. (2002). Net CO_2 exchange rates in three different successional stages of the "Dark Taiga" of central Siberia. *Tellus B-* **54**: 642–654.

Sarmiento, J.L. and Sundquist, E.T. (1992) Revised budget for the oceanic uptake of anthropogenic carbon dioxide. *Nature* **356**: 589–593.

Scarascia-Mugnozza, G., Bauer, G.A., Persson, T., Matteucci, G. and Masci, A. (2000) Tree biomass, growth and nutrient pools. In: Schulze, E.-D. (ed) *Carbon and Nitrogen Cycling in European Forest Ecosystems*, 142 pp. 49–62. Ecological Studies, Springer-Verlag. Heidelberg.

Schelhaas, M.J. and Nabuurs, G. (2001) Spatial distribution of regional whole tree carbon stocks and fluxes of forests in Europe. Alterra Report 300, Alterra, Wageningen.

Schelhaas, M.J. and Schuck, V.S. (2001) Database on forest disturbances in Europe. Joensuu, Finland, European Forest Institute.

Schimel, D., Melillo, J., Tian, H., McGuire, A.D., Kicklighter, D., Kittel, T. *et al.* (2000) Contribution on increasing CO_2 and climate to carbon storage by ecosystems in the United States. *Science* 287: 2004–2006.

Schimel, D.S., House, J.I., Hibbard, K.A., Bousquet, P., Ciais, P., Peylin, P. *et al.* (2001) Recent patterns and mechanisms of carbon exchange by terrestrial ecosystems. *Nature* **414**: 169–172.

Schulze, E.-D. and Heimann, M. (1998) Carbon and water exchange of terrestrial ecosystems. In: Galloway, J.N. and Melillo, J. (eds) *Asian Change in the context of global Change*, pp. 145–161. Cambridge University Press.

Schulze, E.-D., Lloyd, J., Kelliher, F.M., Wirth, C., Rebmann, C., Luhker, B. *et al.* (1999) Productivity of forests in the Eurosiberian boreal region and their potential to act as a carbon sink – a synthesis. *Global Change Biology* **5:** 703–722.

Schulze, E.-D., Högberg, P., Van Oene, H., Persson, T., Harrison, A.F., Read, D., Kjoller, A. and Matteucci, G. (2000) Interactions between the carbon and nitrogen cycle and the role of biodiversity: A synopsis of study along a north–south transect through Europe. In: Schulze, E.-D. (ed) *Carbon and Nitrogen Cycling in European Forest Ecosystems*, 142 pp. 468–492. Ecology Studies, Springer-Verlag. Heidelberg.

Schulze, E.-D., Vygodskaya, N.N., Tchebakova, N.M., Czimczik, C.I., Kozlov, D.N., Lloyd, J., Mollicone, D., Parfenova, E., Sidorov, K.N., Varlagin, A.V. and Wirth, C. (2002) The Eurosiberian Transect: an introduction to the experimental region. *Tellus Series-B* **54:** 421–428.

Shvidenko, A. Z. and Nilsson, S. (2000) Extent, distribution, and ecological role of fire in Russian forests. In: Kasischke, E.S. and Stocks, B.J. (eds) *Fire, Climate Change, and Carbon Cycling in the Boreal Forest*, pp. 32–150. Ecological Studies. Springer. New York.

Shvidenko, A.Z. and Nilsson, S. (2002) Dynamics of Russian forests and the carbon budget in 1961–1988: An assessment based on long-term forest inventory data. *Climatic Change* **55:** 5–37.

Shvidenko, A.Z. and Nilsson, S. (2003) A synthesis of the impact of Russian forests on the global carbon budget for 1961–1998. *Tellus Series B – Chemical and Physical Meteorology* **55:** 391–415.

Shvidenko, A.Z., Nilsson, S., Stolbovoi, V.S., Rozhkov, V.A. and Gluck, M. (2001) Aggregated estimation of basic parameters of biological production and the carbon budget of Russian terrestrial ecosystems: 2. Net primary production. *Russian Journal of Ecology* **32:** 71–77.

Spiecker, H., Mielikäinen, K., Köhl, M. and Skovsgaard, J. (1996) Growth trends in European forests. In: Spiecker, H., Mielikäinen, K., Köhl, M. and Skovsgaard, J. (eds), *European Forest Institute Research Report no. 5*. Springer-Verlag. Berlin.

Stanners, D. and Bourdeau, P. (eds) (1995) Europe's Environment. In: *The Dobris Assessment*. European Environment Agency, Copenhagen.

Steele, J.S., Gower, S.T., Vogel, J.G. and Norman, J.M. (1997) Roots mass, net primary production and turnover in aspen, jack pine and black spruce forests in Saskatchewan and Manitoba. *Canadian Tree Physiology* **17:** 577–587.

Telles, E.C.C., De Camargo, P.B., Martinelli, L.A., Trumbore, S.E., Da Costa, E.S., Santos, J., Higuchi, N. and Oliveira, R.C. (2003) Influence of soil texture on carbon dynamics and storage potential in tropical forest soils of Amazonia. *Global Biogeochemical Cycles* **17(2):** doi:10.1029/2002GB001953.

Thompson, M. V., Randerson, J.T., Malmström, C.M. and Field, C.B. (1996) Change in net primary production and heterotrophic respiration : How much is necessary to sustain the terrestrial carbon sink ? *Global Biogeochemical Cycles* **10:** 711–726.

Tian, H.Q., Melillo, J.M., Kicklighter, D.W., McGuire, A.D., Helfrich, J., Moore, B. and Vorosmarty, C.J. (1998). Effect of interannual climate variability on carbon storage in Amazonian ecosystems. *Nature* **396**: 664–667, doi:10.1038/25328.

Trumbore, S.E. (1993) Comparison of carbon dynamics in two soils using measurements of radiocarbon in pre-and post-bomb soils. *Global Biogeochemical Cycles* **7**: 275–290.

Trumbore, S. E., Davidson, E.A., Camargo, P.B., Nepstad, D.C. and Martinelli, L.A. (1995) Belowground cycling of carbon in forests and pastures of eastern Amazonia. *Global Biogeochemical Cycles* **9**: 515–528.

Usoltsev, V.A. (2001) *Forest biomass of Northern Eurasia*. Russian Academy of Sciences. Yekaterinburg. 707 pp. [In Russian.]

van Oene, H., Berendse, F., Persson, T., Harrison, A.F., Schulze, E.-D. *et al.* (2000) 20 Model analysis of carbon and nitrogen cycling in *Picea* and *Faqus* forests. In *Carbon and Nitrogen cycling in European Forest Ecosystems*, (ed) E.–D. Schulze. Springer-Verlag Berlin

Victoria, R.L., Martinelli, L.A., Moraes, J.M., Ballester, M.V., Krusche, A.V., Pellegrino, G., Almeida, R.M.B and Richey, J.E. (1998) Surface air temperature variations in the Amazon region and its borders during this century. *Journal of Climate* **11**: 1105–1110.

Vleeshouwers, L.M. and Verhagen, A. (2002) Carbon emission and sequestration by agricultural land use: a model study for Europe. *Global Change Biology* **8**: 519.

Vogt, K.A., Vogt, D.J., Palmiotto, P.A., Boon, P., O'Hara, J. and Asbjornsen, H. (1996) Review of root dynamics in forest ecosystems grouped by climate, climatic forest types and species. *Plant and Soil* **187**: 159–219.

Vygodskaya, N.N., Schulze, E.-D., Tchebakova, N.M., Karpachevskkii, L.O., Kozlov, D., Sidorov, K.N. *et al.* (2002) Climatic control of stand thinning in unmanaged spruce forests of the southern Tauga in European Russia. *Tellus Series B – Chemical and Physical Meteorology* **54**: 443–461.

Wang, Y.P. and Polglase, P.J. (1995) Carbon balance in the tundra, boreal forest and humid tropical forest during climate change- scaling up from leaf physiology and soil carbon dynamics. *Plant Cell & Environment* **18**: 1226–1244.

Wirth, C.E.D. and Schulze, E.D. (2002) Comparing the influence of site quality, stand age, fire and climate on aboveground tree production in Iberian Scots pine forests. *Tree Physiology* **22**: 537–552.

Wirth, C., Czimczik, C.I. and Schulze, E.-D. (2002) Beyond annual budgets: carbon flux at different temporal scales in fire-prone Siberian Scots pine forests. *Tellus Series B – Chemical and Physical Meteorology* **54**: 611–630.

Wirth, C., Schulze, E.-D. and Weller, E. (2003) Carbon stocks and stock changes in Thrinigian state forest. *Tree Physiology*, in preparation.

Wullschleger, S. D., Post, W.M. and King, A.W. (1995) On the potential for a CO_2 fertilization effect in forests: estimates of the biotic growth factor based on 58 controlled-exposure studies. In: Woodwell, G.M. and Mackenzie, F.T. (eds) *Biotic Feedbacks in the Global Climatic System: Will Warming Feed the Warming?* pp. 85–107. Oxford University Press, New York.

8

Measurement of CO_2 exchange between Boreal forest and the atmosphere

T. Andrew Black, David Gaumont-Guay, Rachhpal S. Jassal, Brian D. Amiro, Paul G. Jarvis, Stith T. Gower, Frank M. Kelliher, Allison Dunn and Steven C. Wofsy

1. Introduction

The Boreal forest occupies much of the circumpolar region between 50° and 70° N. It is the world's second largest forest biome with an area of 12.2 million km² (Landsberg and Gower, 1997) (or 20 million km² from a classification by DeFries and Townshend (1994)) accounting for about 21% of the forested land surface (Whittaker and Likens, 1975). Of this area, 43% is in Siberia, 21% is in Europe, and 36% is in North America (Schulze *et al.*, 1999). This biome accounts for 43% of the Earth's carbon stored in soil and 13% of the carbon stored in biomass (Schlesinger, 1991). The Boreal forest has warmed significantly over the past several decades and is likely to warm more than other forest biomes in response to changing climate (IPCC, 2001). Because of the quantity of carbon stored in this biome, especially in the soil, its sensitivity to variations in climate and its influence on the climate of the Northern Hemisphere, it is important to understand the processes controlling the exchange of carbon between the Boreal forest and the atmosphere (Sellers *et al.*, 1997).

Several studies on atmospheric CO_2 concentration and isotope analyses (Ciais *et al.*, 1995, 2000; Fan *et al.*, 1998; Gurney *et al.*, 2002; Sarmiento and Gruber, 2002; Chapter 7, this volume) have strongly suggested that the Northern Hemisphere continents have been a large sink for anthropogenic CO_2. For example, using $^{13}C/^{12}C$ measurements of atmospheric CO_2, Ciais *et al.* (1995) found that northern temperate and Boreal ecosystems between 35° and 65° N were a major sink of 3.5 ± 1.0 Pg (C) per year in 1992–1993, accounting for roughly half of the global fossil fuel emissions during those years. The question of how important the Boreal forest is in determining the strength of the global carbon sink was one of the reasons for the Boreal Ecosystem–Atmosphere Study (BOREAS) of the mid-1990s (Sellers *et al.*, 1997). BOREAS was unable to resolve this question fully because its intensive and multidisciplinary field measurements were focused mainly on the growing season. BOREAS stressed the importance of process-based research and recognized strong coupling between the physical system and the carbon cycle (Sellers *et al.*, 1997). At the time of BOREAS, it was recognized that it would require an average net uptake of 80 g (C) m^{-2} per year for the Boreal forest to account for a global sink of 1 Pg (C) per year. Certainly this is in the range of the values obtained by year-round eddy covariance measurements (see below), but it will require careful scaling up over the very heterogeneous Boreal landscape to determine the overall sink strength of the Boreal forest biome. Furthermore, careful accounting for year-to-year climate variability and disturbance resulting from fire, harvest, insect attack, disease, and windthrow will be necessary to determine how the carbon balance of the Boreal forest biome changes over time.

Landsberg and Gower (1997) provided a good description of climate, soils and vegetation of the Boreal forest biome and Baldocchi *et al.* (2000) discussed climate and vegetation controls on energy exchanges with the atmosphere. Malhi *et al.* (2000) analysed carbon flux measurements made at a tower flux site in each of Boreal, temperate, and tropical forests, and compared different components of the carbon balance for the three sites. The objectives of this chapter are: (i) to summarize recent work using eddy covariance to measure the net exchange of CO_2 between the atmosphere and the Boreal forest; (ii) to review progress in understanding controls on the net ecosystem productivity (NEP); (iii) to report on recent studies aimed at determining effects of disturbance and stand age on NEP; and (iv) to review methods to upscale forest stand CO_2 fluxes to regional scale. The last includes evaluating the carbon balance on a regional or biome scale, accounting for spatial and temporal variability as well as the consequences of disturbance, to obtain the net biome productivity (NBP) (see Chapters 7 and 15, this volume). We end by commenting on some knowledge gaps and future research needs.

2. Boreal landscape characteristics

The Boreal landscape is a remarkable mosaic of forests of different ages reflecting the past history of fires. There are relatively few tree species because of the relatively short time since the last glaciation and adaptation to the harsh, cold environment. The main overstorey genera present include spruce (*Picea*), pine (*Pinus*), larch (*Larix*), fir (*Abies*), hemlock (*Tsuga*), poplar (*Populus*), birch (*Betula*), willow (*Salix*), and alder (*Alnus*) (*Table 1*). Soils include poorly drained organic soils supporting black spruce (*Picea mariana* (Mill.) BSP), moderately well-drained fine-textured soils supporting white spruce (*Picea glauca* (Moench) Voss) and trembling aspen (*Populus tremuloides* Mich.), and droughty, sandy soils supporting jack pine (*Pinus banksiana* Lamb.). The understoreys range from hazelnut in aspen stands, moss in spruce stands to lichen in pine stands.

Climatically, the Boreal zone can be divided into maritime, continental, and high continental sub-zones, of which the continental area is the largest. *Table 2* summarizes the mean monthly air temperature and annual precipitation in the three sub-zones. Soil temperatures at a depth of 5 cm lie between –3 and 15 °C in the maritime and continental sub-zones, while permafrost occurs in the winter months in the high continental sub-zone.

Table 1. *Dominant tree species and extent of their approximate land cover in different regions of the Boreal forest biome.*

Region	Species	% Land cover
North America [a]	*Picea (mariana, glauca)*	39
	Pinus banksiana	14
	Abies balsamea	11
	Populus & Betula	21
Scandinavia[b]	*Picea abies*	26
	Pinus sylvestris	65
	Populus & Betula	8
Siberia[c]	*Larix gmelinii*	35
	Pinus sylvestris	17
	Picea abies	11
	Populus & Betula	17

[a]Compiled from Canada Centre for Remote Sensing data (G. Pavlic and W. Chen, personal communication).
[b]Complied from Finnish National Forest Inventory data available at http://www.metla.fi/ohjelma/vmi/nfi-resu.htm.
[c]Nilsson *et al.* (2000).

Table 2. *Climate of the Boreal forest biome.*

Region	Mean temperature of coldest month (°C)	Mean temperature of warmest month (°C)	Annual precipitation (mm)
Maritime	−5 to 2	10 to 15	400–1000
Continental	−20 to −10	15 to 20	400–600
High Continental	−35 to −25	10 to 20	300–400

In the maritime sub-zone, the range of climatic extremes is relatively small. Winters are comparatively mild and summers are often cool. The mean temperature of the warmest month is 10 to 15 °C and of the coldest month −5 to 2 °C. The annual precipitation, much of it as snow, ranges from 400 to 800 mm but can be 1000 mm or more in western Norway and Newfoundland.

Continental winters are long and cold and there is abundant snow for five to seven months in most of the Canadian Boreal forest. Considerable variation of the monthly mean temperature, especially in winter, is a marked feature of the zone, requiring adaptability on the part of trees. Desiccating winds and temperatures of −20 to −40 °C can be lethal for trees in the northern parts of the sub-zone. Warming in the spring is rapid; however, there is large variation in the time at which photosynthesis begins. Summer weather is comparatively warm but can be very changeable. The vegetation period, measured by the number of days when the mean daily temperature is more than 6 °C, varies from 100 to 150 days. The mean temperature of the warmest month ranges from 15 to 20 °C. The annual precipitation varies from 400 to 600 mm, with the major part of it falling during the summer months.

In the high continental sub-zone, as in eastern Siberia for example, the winter is very long, extremely cold and dry. The mean annual temperature varies from −7 to −10 °C.

The range of mean monthly temperatures can be more than 40 °C with the lowest temperature around –50 to –60 °C. The mean temperature of the coldest month can be below –25 °C. Spring comes rapidly, the summer is short and comparatively warm but frost is possible every night. Annual precipitation of around 300 to 400 mm is generally lower than in the other sub-zones. The major part of the precipitation falls during the growing season but the precipitation/evaporation ratio is less than unity in warm summer months.

3. Importance of eddy covariance measurements in Boreal carbon balance studies

The eddy covariance (EC) method of measuring the exchange of CO_2 between the atmosphere and the Earth's surface is based on measurement of the fluctuations of vertical wind speed and CO_2 concentration in the surface layer. In the case of terrestrial ecosystems, measurement sensors are placed on towers above the vegetation canopy. Because forests are aerodynamically rough surfaces with very small vertical concentration gradients above, making it difficult for the gradient approach to work, the EC method has come to play an important role in the measurement of fluxes over forests. With the recent availability of fast response sonic anemometers and infrared gas analysers, this technique has become widely accepted during the past decade for the measurement of CO_2 exchange between the atmosphere and terrestrial ecosystems Valentini (2003). Baldocchi (2003) has presented a recent overview of the EC technique describing its theoretical basis and implementation in evaluating carbon exchange rates of ecosystems.

The EC method enables flux measurements to be made continuously for periods of hours to years with minimal disturbance to the vegetation, largely unattended and in climates as harsh as in the Boreal zone. Consequently, the EC method has contributed substantially to the progress that has been made in understanding the ecophysiological processes that control ecosystem carbon exchange. As generally used, EC provides half-hourly and hourly values of net exchange of CO_2 between the ecosystem and the atmosphere (net ecosystem exchange: NEE) with positive values (upward fluxes) corresponding to losses and negative values to gains of carbon from the atmosphere by the ecosystem. The measurements of eddy flux above the canopy are combined with the change in CO_2 storage in the air column beneath the eddy flux sensors to obtain NEE. NEE with a change in sign is a good approximation of net ecosystem productivity (NEP). The relatively small difference results from the gain or loss of dissolved organic carbon as a result of water leaving or passing through the ecosystem and losses of other carbon containing volatile organic compounds (VOCs). The method has proved to be a particularly effective tool in forest carbon exchange research, as it provides estimates of NEE at the scale of the stand, i.e., estimates integrated over a footprint extending about 500 m around the measuring point. It also provides invaluable data for testing soil–vegetation–atmosphere transfer (SVAT) models.

The first major initiatives to measure CO_2 exchange between Boreal forests and the atmosphere using EC occurred in the early 1990s with BOREAS in North America, NOPEX (Northern Hemisphere climate processes land-surface experiment) in Sweden, and several campaigns in Siberia.

Energy balance closure has been suggested as a means to assess confidence in EC fluxes, but it is still controversial as to whether the lack of energy balance closure can be used to correct CO_2 fluxes. Generally, measurements during day-time have high reliability, but night-time measurements are often difficult to interpret. For most sites, when night-time conditions are calm the eddy-flux approach is uncertain and most workers reject these measurements. Corrections using the relationship between night-time fluxes and soil temperature, when friction velocity (u_*), a measure of turbulent mixing, exceeds some threshold value, are frequently used to replace values of NEE obtained at low values of u_* (see, for example, Black *et al.*, 1996; Wofsy *et al.*, 1993). This is based on the observation that ecosystem CO_2 efflux is not very sensitive to u_* above the threshold value.

While providing direct measurements of NEE, EC also enables ecosystem respiration (R) and gross ecosystem productivity or photosynthesis (GEP, P_g) to be estimated. Night-time values of NEE at times with adequate mixing provide a good estimate of night-time R. GEP can be estimated by adding day-time NEP (P_e) to day-time R, with the latter obtained from the relationship between night-time R and temperature, i.e., $P_g = P_e + R$. A good indication of day-time R can also be obtained by extrapolating day-time NEE values to zero photosynthetic photon flux density (PPFD), using the rectangular hyperbolic light response relationship (Griffis *et al.*, 2003) or the non-rectangular hyperbolic light response function (Johnson and Thornley, 1984).

With independent estimates of heterotrophic respiration (R_h) for Boreal forest stands becoming available (see, for example, Högberg *et al.*, 2001, and Chapter 12, this volume) it is possible to estimate NPP (P_n) from NEP, i.e., $P_n = P_e + R_h$ (Griffis *et al.*, 2004). This can provide important confirmation of NPP estimates obtained from measurements of above-ground and below-ground biomass production (Gower *et al.*, 1997, 2001). With both NPP and GEP known, it is possible to estimate autotrophic respiration (R_a), i.e., $R_a = P_g - P_n$. NEP upscaled to a region, taking into account spatial variability resulting from stand species, age, and disturbance, can then be used to provide an estimate of net biome productivity, NBP.

4. Tower flux measurements

The EC measurements described here are largely tower based, but aircraft-based EC measurements will be briefly discussed later. Tower-based EC measurements have been made at 38 sites in Boreal forest: 21 in North America, 5 in North Europe and 12 in Siberia (*Figure 1* and *Table 3*). The sites include evergreen conifers (black spruce, jack pine, Norway spruce (*Picea abies* (L.) Karst), Scots pine (*Pinus sylvestris* L.)), deciduous conifers (Dahurian larch (*Larix gmelinii* (Rupr.) Kuzeneva)), deciduous broadleaves (trembling aspen, aspen-poplar (*Populus balsamifera* L.), silver birch (*Betula pendula* Roth.), white birch (*B. papyrifera* (Marsh.)) and fens and bogs. These are stands of various ages (i.e., *chronosequences*, described below). Maximum daily NEP values ranged from 1 g (C) m^{-2} d^{-1} for jack pine to 6 g (C) m^{-2} d^{-1} for aspen. Many studies were for periods of a several days or a few weeks. Some, like the BOREAS project in Canada, extended over the entire growing season from early spring until late fall. At nine sites, the studies have evolved into long-term, year-round flux measurement programmes.

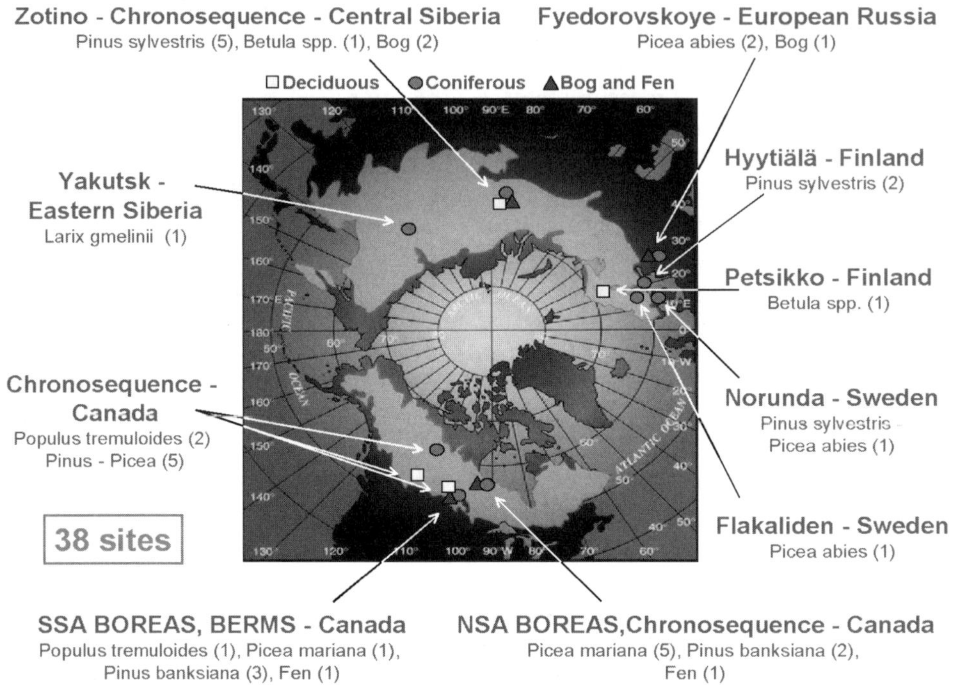

Figure 1. Map of circumpolar Boreal forest showing distribution of tower flux measurement sites in North America, northern Europe, Siberia, and European Russia.

4.1 Short-term eddy covariance studies

Short-term studies through the growing season and diurnal cycles of NEP have been important in determining the effects of environmental variables on canopy CO₂ exchange and have provided critical data for evaluating ecophysiological models. Comparative studies on different species have proved useful in understanding ecophysiological processes. Here, we summarize the estimated maximum gross ecosystem productivity ($P_{g,max}$) at different Boreal sites. $P_{g,max}$ was estimated by adding maximum midday (mean for 10.00–14.00 h) NEP ($P_{e,max}$) to maximum night-time respiration ($R_{n,max}$), obtained on nights during the same week as maximum NEP was measured. These are approximate numbers but give a good indication of photosynthetic capacity and ecosystem respiration in mid-season for different Boreal ecosystems. Values of $P_{g,max}$ (*Table 4*) vary from a low of 6 μmol (C) m⁻² s⁻¹ for stands of larch, followed by bogs, fens, pines, spruce, and birch, to a high of about 30 μmol (C) m⁻² s⁻¹ for aspen. These values reflect stand structure, presence or absence of understorey, age, and environmental conditions, as well as species. Notably, the value for the very old (215-year-old) Scots pine was only 6 μmol (C) m⁻² s⁻¹. Values of $R_{n,max}$ ranged from 1 to 3 μmol (C) m⁻² s⁻¹ for fens and bogs, 3 to 5 μmol (C) m⁻² s⁻¹ for pines, birch and larch, to 7 μmol (C) m⁻² s⁻¹ for aspen and spruce stands. There tends to be higher $R_{n,max}$ for stands with higher $P_{g,max}$ (*Figure 2*) as respiratory losses are likely to

Table 3. Site characteristics (site name (symbol), number of sites, location, species, age, elevation, LAI, period of measurements and references) for eddy covariance (EC) sites in North American, north European, and Siberian Boreal forests.

Site	Number of sites	Location	Species	Age (years)	Elevation (m)	LAI (m^{-2})	Period	Reference
North America								
Saskatchewan, Canada (SOA)	1	53°37′ N 106°11′ W	trembling aspen, hazelnut	85	601	2	1994, 1996–2002	Black *et al.*, 1996 Chen *et al.*, 1999 Black *et al.*, 2000
Saskatchewan, Canada (SOBS)	1	53°59′ N 105°70′ W	black spruce	130	629	4.2	1994, 1996, 1999–2002	Jarvis *et al.*, 1997 Pattey *et al.*, 1997 Swanson and Flanagan, 2001 Rayment *et al.*, 2002
Saskatchewan, Canada (SOJP)	1	53°54′ N 104°41′ W	jack pine	80	579	2	1994, 1999–2002	Baldocchi *et al.*, 1997
Saskatchewan, Canada (SYJP)	1	53°88′ N 104°65′ W	jack pine	25	534		1994	
Saskatchewan, Canada (SCJP)	1	53°90′ N 104°60′ W	jack pine	8	580		2001, 2002	
Saskatchewan, Canada (Sfen)	1	53°57′ N 105°57′ W	fen		525		1994, 1995	Suyker *et al.*, 1997
Manitoba, Canaca (NOBS)	1	55°52′ N 98°29′ W	black spruce	130	259	4.8	1993–2002	Goulden *et al.*, 1997 Goulden *et al.*, 1998 Hirsch *et al.*, 2002
Manitoba, Canada (NOJP)	1	55°93′ N 98°62′ W	jack pine	80	255		1994, 1996	
Manitoba, Canada (NYJP)	1	55°89′ N 98°29′ W	jack pine	25	249	1.2	1994, 1996	McCaughey *et al.*, 1997
Manitoba, Canada (Nfen)	1	55°91′ N 98°42′ W	fen		211		1994, 1996	Lafleur *et al.*, 1997
Manitoba, Canada (BurnNBS)	4	55°51′ N 55°54′ N 98°21′ W 98°58′ W	black spruce	11–70		0–2.9	1999, 2000	Litvak *et al.*, 2003
NWT, Alberta, Saskatchewan, Canada (BurnNM)	7	54°05′ N 61°35′ N	trembling aspen, poplar, jackpine,	1–80			1998, 1999	Amiro, 2001 Amiro *et al.*, 2003

Table 3. continued

Site	Number of sites	Location	Species	Age (years)	Elevation (m)	LAI (m^{-2})	Period	Reference
		105°53′ W 118°20′ N	spruce					
Northern Europe Petsikko, Finland (PET)	16	9°28′ N 27°14′ E	mountain birch		280	2.5	1996	Aurela *et al.*, 2001
Hyytiälä, Finland (HYT)	2	61°51′ N 24°17′ E	Scots pine	5, 38		3	1997, 2000	Rannik *et al.*, 2002
Norunda, Sweden (NOR)	1	60°05′ N 17°28′ E	Scots pine, Norway spruce	80–100	45	5	1994–1998	Lindroth *et al.*, 1998 Cienciala *et al.*, 1998 Constantin *et al.*, 1999 Levy *et al.*, 1999; Halldin *et al.*, 1999; Moren and Lindroth, 2000 Widen and Majdi, 2001 Widen, 2002
Flakaliden, Sweden (FLA)	1	64°07′ N 19°27′ W	Norway spruce	35	310	2	1997, 1998	Wallin *et al.*, 2001
Siberia and European Russia Zotino, Central Siberia (ZOBI)	1	60°59′ N 89°43′ W	white birch	50	200		1999	Meroni *et al.*, 2002
Zotino, Central Siberia (ZOP)	5	60°45′ N 89°23′ E	Scots pine	7–215	160	1.5	1996, 1998–2000	Kelliher *et al.*, 1999 Schulze *et al.*, 1999 Lloyd *et al.*, 2002
Zotino, Central Siberia (ZOBog)	2	60°44′ N 89°09′ E	sphagnum bog				1996, 1998–2000	Schulze *et al.*, 1999 Arneth *et al.*, 2002
Yakutsk, Eastern Siberia (YAK)	1	60°51′ N 128°16′ E	Dahurian larch	130	348	1.4	1993, 1994, 1996	Hollinger *et al.*, 1995 Hollinger *et al.*, 1998 Schulze *et al.*, 1999
Fyedorovskoye, European Russia (FYS)	2	56°27′ N 32°55′ E	Norway spruce	2, 150	220	3.5	1996, 1998–2000	Schulze *et al.*, 1999 Knohl *et al.*, 2002 Milyukova *et al.*, 2002
Fyedorovskoye, European Russia (FYBog)	1	56°27′ N 32°55′ E	sphagnum bog				1996, 1998–2000	Schulze *et al.*, 1999 Arneth *et al.*, 2002

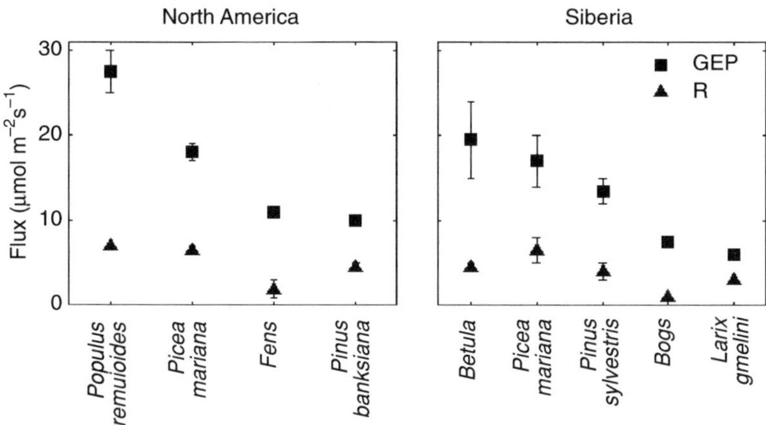

Figure 2. *Comparison of mean day-time maximum GEP and mean night-time maximum R at some North American and Siberian sites shown in Table 4. Vertical bars indicate range of values.*

be related to productivity. These results indicate high capacity to assimilate carbon during the relatively short growing season of most Boreal stands. However, unlike photosynthesis, respiration continues throughout the year although at a lower rate.

Short-term studies have also shown how NEP of the Boreal forest stands is related to PPFD and atmospheric water vapour saturation deficit (D). For example, Hollinger *et al.* (1998) found that the NEP of a larch stand in eastern Siberia increased with increasing PPFD and decreased as D increased during the summer: NEP saturated at 6 μmol (C) m^{-2} s^{-1} for PPFD > 500 μmol (C) m^{-2} s^{-1} and decreased with increasing D. They also found that day-time NEP increased immediately after rain, but on summer afternoons, the ecosystem became a source of CO_2, probably because of increased soil respiration and reduced photosynthetic CO_2 uptake resulting from increased atmospheric saturation deficit. Although PPFD-saturated CO_2 uptake was not very responsive to air temperature or D, Jarvis *et al.* (1997) found that there was a net loss of CO_2 on 31 days of high temperature (and consequently also high D) or low PPFD (or both) over the 120 days of the 1994 growing season, at the southern BOREAS black spruce site (SOBS).

Year-round measurements are necessary to characterize how NEP varies with the seasons, including the very cold winters when the soil is frozen and CO_2 exchanges are expected to be small. *Table 5* compares NEP for the growing season (April–October) and winter (November–March) at the SOBS and shows that winter losses were 56, 49, and 83% of growing season NEP during the years 2000, 2001, and 2002, respectively. These are relatively large proportions of the annual carbon gain and emphasize the variable impact of winter weather conditions on annual NEP.

Several such short-term studies have provided useful empirical relationships between R, P_e, and environmental variables (Sellers *et al.*, 1997), as well as seasonal estimates of carbon exchange. Results from BOREAS suggested that Boreal ecosystems range from moderate sinks to weak carbon sources and that inter-annual variability in carbon dynamics is strongly influenced by timing of snow melt and

Table 4. *Growing season estimates of maximum gross ecosystem productivity (GEP) and maximum night-time ecosystem respiration (R) obtained from eddy covariance (EC) measurements made in North American, European, and Siberian Boreal forests. Maximum GEP was estimated using GEP = maximum day-time NEP (data not presented) + maximum night-time R.*

Vegetation type, site	Species	Age	Maximum GEP (μmol m^{-2} s^{-1})	Maximum R (μmol m^{-2} s^{-1})	Reference
North America					
Saskatchewan, Canada (SOA)	trembling aspen, hazelnut	80	30	7	Black *et al.*, 1996 Chen *et al.*, 1999 Black *et al.*, 2000
Alberta, Canada (BurnNM) (deciduous)	trembling aspen, poplar	80	25	7	Amiro *et al.*, 2003
		1	10	7.5	
Saskatchewan, Canada (SOBS)	black spruce	130	19	6	Jarvis *et al.*, 1997 This study
Manitoba, Canada (NOBS)	black spruce	130	17	7	Goulden *et al.*, 1997
Manitoba, Canada (BurnNBS) (coniferous)	black spruce	70			Litvak *et al.*, 2003
		36			
		19			
		11			
Saskatchewan, Canada (SOJP)	jack pine	80	10	4.5	Baldocchi *et al.*, 1997
Manitoba, Canada (NOJP)	jack pine	80			
Manitoba, Canada (SYJP)	jack pine	25			
Manitoba, Canada (NYJP)	jack pine	25	12	3.5	McCaughey *et al.*, 1997
NWT, Saskatchewan, Canada (BurnNM)	jack pine, black spruce	80	10	4	Amiro *et al.*, 2003
	jack pine, spruce, aspen, poplar	50	19	8	
		10	11	3	
Saskatchewan, Canada (Sfen)	fen		12	0.8	Suyker *et al.*, 1997
Manitoba, Canada (Nfen)	fen		10	3	Lafleur *et al.*, 1997
Northern Europe					
Petsikko, Finland (PET)	mountain birch		15	4	Aurela *et al.*, 2001
Flakaliden, Sweden (FLA)	Norway spruce	35	20	8	Wallin *et al.*, 2001
Hyytiälä, Finland (HYT)	Scots pine	38	15	5	Rannik *et al.*, 2002
		5	5	4	
Norunda, Sweden (NOR)	Scots pine, Norway spruce	80–100	14	5	Lindroth *et al.*, 1998
Siberia and European Russia					
Zotino, Central Siberia (ZOBI)	white birch	50	24	5	Meroni et al., 2002
Zotino, Central Siberia (ZOP)	Scots pine	215	6	2	Schulze *et al.*, 1999
		67	12	3	
		13	4	2	
		7	3	1.5	
Yakutsk, Eastern Siberia (YAK)	Dahurian larch	130	6	3	Hollinger *et al.*, 1998 Schulze *et al.*, 1999
Fyedorovskoye, European Russia (FYS)	Norway spruce	150 2			Knohl *et al.*, 2002
Zotino, Central Siberia (ZOBog)	sphagnum bog		7	1	Arneth *et al.*, 2002
Fyedorovskoye, European Russia (FYBog)	sphagnum bog		8	1	Arneth *et al.*, 2002

Table 5. *Seasonal variation in NEP (g (C) m^{-2} per year) at the southern old black spruce (SOBS) site in Saskatchewan, Canada.*

Season	2000	2001	2002
April–October	112	153	72
November–March	–62	–75	–56
November–March temperature at 2 cm depth (°C)	–3	–2.5	–3.7

summer temperatures. However, it is clear that long-term flux measurements are necessary to quantify annual NEP and to understand fully the processes controlling carbon sequestration.

4.2 *Long-term eddy covariance studies*

There have been few long-term studies covering the full 12 months over several years. Such studies are particularly important because they provide temporally integrated, annual net total carbon sequestration, together with valuable information on the effects of inter-annual climate variability on carbon gains and losses. Such data also provide clues as to the likely future effects of changing climate and can be used to test carbon exchange models.

Long-term flux measurements have been made at nine Boreal sites. *Table 6* lists the annual values of NEP, *R*, and GEP obtained from EC measurements for several years at each of the nine sites. Annual estimates of GEP of Boreal forest stands varied from about 540 to 1460 g (C) m^{-2} per year, and that of *R* from about 410 to 1790 g (C) m^{-2} per year, depending on species, age, and climate and weather conditions at the site. Accordingly, the ratio of *R* to GEP, an indicator of whether the site is a source or a sink for CO_2, varied from 0.72 to 1.25. *Figure 3* shows mean annual NEP at the nine long-term Boreal sites, and indicates that NOR (Norunda) (*P. abies* and *P. sylvestris*) and FYS (Fyedorovskoye) (*P. abies*) sites were net sources, NOBS (Northern Old Black Spruce) near neutral and all other sites were weak to moderate carbon sinks. High *R* (*Figure 4*) explains why the NOR and FYS sites were sources of CO_2.

Flux measurements at the 130-year-old NOBS site have been made since 1994, the beginning of BOREAS, and are ongoing. Mean NEP at this site is near zero, varying from –71 g (C) m^{-2} per year to 36 g (C) m^{-2} per year (*Figure 5*). Although NPP of the trees and moss is roughly 220 g (C) m^{-2} per year (Gower *et al.*, 1997), the average loss of 30 g (C) m^{-2} per year measured by EC implies a heterotrophic respiratory loss of about 250 g (C) m^{-2} per year. From 1994 to 1998, the site was losing carbon. Much of this loss appears to result from decomposition of old carbon deep in the soil profile because of high soil temperatures. This net efflux of carbon has been attributed to climate warming, particularly enhanced thaws (Goulden *et al.*, 1998). The hypothesis of release of carbon deep in the soil profile has recently been confirmed through analysis of soil CO_2 concentration at various depths at the site (Hirsch *et al.*, 2002). However, this site sequestered carbon during 2001, 2002 and 2003. Current research

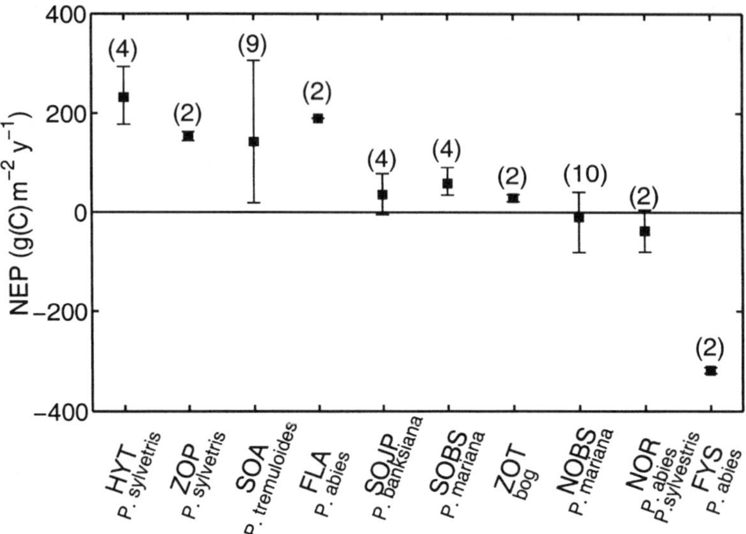

Figure 3. *Mean annual NEP at the ten long-term flux measurement Boreal sites (see Figure 1 and Table 3 for site symbols and description). Numbers in parentheses refer to the number of years. Vertical bars indicate range of values. Note the high variation in the only deciduous site, SOA.*

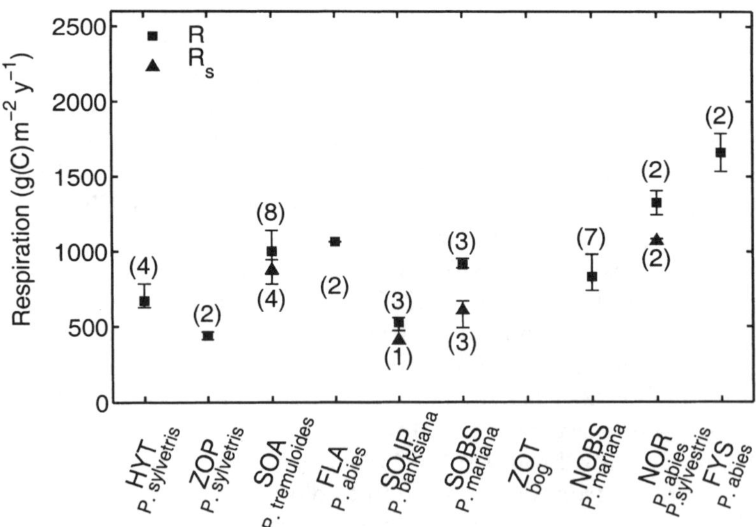

Figure 4. *Mean annual soil and ecosystem respiration at nine long-term Boreal sites shown in Figure 3. Numbers in parentheses refer to the number of years. Vertical bars indicate range of values.*

Table 6. *Annual estimates of net ecosystem productivity (NEP), ecosystem respiration (R) and gross ecosystem productivity (GEP) obtained from eddy covariance (EC) measurements made in North American, European, and Siberian Boreal forests. GEP was estimated using GEP = NEP + R.*

Site	Year	NEP g (C) m^{-2} yr^{-1}	R g (C) m^{-2} yr^{-1}	GEP g (C) m^{-2} yr^{-1}	R/GEP	References
Saskatchewan, Canada (SOA)	1994	144	1140	1284	0.89	This study
	1996	19	993	1012	0.98	
	1997	105	1148	1254	0.88	
	1998	268	1069	1337	0.81	
	1999	115	1115	1229	0.90	
	2000	148	1086	1234	0.88	Kljun *et al.*, 2004
	2001	360	1026	1386	0.74	
	2002	139	880	1019	0.86	
	2003	95	944	1039	0.91	
Saskatchewan, Canada (SOBS)	2000	66	813	879	0.92	Kljun *et al.*, 2004
	2001	68	826	894	0.92	
	2002	21	744	765	0.96	
	2003	62	813	875	0.93	
Saskatchewan, Canada (SOJP)	2000	78	605	684	0.89	Kljun *et al.*, 2004
	2001	41	676	717	0.94	
	2002	−23	626	602	1.04	
	2003	29	624	653	0.96	
Manitoba, Canada (NOBS)	1995	−76	692	616	1.12	This study
	1996	−49	699	650	1.07	
	1997	−56	666	610	1.09	
	1998	−19	746	727	1.03	
	1999	−2	714	712	1.00	
	2000	−7	681	713	0.96	
	2001	23	711	773	0.92	
	2002	26	581	607	0.96	
	2003	28	664	692	0.96	
Norunda, Sweden (NOR)	1996	5	1245	1250	1.00	Janssens *et al.*, 2001
	1997	−80	1405	1325	1.06	
Flakaliden, Sweden (FLA)	1997	190	1065	1255	0.85	Janssens *et al.*, 2001
Hyytiälä, Finland (HYT)	1997	252	639	891	0.72	Vesala, 2003
	1998	293	629	921	0.68	
	1999	206	634	840	0.75	
	2000	178	785	963	0.82	
Zotino, Central Siberia (ZOP)	1999	145	414	559	0.74	Lloyd *et al.*, 2002
	2000	163	464	627	0.74	
Fyedorovskoye, European Russia (FYS)	1999	−324	1788	1464	1.22	Milyukova *et al.*, 2002
	2000	−312	1536	1224	1.25	

at NOBS shows that the water regime is another important factor controlling the carbon balance of the site. Large growing season rainfall decreases water table depth, reduces respiratory losses from the soil and thus increases NEP. This would explain positive NEP values at the site in recent years. Similarly, small rainfall during the early years of the NOBS study helps to explain the negative NEP values during those years. The results further confirm the importance of long-term studies to identify processes important in controlling carbon exchange and to identify sites and ecosystems as likely future sinks or sources.

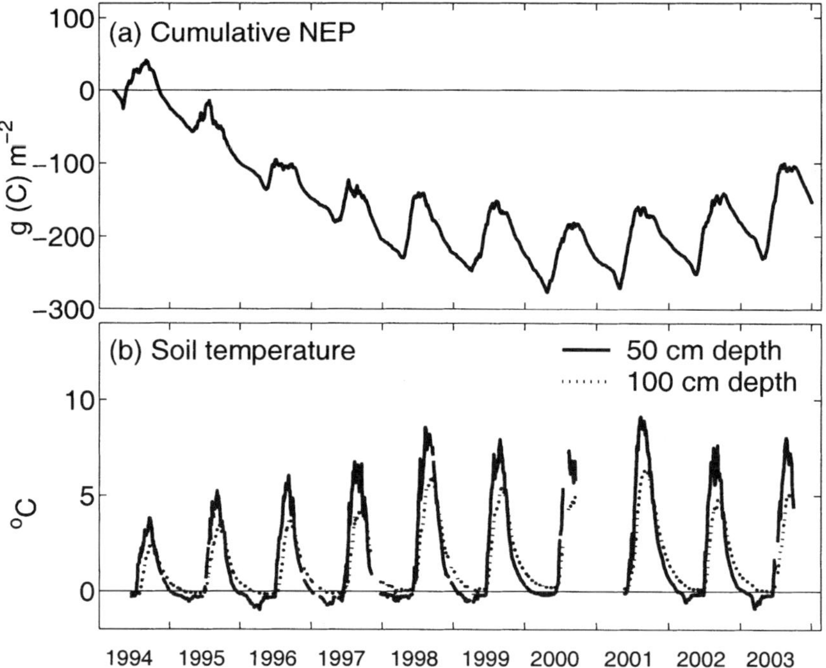

Figure 5. *Time series since 1994 of cumulative NEP (a) and soil temperatures (b) at the northern old black spruce (NOBS) site in Manitoba, Canada.*

Long-term studies at the southern old aspen site (SOA) tell us much about the response of a Boreal deciduous forest to inter-annual variability. As can be seen in *Figure 6*, there has been considerable variation in annual NEP over the past 10 years. NEP varied from 80 g (C) m^{-2} per year in 1996 to 361 g (C) m^{-2} per year in 2001, a moderate carbon sink with an average of 150 g (C) m^{-2} per year. Date of leaf emergence varied from as early as 26 April to 19 May (Barr *et al.*, 2004). Annual NEP was highly correlated with date of leaf emergence (*Figure 7*), which in turn was highly correlated with spring temperature (*Figure 8*) (Barr *et al.*, 2002; Black *et al.*, 2000; Chen *et al.*, 1999). The earlier the start of the growing season, the higher was NEP. However, spring temperature was only weakly correlated with mean annual temperature over these years. In the past three years (2001–2003), the site has been subjected to the effects of low growing season soil moisture as a result of a major drought in the western plains of North America. Surprisingly, annual NEP in 2001 was the highest in the history of measurements at this site, and was probably the result of the early leaf emergence and a short-term effect of the initial stage of the drought. In that year, GEP was largely unaffected probably because roots were extracting moisture from deep in the soil; however, R was significantly reduced. This may have been the result of reduced heterotrophic respiration because of low moisture in the forest floor and shallow soil mineral layers (Griffis *et al.*, 2004). However, Lavigne *et al.* (2004) showed that in a stand of New Brunswick balsam fir (*Abies balsamea* (L.) Mill.), root respiration was more sensitive than microbial respiration to water stress in spring, although by early fall, roots had acclimated to the water stress. In 2002, the combination of low spring temperatures, causing late leaf emergence, and periodic low soil moisture during the growing season, reduced NEP to below the average for the past eight years. In 2003, spring was warm and soil moisture was very similar to that in 2002, resulting in a slight further reduction in NEP, giving rise to the second lowest value (97 g (C) m^{-2} per year) at SOA in the 10 years of flux measurements there.

The current BERMS (Boreal ecosystems research and monitoring sites) programme, which includes long-term flux monitoring in three mature forest stands, provides a unique opportunity to compare the responses of aspen, black spruce and jack pine ecosystems to seasonal and inter-annual climate variability. These three stands (SOA, SOBS, and SOJP (southern old jack pine)) are within 100 km of each other in the southern Boreal forest of Saskatchewan. There are also chamber systems in each stand monitoring soil CO_2 efflux year round. Measurements at SOBS and SOJP were restarted in 1999 after being inactive since BOREAS ended in 1996. SOA is on a mineral soil with moderate water holding capacity, whereas SOBS is in a basin with a 30 cm surface layer of organic matter and a high water table, and SOJP is on deep sand with low water holding capacity. The results over the four years (2000–2003) (*Figure 9*) are instructive in comparing the responses of deciduous SOA

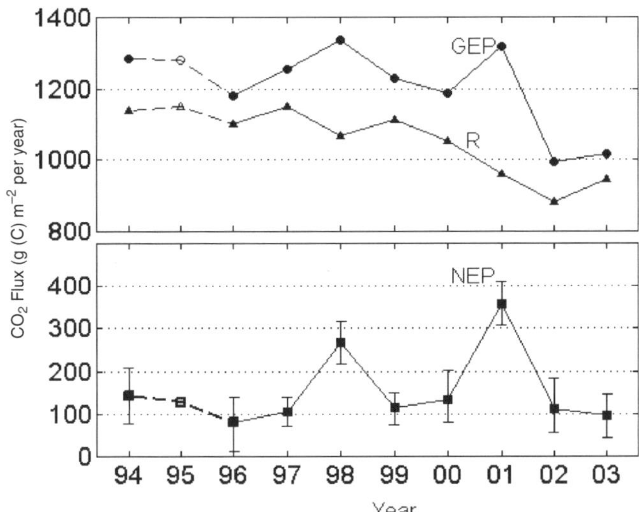

Figure 6. *Long-term variation in the carbon balance components at the southern old aspen (SOA) site in Saskatchewan, Canada. Values for 1995 were estimated using the C-CLASS model (Arain et al., 2002). Vertical bars represent the estimated uncertainty.*

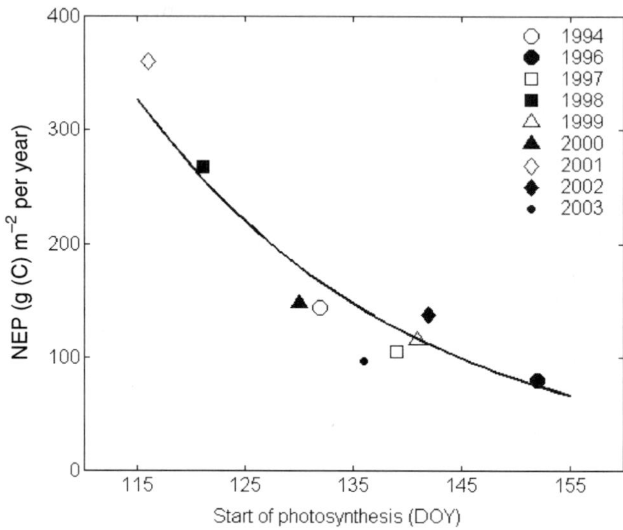

Figure 7. Relationship between NEP and start of photosynthesis for different years at the southern old aspen (SOA) site in Saskatchewan, Canada.

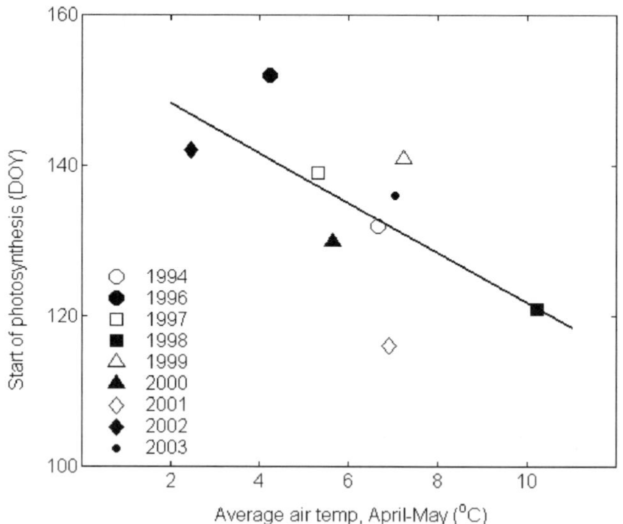

Figure 8. Relationship between start of photosynthesis and average spring-season (April–May) air temperature for different years at the southern old aspen (SOA) site in Saskatchewan, Canada. DOY, day of year.

with wet (SOBS) and dry (SOJP) coniferous sites. The weather of the first year, 2000, was relatively normal, with 400 mm of precipitation and a mean spring (April–May) temperature of 7 °C. NEP was 148, 66, and 78 g (C) m^{-2} per year for SOA, SOBS, and SOJP, respectively. In 2001 both stands of conifers showed little response to the warm spring that greatly increased photosynthesis at SOA, as mentioned above. In contrast, the cold spring in 2002 delayed soil thaw and, consequently, significantly decreased annual NEP in both stands of conifers, just as it did at SOA as a result of late leaf emergence. The drought of 2001–2003 had much smaller impact on the two conifer stands than on the deciduous broadleaf aspen stand, because rainfall during those three years was somewhat higher at SOBS and SOJP than at SOA. In addition, because of its topographically low position, there was almost no reduction in soil moisture at depths below 30 cm at SOBS. However, in the worst year of the drought, 2003, there was a noticeable reduction in GEP in both stands of conifers during the growing season (Kljun *et al.*, 2004).

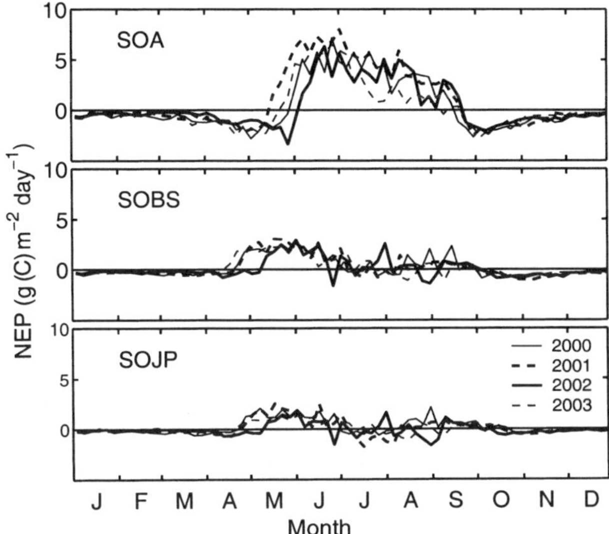

Figure 9. *Comparison of seasonal and interannual variation in NEP at southern old aspen (SOA), southern old black spruce (SOBS) and southern old jack pine (SOJP) sites in Saskatchewan, Canada, during 2000–2003.*

Figure 10 shows the relationship between annual NEP and mean air temperature during spring (April–May) for SOA, SOBS and SOJP during the four years, 2000–2003. With the exception of SOA in 2003, when soil moisture was particularly low, NEP increased with spring temperature in all three stands, more strongly in the deciduous stand. These results extend earlier findings (Barr *et al.*, 2002; Black *et al.*, 2000) that warm springs result in higher NEP in Boreal deciduous broadleaf stands compared to Boreal conifer stands, and support the hypothesis that climate warming may increase Boreal forest productivity (Bergh and Linder, 1999; Grace and Rayment, 2000; Jarvis and Linder, 2000; Mellilo *et al.*, 2002), provided that there is no consequent shortage in soil water. Bergh and Linder (1999) found that soil warming by 5 °C at 10 cm depth during the growing season at Flakaliden in northern Sweden increased stem growth of Norway spruce trees by more than 50%. Jarvis and Linder (2000) concluded that this effect resulted from enhanced capture of CO_2 by a larger leaf area per tree, stimulated by nutrients released from increased decomposition of the readily decomposable soil organic matter fraction (SOM). Based on the observation that decomposition of SOM in Finnish soil was not very sensitive to temperature over long periods of time (Liski *et al.*, 1999), Grace and Rayment (2000) argued that an increase in photosynthesis with such a rise in temperature would override an increase in respiration.

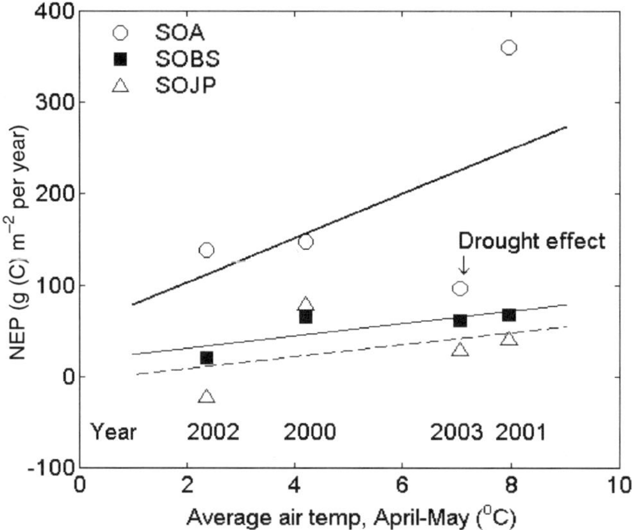

Figure 10. Relationship between NEP and mean air temperature during spring season at southern old aspen (SOA), southern old black spruce (SOBS) and southern old jack pine (SOJP) sites in Saskatchewan, Canada, during 2000–2003.

It is useful to compare NEP at SOBS with that at NOBS, which is situated about 600 km northeast of SOBS. The discontinuous permafrost at NOBS does not occur at SOBS. Measurements made in 1994 (Jarvis *et al.*, 1997) and 1996 (Jarvis, P.G., unpublished observations) and those made during the past three years as part of the BERMS programme show that SOBS is likely to be a weak sink for carbon. GEP and R for SOBS were larger than for NOBS by 12 and 5%, respectively (*Figure 11*). There is a distinct similarity in the seasonal course of NEP at the two sites; the mid-summer drop in NEP is the result of R rising in mid-summer to almost balance ecosystem photosynthesis. This mid-summer drop in NEP seems to be a character-istic of Boreal coniferous stands, whether on wet (SOBS and NOBS) or dry (SOJP) soils with contrasting SOM (see *Figure 9*).

The role of soil moisture in the Boreal forests is further illustrated by the long-term flux measurement at Norunda in central Sweden in the NOPEX and EUROFLUX pro-grammes (Valentini *et al.*, 2000). As mentioned earlier, at this site, on average R exceeds GEP and NEP is negative (Janssens *et al.*, 2001; Lindroth *et al.*, 1998; Valentini *et al.*, 2000). Surface water drainage some 20 years ago has led to increased rates of decomposi-tion of SOM and thus enhanced emission of CO_2. At this site, R shows high inter-annual variability, whereas at SOA, for example, GEP shows more inter-annual variability than R, and thus drives the inter-annual variability of NEP there.

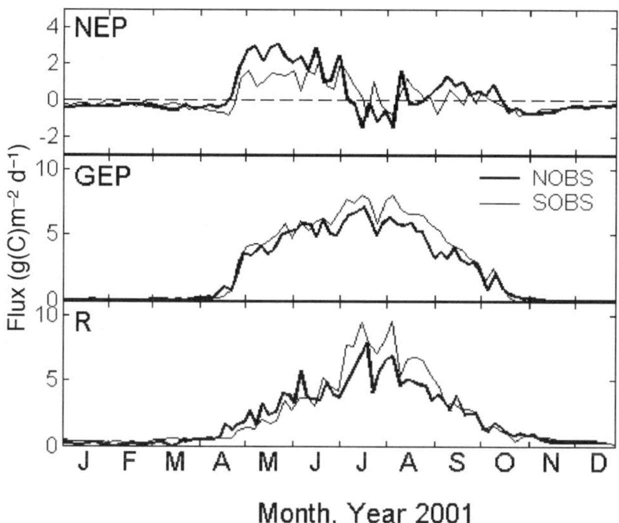

Figure 11. Comparison of carbon balance components at the Northern Old Black Spruce (NOBS) site in Manitoba, Canada and the southern old black spruce (SOBS) site in Saskatchewan, Canada.

Long-term flux monitoring is also underway at Flakaliden in Sweden (Grelle and Lindroth, 1996; Wallin *et al.*, 2001), Hyytiälä in Finland (Vesala *et al.*, 1998), Zotino in central Siberia (Lloyd *et al.*, 2002), and Fyedorovskoye in European Russia (Milyukova *et al.*, 2002; Schulze *et al.*, 1999). The Scots pine stand in Hyytiälä is a moderately strong sink (228–264 g (C) m^{-2} per year). At Flakaliden, higher respiration as a result of abnormally high temperature and reduced soil moisture during the growing season led to the young stand of Norway spruce switching from carbon uptake in 1994 to carbon loss in 1995. At Zotino, the lowest *R* among the nine long-term EC sites led to carbon uptake in spite of low GEP. At Fyedorovskoye, Milyukova *et al.* (2002) showed that photosynthesis and respiration of a Norway spruce stand increased markedly with increase in temperature, but temperature sensitivity of respiration was higher. They also showed that decomposition of windthrown trees played an important role in the negative carbon balance of the stand.

5. Ecophysiological stand-scale flux measurements

5.1 *Chamber measurements*

Measurements of the components of stand carbon budgets using chambers contribute to understanding carbon exchange processes between the atmosphere, soil and vegetation. For example, chambers of various sizes, designs, and operating principles have been used to measure CO_2 exchanges between the atmosphere and foliage, branches, and tree boles at different heights in a canopy and between the soil surface, soil horizons and roots at different depths in the soil. However, scaling up from chamber measurements to the stand scale is a challenge because of the very considerable spatial variability within a stand and the few chambers it is practical to operate. Several short-term chamber studies were made, for example at BOREAS (Lavigne *et al.*, 1997; Rayment and Jarvis, 2000; Rayment *et al.*, 2002), at NOPEX (Moren and Lindroth, 2000), and at other Swedish (Widen, 2002) and Siberian (Hollinger *et al.*, 1998; Kelliher *et al.*, 1999) Boreal forest sites. These have provided valuable information on diurnal and seasonal variation in photosynthesis and respiration rates and the influence of environmental variables on ecophysiological processes. For example, year-round measurements of soil CO_2 efflux using automated chamber systems have shown that mean annual rates of soil respiration (R_s) at SOA, SOBS, and SOJP of 880, 625, and 412 g (C) m^{-2} per year, respectively, are about 80, 67, and 83% of total ecosystem respiration, R, respectively (*Figure 4*). *Figure 12* compares the seasonal variations of R and R_s at SOBS in 2002. During the winter non-growing season, R and R_s were almost equal. Early in the growing season, R_s accounted for only about 50% of R, but as both R and R_s rose with the increase in temperature as the season progressed, this proportion also increased to about 70% so that, as noted earlier, in mid-season R came close to equalling GEP. Soil moisture stress in August resulted in decreases in both R and R_s. Innovative research approaches of stem girdling (Högberg *et al.*, 2001, and Chapter 12, this volume) and long-term soil warming experiments (Bergh and Linder, 1999; Melillo *et al.*, 2002) show promise for partitioning R_s into R_a and R_h, and thus of determining component responses of ecosystems to climate warming.

5.2 *Biometric carbon budgeting*

Biometric carbon budgeting involves full accounting of the changes in above-ground and below-ground carbon storage to estimate NEP. Whereas biometric measure-

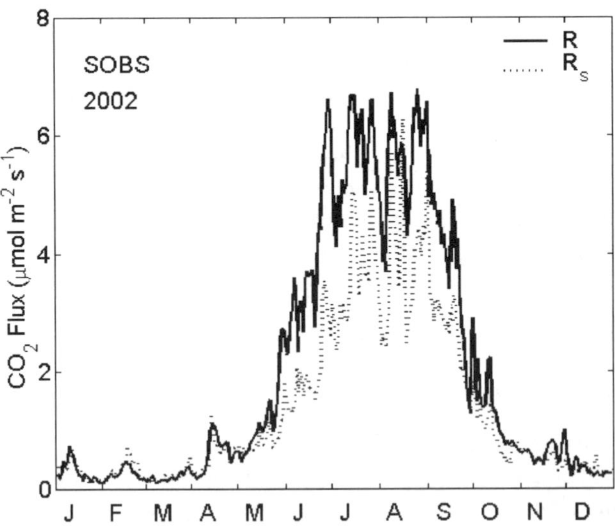

Figure 12. *Comparison of time series of ecosystem respiration (R) and soil CO$_2$ efflux (soil respiration, R$_s$) at the southern old black spruce (SOBS) site in Saskatchewan, Canada in 2002. R$_s$ is an average of six automatic chambers in summer and three or four in the winter.*

ments have been used to estimate NEP of tropical forests (Phillips *et al.* 1998; Chapter 10, this volume) and temperate forests (Barford *et al.*, 2001; Schulze *et al.*, 2000; Chapters 4 and 15, this volume), this approach does not seem to have been used to estimate NEP of Boreal forests. Although the studies referenced above promise agreement between biometric and EC approaches, the use of the former remains an important challenge in Boreal forests where growth rates are extremely low.

A related approach is to estimate NEP by subtracting estimates of R_h from NPP measurements. Gower *et al.* (2001), using complete NPP budgets, determined NPP of 24 Boreal forest stands to lie in the range 220–910 g (C) m^{-2} per year. Bond-Lamberty *et al.* (2004) used the NPP-based approach, i.e., $(P_n - R_h)$, to calculate the NEP of seven black spruce stands comprising a Boreal forest chronosequence in northern Manitoba, Canada (see below). However, our attempts to calculate NEP of the stands at SOA, SOBS, and SOJP from NPP and estimates of R_h (from root exclusion plots) suggested that either the NPP values were underestimated or the R_h values were overestimated. Eventually, it will be necessary for biometric estimates and EC estimates of NEP to converge to give confidence in our estimations of carbon exchange and our understanding of the processes controlling it in Boreal forests.

6. Effects of disturbance

Disturbances are part of the natural cycle, often releasing nutrients back to the forest to continue supporting the next generation of vegetation, and dictating stand renewal in most of the Boreal forest. The major disturbance processes are fire, insect herbivory, disease, harvesting and windthrow. Estimates of the extent of these disturbances are given in *Table 7*. Here, we concentrate on severe events that have a well-defined impact on carbon exchange: particularly fire.

Table 7. *Extent of major disturbances in the Boreal forest.*

Area (million ha)	Eurasia	North America	Reference
Boreal forest area	850	350	FAO, 2000 Apps *et al.*, 1993
Typical area burned annually	2–13	1–8	Dixon & Krankina, 1993; Conard *et al.*, 2002; Stocks *et al.*, 2002; Murphy *et al.*, 1999
Area of forest with moderate to severe insect infestation annually		10–25	Volney and Fleming, 2000
Area of forest harvested annually		1	CCFM, 2002

Fire plays a major role in Boreal forests, with an average return period of about 120 years, and thus is the reason why the Boreal forest is a mosaic of different-aged stands of various species. Fires burned an average of 2 million ha (1 hectare (ha) = 10^4 m^2) of Boreal forest annually in Canada from 1959 to 1999, releasing an average 27 Tg (C) per year (Amiro *et al.*, 2001). However, 115 Tg (C) were released in 1995 alone.

Conard *et al.* (2002) estimated that 13 million ha burned in Siberia in 1998, while Conard and Ivanova (1997) reported that 12 million ha is a conservative estimate of Boreal forest burned annually in Russia. In Canada, the area burned has increased during the past four decades (Stocks *et al.*, 2002) and will likely continue to increase with a warming climate (Flannigan *et al.*, 1998). It is likely that climate change will alter the timing, intensity, frequency, and extent of disturbances, including their interactions (Dale *et al.*, 2001; Volney and Fleming, 2000). Several studies have focused on measuring carbon fluxes after fire and harvesting. Flux measurement data after insect infestation, major disease, and windthrow events are largely lacking.

Post-fire CO_2 EC flux measurements along a black spruce fire chronosequence (Burn NBS) in northern Manitoba by Litvac *et al.* (2003) show growing season NEP increasing with stand age, peaking at about 36 years with lower net carbon uptake at 70 and carbon emission at 130 years after fire (*Figure 13*). Using their $(P_n - R_h)$ method for estimating NEP, Bond-Lamberty *et al.* (2004) found that the youngest stands of the same chronosequence were moderate annual carbon sources of about 100 g (C) m^{-2} per year, the middle-aged stands relatively strong sinks of 100–300 g (C) m^{-2} per year, and the oldest stands about neutral.

Figure 14 compares diurnal patterns of NEP of different-aged, post-burn, jack pine stands in the southern Boreal forest of Saskatchewan (Burn NM) during the growing season. Daytime photosynthetic uptake of CO_2 was found to increase with age so that NEP of the 10-year-old stand was very similar to that of the mature stand. However, night-time CO_2 respiration loss also increased with age and consequently on a 24-hour basis, the 10-year-old stand sequestered more CO_2 than the mature stand. Most chamber studies confirm a decrease in both R_a and R_h following fire (Amiro *et al.*, 2003). However, night-time R can be quite high after harvesting of an aspen stand, likely caused by root respiration because aspen roots remain alive and new shoots sucker from them. On the other hand, Rannik *et al.* (2002) reported that night-time respiration was similar in a 5-year-old harvested site and a 38-year-old Scots pine stand in Finland, but that the harvested site was a source whereas the older stand was a sink. A 6-year-old white spruce clear-cut in northern British Columbia could be either a carbon source or sink during the growing season, depending on the year, but annually it was a source (Pypker and Fredeen, 2002). Schulze *et al.* (1999) found that post-fire 7- and 13-year-old pine stands in Siberia were net carbon sources in summer, whereas older stands were sinks (*Figure 13*). They also reported that a site that had experienced severe windthrow two years previously was a strong source of 2 g (C) m^{-2} d^{-1}, but on Canadian Boreal forest sites where rapid natural regeneration occurs one might expect quick conversion to a sink, with the added bonus of a large addition of carbon to the soil pool. This is what distinguishes disturbance by windthrow from fire: windthrow adds organic carbon to the soil pool, whereas fire adds CO_2 to the atmosphere!

7. Upscaling eddy covariance measured fluxes to the biome scale

Several different approaches used for this purpose are briefly described below. It is beyond the scope of this review to summarize all the results obtained by these methods, but the references should provide entry into their application to the Boreal forest biome.

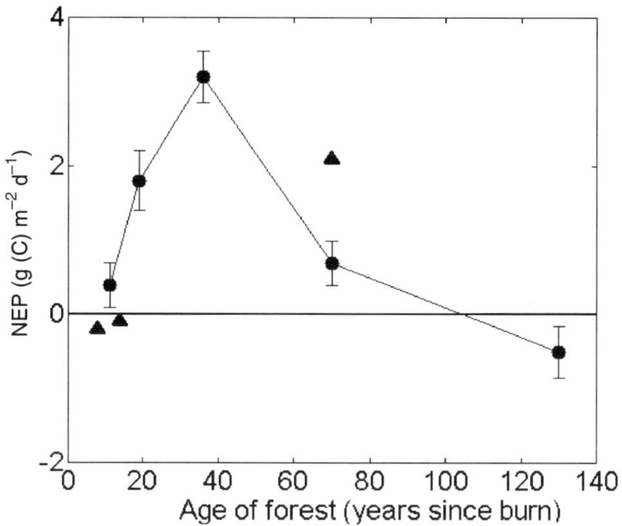

Figure 13. *Effect of stand age since burn on growing season NEP of Boreal forests. Filled circles,* Picea mariana, *Canada (Litvak et al., 2003); filled triangles,* Pinus sylvestris, *Siberia (Schulze et al., 1999).*

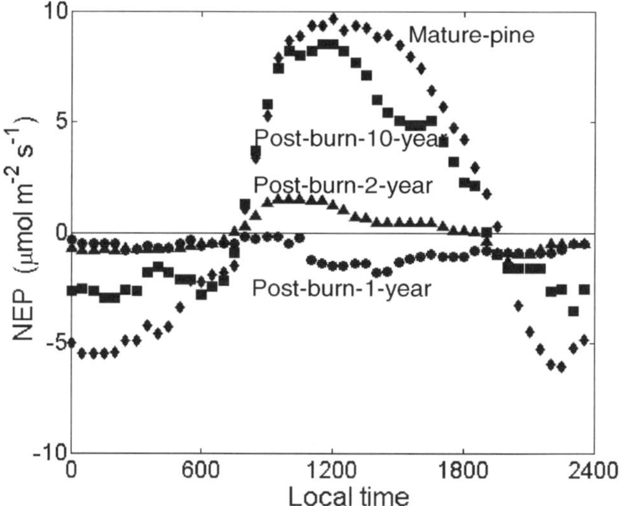

Figure 14. *Summer-time diurnal CO_2 flux density patterns in different-aged post burn jack pine stands in the Southern Canadian Boreal forest. Redrawn from Amiro et al. (2003).*

7.1 Aircraft-based eddy covariance fluxes

Aircraft have been used to measure landscape-scale fluxes using EC. Aircraft cover spatial scales that can readily incorporate a range of stand ages. The measurement length-scale is typically about 2 km so that individual stands can either be separated or a mosaic of stand ages can be averaged. Desjardins *et al.* (1997) using BOREAS measurements developed relationships between aircraft- and tower-based EC CO_2 fluxes measured over Boreal forest stands. They used these relationships to scale up tower-based fluxes to 16 km × 16 km grid squares. Data from aircraft flights along a 500 km transect over stands of known ages since fire showed that fresh burns were carbon sources but that summer day-time canopy carbon fluxes were downward into the canopy just two years after a fire, and recovered with stand age to mature rates within about 15–20 years (Amiro *et al.*, 1999, 2003). Whereas aircraft-borne EC can measure regional-scale fluxes during day-time, these measurements cannot be made at night for safety reasons and are otherwise severely restricted for the time-scale over which they can be made.

7.2 *Estimating net biome productivity*

Estimates of NBP are required for national and continental carbon balances. Flux measurements from all stand species and ages must be considered for integration to biome scale, and transient processes, such as direct releases by fire, must also be taken into account. There have been some recent initiatives to account for disturbance and spatial variability in carbon fluxes across patchworks of stands of various ages to estimate NBP (Schulze *et al.*, 1999; Chapter 15, this volume). They estimated NBP for Siberia to be about 0.2 g (C) m^{-2} per year, up to 2–5 g (C) m^{-2} per year for Scandinavia, and 1 g (C) m^{-2} per year for North America. Current measurements on post-fire and post-harvest chronosequences (see, for example, Amiro *et al.*, 2004) will help to estimate regional-scale NBP with higher precision.

7.3 *Ecosystem process models*

Many ecosystem process models have been used to estimate forest-scale NEP and regional-scale NBP. For example, Cienciala *et al.* (1998) used the FOREST-BGC ecosystem process model (Running and Coughlan, 1988; Running and Gower, 1991) to estimate annual NEP of Scots pine and Norway spruce stands in central Sweden. The model requires data of daily microclimate, key site and vegetation variables, to calculate photosynthesis, growth and maintenance respiration, allocation, litter-fall, and decomposition. They found modelled NEP to vary between 60 and 200 g (C) m^{-2} per year for an LAI range of 3–6. Amiro *et al.* (2003) used remotely sensed high-resolution radiometer satellite measurements to estimate LAI and land cover to model NPP at a 1 km × 1 km resolution with the Boreal ecosystem productivity simulator (Liu *et al.*, 1997, 1999). This estimate of NPP with an average of 273 g (C) m^{-2} per year was consistent with measured NPP for Boreal coniferous forests (Gower *et al.*, 1997). Grant *et al.* (1999) used the *Ecosys* model to describe CO_2 exchange between the SOA stand and the atmosphere, and predicted the stand to be a net sink of 68 g (C) m^{-2} per year compared with 130 g (C) m^{-2} per year measured with EC. Arain *et al.* (2002) incorporated two sub-models – leaf canopy conductance and photosynthesis, and autotrophic and heterotrophic respiration – into the Canadian Land Surface Scheme (CLASS) (Verseghy, 1991) to calculate the productivity of Boreal broadleaf and conifer forest stands. They found that SOA was a weak to moderate carbon sink with considerable inter-annual variability in carbon uptake, while SOBS was a weak carbon sink in cool years and a weak carbon source in warm years. Chen *et al.* (2000) integrated an upscaling algorithm of Farquhar's biochemical model of photosynthesis with the soil carbon–nitrogen cycle of the Century model into the Integrated Terrestrial Ecosystem C-budget model (InTEC), which also incorporates the effects of fire, harvest, and insect damage. Treating the entire forest area of Canada as one unit, their analysis using InTEC, in conjunction with that of Houghton *et al.* (1999), showed that between 1980 and 1996 North America was a weak sink of 0.2–0.5 Pg (C) per year. This is much less than the 1.7 Pg (C) per year estimated by Fan *et al.* (1998), who used an atmospheric inversion model. These ecosystem process models can be tested against stand-scale carbon balances obtained using EC and used to assist partitioning of such EC data into estimates of stand component fluxes. In addition, these models, or models like them, are used to obtain regional-scale carbon balances and as components of general circulation models (GCMs).

7.4 *Convective boundary-layer budgeting approach*

Another approach that provides estimates of carbon exchange over the scale of 100–1000 km^2 is the (CBL) convective boundary-layer budgeting technique. The method uses temporal variations in the CO_2 mixing ratio in the well-mixed CBL, and treats the CBL as a large chamber within which conservation of mass applies (see Chapter 5, this volume). This method has been used with some success to estimate regional scale CO_2 fluxes in Russia (Hollinger *et al.*, 1995) and Sweden (Levy *et al.*, 1999). Using the CBL technique, Lloyd *et al.* (2001) estimated regional day-time NBP of the predominantly conifer forest vegetation in central Siberia to be 3–9 µmol m^{-2} s^{-1} on 15 and 16 July 1996, and found reasonable agreement with tower flux measurements of NEP within the area. This method can, however, only be used on days of fine weather in which, typically, a characteristic CBL develops from an overnight inversion to a known height above the ground. Furthermore, information is required on the composition of the air entrained into the CBL as it grows during the course of the day. Consequently, this method is best used for testing hypotheses rather than for carbon budgeting.

7.5 *Stable isotope methods*

This method of estimating the carbon balance of terrestrial ecosystems is based on measurement of the stable isotopic composition $^{13}C/^{12}C$ of atmospheric CO_2. Because plant photosynthesis discriminates against ^{13}C, the ratio $^{13}C/^{12}C$, in combination with atmospheric CO_2 concentration, can be used to estimate terrestrial biospheric exchanges of CO_2 with the atmosphere at a range of scales. Calculation of the carbon flux using $^{13}C/^{12}C$ isotope ratios at continental scale integrates to give estimates of NBP. The advantage of the approach is that large areas of the world can be analysed as a single unit and their net carbon balances inferred. Using this method with observations from a global network of 43 sites in 1992 and 1993, Ciais *et al.* (1995) found a strong terrestrial biospheric sink for carbon in the temperate latitudes of North America (see also Chapter 7, this volume).

7.6 Atmospheric inverse models

For the past decade, 'inverse modelling' has been one of the primary tools to improve our estimates and understanding of annual regional and global-scale carbon exchanges (Ciais *et al.* 2000; Fan *et al.*, 1998; Gurney *et al.*, 2002; Chapter 7, this volume). In this method, prior estimates of both the fluxes and their uncertainties are specified and atmospheric transport models are used to reproduce best the spatial pattern of atmospheric CO_2. Fluxes are then estimated as *differences* from similarly obtained 'background' fluxes which represent fossil fuel emissions, and are annually balanced. For the period 1985–1995, Ciais *et al.* (2000) obtained a mean annual carbon sink of 0.3, 0.8, and 1.3 Pg (C) per year for North America, Europe, and Siberia, respectively. Gurney *et al.* (2002), using 16 different atmospheric transport models, showed that during 1992–1996, Boreal forest in North America was a small source and that in Eurasia a large sink of CO_2. However, with the present large estimated uncertainties in the magnitude of the quantities, one can go little further than to define whether a region is a sink or source. Further improvements in atmospheric transport models and expansion of the CO_2 observation network are required to obtain accurate quantitative estimates using the inverse modelling methodology.

8. Conclusions

8.1 Achievements

EC measurements have provided valuable data on diurnal, seasonal and annual CO_2 fluxes over a range of Boreal ecosystems. This has helped in understanding the role weather conditions, microclimate, and ecophysiology play in determining CO_2 exchange at the ecosystem scale. Date of the start of the growing season plays a major role in determining annual NEP of Boreal broadleaf stands, whereas summer respiration significantly suppresses NEP in conifer stands. Hydrology plays an important role in the carbon balance of Boreal ecosystems. Many Boreal forests and bogs are weak to moderate carbon sinks; some are weak sources. Inter-annual variability in carbon sequestration of Canadian Boreal forest stands appears to be largely controlled by spring temperature and growing season soil water regime.

Chamber measurements have proved to be valuable in determining the contributions of various ecosystem components to total carbon exchange and in understanding process and factors controlling ecosystem CO_2 exchange. Biometric measurements provide an important, independent method of estimating NEP. In addition, independent measurements of NPP and heterotrophic respiration can provide an additional means of estimating NEP.

Studies on the influence of fire – a major disturbance in the Boreal forest – have shown that it takes about 10 years for a Boreal black spruce stand to restart sequestering carbon on a growing season basis after fire, and growing season NEP reaches its maximum at 30 to 40 years after fire. EC measurements have provided valuable data for ecophysiological models that have increased our understanding of the processes controlling ecosystem NEP.

Regional-scale NBP estimates of Boreal forest in Siberia, Scandinavia, and North America have been made. Aircraft EC fluxes, convective boundary layer budgets, stable isotopic analysis and inverse modelling have proved to be valuable tools in validating upscaled CO_2 fluxes, but convergence among these methods still leaves much to be desired.

8.2 *Where next?*

Despite significant progress made in measuring stand-scale CO_2 exchange and inferring GEP and R over the past 10 years, there still remain many gaps in our knowledge. Uppermost among these is the need to resolve discrepancies between estimated and EC-measured NEP values. NEP is a small difference between two large terms, carbon gain (P_g) and carbon loss (R). More work is required to refine estimates of R and to test alternative means of estimating day-time respiration and partitioning R into R_a (above and below ground) and R_h. This may be possible with new, innovative process-scale studies on soil and ecosystem respiration and photosynthesis. Means to account for underestimation of EC nocturnal fluxes when light winds decrease vertical mixing (Baldocchi, 2003; Lee, 1998) need to be further refined. These include careful accounting of CO_2 storage in the soil and air column beneath the EC measurement height, and its advective transport. The use of the nocturnal boundary-layer budget technique proposed (and tested at SOBS) by Pattey *et al.* (2002) is promising. More EC measurement sites, representative of different regions of the Boreal forest, should be involved in long-term investigations of the influences of hydrology and expected changes in climate in the near future. To gain confidence in our estimates of CO_2 exchange between Boreal forest and the atmosphere, there is an urgent need to achieve agreement between EC and biometric estimates of NEP.

Upscaling fluxes from stand scale to regional scale still remains a daunting task, and confirmation of results difficult to obtain. Process-based models and remote sensing techniques should be increasingly developed for this purpose. More emphasis should be placed on long-term biometric, (i.e., carbon stock) studies, involving more sites, and results compared with EC estimates. Estimating annual and long-term effects of disturbance, both managerial and 'natural', e.g., thinning, harvesting, insect infestation, windthrow, and fires, on stand, ecosystem, and regional carbon balances, remains a challenge. This will require year-round EC measurements in stands of increasing age since establishment following disturbance. Such measurements are needed to determine the number of years it takes for these stands to resume carbon sequestration on an annual basis.

There has been lack of consistency in results from numerous models as well as from stable isotopic studies, and these need further refinement. There is much scope for making advances in accurately estimating the carbon balance at stand, biome, and global scales by combining measurements with modelling and remote sensing techniques.

9. Summary

The Boreal forest is the world's second largest forested biome occupying the circumpolar region between 50° N and 70° N. This heterogeneous biome stores about 25% of all terrestrial carbon. We have reviewed EC measurements of CO_2

exchange between the atmosphere and Boreal forests, and assessed progress in understanding the controlling processes. We have assessed net ecosystem productivity, the net balance between net primary productivity and heterotrophic respiration, measured using the EC method, for 38 Boreal forest sites. Gross ecosystem productivity has been estimated by adding day-time EC-measured CO_2 fluxes to respiration estimated from night-time relationships between respiration and temperature.

Maximum midday values of gross ecosystem productivity vary from 33 µmol m^{-2} s^{-1} for aspen to 6 µmol m^{-2} s^{-1} for larch stands. Long-term EC flux measurements, ongoing at nine Boreal sites, have shown the strong impact of spring weather and growing season water balance on annual net ecosystem productivity.

Estimation of net biome production, incorporating the effects of disturbance resulting from forest fires and logging, has progressed significantly in recent years. After disturbance, summer measurements in Boreal chronosequences suggest that it takes about 10 years before growing season carbon uptake offsets the decomposition emissions.

Small-scale exchange rate measurements using chambers and manipulative experiments such as stem girdling and soil heating help to understand the processes and mechanisms playing major roles in the carbon balance of terrestrial ecosystems.

Aircraft EC flux measurements, convective boundary layer carbon budgets, and $^{13}C/^{12}C$ changes in the atmosphere play an important role in validating estimates of regional carbon exchange based on scaled up EC measurements. Atmospheric inverse models are an important approach to studying regional and global carbon balance but need further improvement to yield reliable quantitative results.

Acknowledgements

We acknowledge the support of BERMS research by the Natural Science and Engineering Research Council of Canada, the Canada Foundation for Climate and Atmospheric Science, the Meteorological Service of Canada, the Canadian Forestry Service, and Parks Canada. We sincerely thank Alan Barr, Harry McCaughey, Natascha Kljun, and Kai Morgenstern for assistance with BERMS data. We acknowledge CarboEurope for the use of flux data from several of its northern forest sites. T.A.B. gratefully acknowledges the travel funds enabling him to attend the Carbon Balance of Forest Biomes Symposium as part of the SEB Annual Meeting in Southampton, 1–4 April 2003.

References

Amiro, B.D. (2001) Paired-tower measurements of carbon and energy fluxes following disturbance in the boreal forest. *Global Change Biology* **7**: 253–268.

Amiro, B.D., MacPherson, J.I. and Desjardins, R.L. (1999) BOREAS flight measurements of forest-fire effects on carbon dioxide and energy fluxes. *Agricultural and Forest Meteorology* **96**: 199–208.

Amiro, B.D., MacPherson, J.I., Desjardins, R.L., Chen, J.M. and Liu, J. (2003) Post-fire carbon dioxide fluxes in the western Canadian boreal forest: evidence from towers, aircraft and remote sensing. *Agricultural and Forest Meteorology* **115**: 91–107.

Amiro, B.D., Barr, A.G., Black, T.A., Iwashita, H., Kljun, N., McCaughey, J.H., Morgenstern, K., Murayama, S., Nesic, Z., Orchansky, A.Z. and Saigusa, N. (2004) Carbon, energy and water fluxes at mature and disturbed forest sites, Saskatchewan, Canada. *Agricultural and Forest Meteorology.* Submitted.

Amiro, B.D., Todd, J.B., Wotton, B.M., Logan, K.A,. Flannigan, M.D., Stokes, B.J., Mason, J.A., Martell, D.L. and Hirsch, K.G. (2001) Direct carbon emissions from Canadian forest fires, 1959 to 1999. *Canadian Journal of Forest Research* **31:** 512–525.

Apps, M.J., Kurz, W.A., Luxmoore, R.J., Nilsson, L.O., Sedjo, R.A., Schmidt, R., Simpson, L.G. and Vinson, T.S. (1993) Boreal forests and tundra. *Water, Air and Soil Pollution* **70:** 39–53.

Arain, M.A., Black, T.A., Barr, A.G., Jarvis, P.G., Massheder, J.M., Verseghy, D.L. and Nesic, Z. (2002) Effects of seasonal and interannual climate variability on net ecosystem productivity of boreal deciduous and conifer forests. *Canadian Journal of Forest Research* **32:** 878–891.

Arneth, A., Kubratova, J., Kolle, O., Shibistova, O.B., Lloyd, J., Vygodskaya, N.N. and Shulze, E.D. (2002) Comparative ecosystem–atmosphere exchange of energy and mass in a European Russian and a central Siberian bog. II. Interseasonal and interannual variability of CO_2 fluxes. *Tellus Series B – Chemical and Physical Meteorology* **54:** 514–530.

Aurela, M., Tuovinen, J.P. and Laurila, T. (2001) Net CO_2 exchange of a sub-arctic mountain birch ecosystem. *Theoretical and Applied Climatology* **70:** 135–148.

Baldocchi, D. (2003) Assessing the eddy covariance technique for evaluating carbon exchange rates of ecosystems: past, present and future. *Global Change Biology* **9:** 479–492.

Baldocchi, D.D., Vogel, C.A. and Hall, B. (1997) Seasonal variation of carbon dioxide exchange rates above and below a boreal jack pine forest. *Agricultural and Forest Meteorology* **83:** 147–170.

Baldocchi, D.D., Kelliher, F.M. and Black, T.A. (2000) Climate and vegetation controls on boreal zone energy exchange. *Global Change Biology* **6:** 69–83.

Barford, C.C., Wofsy, S.C., Goulden, M.L., Munger, J.W., Pyle, E.H., Urbanski, S.P., Hutyra, L., Saleska, S.R., Fitzjarrald, D. and Moore, K. (2001) Factors controlling long and short-term sequestration to atmospheric CO_2 in a mid-latitude forest. *Science* **294:** 1688–1691.

Barr, A.G., Griffis, T.A., Black, T.A., Lee, X., Staebler, R.M., Fuentes, J.D., Chen, Z. and Morgenstern, K. (2002) Comparing the carbon budgets of boreal and temperate deciduous forest stands. *Canadian Journal of Forest Research* **32:** 813–822.

Barr, A.G., Black, T.A., Hogg, E.H., Kljun, K., Morgenstern, K. and Nesic, Z. (2004) Variability in the leaf area index of a Boreal aspen–hazelnut forest in relation to net ecosystem production. *Agricultural and Forest Meteorology.* **126:** 237–255.

Bergh, J. and Linder, S. (1999) Effects of soil warming during spring on photosynthetic recovery in boreal Norway spruce stands. *Global Change Biology* **5:** 245–253.

Black, T.A., den Hartog, G., Neuman, H., Neuman, H.H., Blanken, P.D., Yang, P.C. *et al.* (1996) Annual cycles of water vapour and carbon dioxide fluxes in and above a boreal aspen forest. *Global Change Biology* 2: 219–229.

Black, T.A., Chen, W.J., Barr, A.G., Arain, M.A., Chen, Z., Nesic, Z., Hogg, E.H., Neuman, H.H. and Yang, P.C. (2000) Increased carbon sequestration by a boreal deciduous forest in years with a warm spring. *Geophysical Research Letters* 27: 1271–1274.

Bond-Lamberty, B., Wang, C. and Gower, S.T. (2004) Net primary production and net ecosystem production of a boreal black spruce wildfire chronosequence. *Global Change Biology* 10: 473–487.

CCFM (Canadian Council of Forest Ministers) (2002) *Compendium of Canadian Forestry Statistics.* Natural Resources Canada, Ottawa, Canada.

Chen, W.J., Black, T.A., Yang, P.C., Barr, A.G., Nueman, H.H., Nesik, Z. *et al.* (1999) Effects of climatic variability on the annual carbon sequestration by a boreal aspen forest. *Global Change Biology* 5: 41–53.

Chen, C., Chen., W., Liu, J. and Cihlar, J. (2000) Annual carbon balance of Canada's forests during 1895–1996. *Global Bigeochemical Cycles* 14: 839–849.

Ciais, P., Tans, P., Trolier, M., White, J. and Francey, R. (1995) A large northern hemisphere terrestrial CO_2 sink indicated by $^{13}C/^{12}C$ ratio of atmospheric CO_2. *Science* 269: 1098–1102.

Ciais, P., Peylin, P. and Bousquet, P. (2000) Regional biospheric carbon fluxes as inferred from atmospheric CO_2 measurements. *Ecological Applications* 10: 1574–1589.

Cienciala, E., Running, S.W., Lindroth, A., Grelle, A. and Ryan, M.G. (1998) Analysis of carbon and water fluxes from the NOPEX boreal forest: comparison of measurements with FOREST–BGC simulations. *Journal of Hydrology* 213: 62–78.

Conard, S.G. and Ivanova, G.A. (1997) Wildfire in Russian boreal forests – Potential impacts of fire regime characteristics on emissions and global carbon balance estimates. *Environmental Pollution* 98: 305–313.

Conard, S.G., Sukhinin, A.I., Stocks, B.J., Cahoon, D.R., Davidenko, E.P. and Ivanova, G.A. (2002) Determining effects of area burned and fire severity on carbon cycling and emissions in Siberia. *Climatic Change* 55: 197–211.

Constantin, J., Grelle, A., Ibrom, A. and Morgenstern, K. (1999) Flux partitioning between understorey and overstorey in a boreal spruce/pine forest determined by the eddy covariance method. *Agricultural and Forest Meteorology* 98–99: 629–643.

Dale, V.H., Joyce, L.A., McNulty, S., Neilson, R.P., Ayres, M.P., Flannigan, M.D. *et al.* (2001) Climate change and forest disturbances. *BioScience* 51: 723–734.

DeFries, R.S. and Townshend, J.R.G. (1994) NDVI-derived land cover classification at a global scale. *International Journal of Remote Sensing* 15: 3567–3586.

Desjardins, R.L., MacPherson, J.I., Mahrt, L., Schuepp, P., Pattey, E., Nuemann, H. *et al.* (1997) Scaling up flux measurements for the boreal forest using aircraft-tower combinations. *Journal of Geophysical Research* 102: 29125–29133.

Dixon, R.K. and Krankina, O.N. (1993) Forest fires in Russia: carbon dioxide emissions to the atmosphere. *Canadian Journal of Forest Research* 23: 700–705.

Fan, S., Gloor, M., Mahlman, J., Pacala, S., Sarmiento, J., Takahashi, T. and Tans, P. (1998) A large terrestrial carbon sink in North America implied by atmospheric and oceanic carbon dioxide data and models. *Science* 282: 442–446.

FAO (Food and Agricultural Organization of the United Nations) (2000) Food Resources of Europe, CIS, North America, Japan and New Zealand (industrialized temperate/boreal countries), ECE/TIM/SP/17. United Nations, Geneva, Switzerland.

Flannigan, M.D., Bergeron, Y., Engelmark, O. and Wotton, B.M. (1998) Future wildfire in circumboreal forests in relation to global warming. *Journal of Vegetation Science* **9**: 469–476.

Goulden, M.L., Munger, J.W., Fan, S.M., Sutton, D.J., Bazzaz, A., Munger, J.W. and Wofsy, S.C. (1997) Physiological responses of a black spruce forest to weather. *Journal of Geophysical Research* **102**: 28987–28996.

Goulden, M.L., Wofsy, S.C., Harden, J.W., Trumbore, S.E., Crill, P.M., Gower, S.T. *et al.* (1998) Sensitivity of boreal forest carbon balance to soil thaw. *Science* **279**: 214–217.

Gower, S.T., Vogel, J., Stow, T.K., Norman, J.M., Steele, S.J. and Kucharik, C.J. (1997) Carbon distribution and aboveground net primary production in aspen, jack pine, and black spruce stands in Saskatchewan and Manitoba, Canada. *Journal of Geophysical Research*. **102**: 29029–29041.

Gower, S.T., Krankina, O., Olson, R.J., Apps, M., Linder, S. and Wang, C. (2001) Net primary production and carbon allocation patterns of boreal forest ecosystems. *Ecological Applications* **11**: 1395–1411.

Grace, J. and Rayment, M. (2000) Respiration in the balance. *Nature* **404**: 819–820.

Grant, R.F., Black, T.A., Hartog, G., Berry, J.A., Neumann, H.H., Blanken, P.D., Yang, P.C., Russell, C. and Nalder, I.A. (1999) Diurnal and annual exchange of mass and energy between an aspen–hazelnut forest and the atmosphere: testing the mathematical model Ecosys with data from the BOREAS experiment. *Journal of Geophysical Research* **104**: 27699–27717.

Grelle, A. and Lindroth, A. (1996) Eddy correlation system for long-term monitoring of fluxes of heat, water vapour and CO_2. *Global Change Biology* **2**: 297–307.

Griffis, T.J., Black, T.A., Morgenstern, K., Barr, A.G., Nesic, Z., Drewitt, G.B., Gaumont-Guay, D. and McCaughey, J.H. (2003) Ecophysiological controls on the carbon balances of three southern boreal forests. *Agricultural and Forest Meteorology* **117**: 53–71.

Griffis, T.J., Black, T.A., Gaumont-Guay, D., Drewitt, G.B., Nesic, Z., Barr, A.G., Morgenstern, K. and Kljun, N. (2004) Seasonal variation and partitioning of ecosystem respiration in a southern boreal aspen forest. *Agricultural and Forest Meteorology*. **125**: 207–233.

Gurney, K.R., Law, R.M., Denning, A.S., Rayner, P.J., Baker, D., Bousquet, P. *et al.* (2002) Towards robust regional estimates of CO_2 sources and sinks using atmospheric transport models. *Nature* **415**: 626–630.

Halldin, S., Gryning, S.E., Gottschalk, L., Jochum, A., Lundin, L.C. and VanDeGriend, A.A. (1999) Energy, water and carbon exchange in a boreal forest landscape – NOPEX experiences. *Agricultural and Forest Meteorology* **98–99**: 5–29.

Hirsch, A.I., Trumbore, S.E. and Goulden, M.L. (2002) Direct measurement of the deep soil respiration accompanying seasonal thawing of a boreal forest soil. *Journal of Geophysical Research* **108**: 1–10.

Högberg, P., Nordgren, A., Buchmann, N., Taylor, A.F.S., Ekbald, L., Högberg, M.N., Nyberg, G., Ottosson-Lofvenius, M. and Read, D.J. (2001) Large-scale forest girdling shows that current photosynthesis drives soil respiration. *Nature* 411: 789–792.

Hollinger, D.Y., Kelliher, F.M., Schulze, E.D., Vygodskaya, N.N., Varlagin, A., Milukova, I. *et al.* (1995) Initial assessment of multi-scale measures of CO_2 and H_2O flux in the Siberian taiga. *Journal of Biogeography* 22: 425–431.

Hollinger, D.Y., Kelliher, F.M., Schulze, E.D., Bauer, G., Arneth, A., Byers, J.N. *et al.* (1998) Forest – atmosphere carbon dioxide exchange in eastern Siberia. *Agricultural and Forest Meteorology* 90: 291–306.

Houghton, R.A., Hackler, J.L. and Lawrence, K.T. (1999). The U.S. carbon budget: contributions from land-use change. *Science* 285: 574–578.

IPCC (2001) *Climate Change 2001: The Scientific Basis.* Cambridge University Press, New York.

Janssens, I.A., Lankreijer, H., Matteucci, G., Kowalski, A.S., Buchman, N., Epron, D. *et al.* (2001) Productivity overshadows temperature in determining soil and ecosystem respiration across European forests. *Global Change Biology* 7: 269–278.

Jarvis, P.G. and Linder, S. (2000) Constraints to growth of boreal forests. *Nature* 405: 904–905.

Jarvis, P.G., Massheder, J.M., Hale, S.E., Moncrieff J.B., Rayment, M. and Scott, S.L. (1997) Seasonal variations of carbon dioxide, water vapour, and energy exchanges of boreal black spruce. *Journal of Geophysical Research* 102: 28953–28966.

Johnson, I.R. and Thornley, J.H.M. (1984) A model of instantaneous and daily canopy photosynthesis. *Journal of Theoretical Biology* 107: 531–541.

Kelliher, F.M., Lloyd, J., Arneth, A., Luhker, B., Byers, J.N., McSeveny, T.M. *et al.* (1999) Carbon dioxide efflux density from the floor of a central Siberian pine forest. *Agricultural and Forest Meteorology* 94: 217–232.

Kljun, N., Black, T.A., Barr, A.G., Gaumont-Guay, D., Griffis, T.J., Morgenstern, K., McCaughey, J.H. and Nesic, Z. (2004) Net carbon exchange of three boreal forests during a drought. Paper 12.5, 26th Conference on Agricultural and Forest Meteorology, pp. 4.

Knohl, A., Kolle, O., Minayeva, T.Y., Milyukova, I.M., Vygodskaya, N.N., Fokens, T. and Schulze, E.D. (2002) Carbon dioxide exchange of a Russian boreal forest after disturbance by wind throw. *Global Change Biology* 8: 31–246.

Lafleur, P.M., McCaughey, J.H., Joiner, D.W., Bartlett, P.A. and Jelinski, D.E. (1997) Seasonal trends in energy, water, and carbon dioxide fluxes at a northern boreal wetland. *Journal of Geophysical Research* 102: 29009–29020.

Landsberg, J.J. and Gower, S.T. (1997) *Application of Physiological Ecology to Forest Management.* Academic Press, San Diego, CA. 354 pp.

Lavigne, M.B., Ryan, M.G., Anderson, D.E., Baldocchi, DD., Crill, P.M., Fitzjarrald, D.R. *et al.* (1997) Comparing nocturnal eddy covariance measurements to estimates of ecosystem respiration made by scaling chamber measurements at six coniferous boreal sites. *Journal of Geophysical Research* 102: 28977–28985.

Lavigne, M.B., Foster, R.J. and Goodine, G. (2004) Seasonal and annual changes in soil respiration in relation to soil temperature, water potential and trenching. *Tree Physiology* **24:** 415–424.

Lee, X. (1998) On micrometeorological observations of surface air exchange over tall vegetation. *Agricultural and Forest Meteorology* **91:** 39–49.

Levy, P.E., Grelle, A., Lindroth, A., Molder, M., Jarvis, P.G., Kruijt, B. and Moncrieff, J.B. (1999) Regional-scale CO_2 fluxes over central Sweden by a boundary layer budget method. *Agricultural and Forest Meteorology* **98–99:** 169–180.

Lindroth, A., Grelle, A. and Moren, A.S. (1998) Long-term measurements of boreal forest carbon balance reveal large temperature sensitivity. *Global Change Biology* **4:** 443–450.

Liski, J., Ilvesniemi, H., Makela, A. and Westman, C.J. (1999) CO_2 emissions from soil in response to climatic warming are overestimated – The decomposition of old soil organic matter is tolerant of temperature. *Ambio* **28:** 171–174.

Litvak, M., Miller, S., Wofsy, S.C. and Goulden, M. (2003) Effect of stand age on whole ecosystem CO_2 exchange in the Canadian boreal forest. *Journal of Geophysical Research* **108: D3** 8225. doi:10.1029/2001JD000854.

Liu, J., Chen, J.M., Cihlar, J. and Park, W.M. (1997). A process-based boreal ecosystem productivity simulator using remote sensing inputs. *Remote Sensing Environment* **62:** 158–175.

Liu, J., Chen, J.M., Cihlar, J. and Park, W.M. (1999). Net primary productivity distribution in the BOREAS region from a process based model using satellite and surface data. *Journal of Geophysical Research* **104(D22):** 27735–27754.

Lloyd, J., Francey, R.J., Mollicone, D., Raupach, M.R., Sogachev, A., Arneth, A. *et al.* (2001) Vertical profiles, boundary layer budgets, and regional flux estimates for CO_2 and its $^{13}C/^{12}C$ ratio and for water vapour above a forest/bog mosaic in central Siberia. *Global Biogeochemical Cycles* **15:** 267–284.

Lloyd, J., Shibistova, O., Zolotoukhine, D., Kolle, O., Arneth, A., Wirth, C., Styles, J.M., Tchebakova, N.M. and Schulze, E.D. (2002) Seasonal and annual variations in the photosynthetic productivity and carbon balance of a central Siberian pine forest. *Tellus Series B – Chemical and Physical Meteorology* **54:** 590–610.

Malhi, Y., Baldocchi, D.D. and Jarvis, P.G. (2000) The carbon balance of tropical, temperate and boreal forests. *Plant, Cell and Environment* **22:** 715–740.

McCaughey, J.H., Lafleur, P.M., Joiner, D.W., Bartlett, P.A., Costello, A.M., Jelinski, D.E., and Ryan, M.G. (1997) Magnitudes and seasonal patterns of energy, water, and carbon exchanges at a boreal young jack pine forest in the BOREAS northern study area. *Journal of Geophysical Research* **102:** 28997–29007.

Melillo, J.M., Steudler, P.A., Aber, J.D., Newkirk, K., Lux, H., Bowles, F.P., Catricala, C., Magill, A., Ahrens, T. and Mirrisseau, S. (2002) Soil warming and carbon-cycle feeedbacks to the climate system. *Science* **298:** 2173–2176.

Meroni, M., Mollicone, D., Belelli, L., Manca, G., Rosellini, S., Stivanello, S. *et al.* (2002) Carbon and water exchanges of regenerating forests in central Siberia. *Forest Ecology and Management* **169:** 115–122.

Milyukova, I.M., Kolle, O., Varlagin, A.V., Vygodskaya, N.N., Schulze, E.D. and Lloyd, J. (2002) Carbon balance of a southern taiga spruce stand in European Russia. *Tellus Series B – Chemical and Physical Meteorology* **54:** 429–442.

Moren, A.S. and Lindroth, A. (2000) CO_2 exchange at the forest floor of a boreal forest. *Agricultural and Forest Meteorology* **101**: 1–14.

Murphy, P.J., Stokes, B.J. and Kasischke, E.S. (1999). Historical fire records in the North American boreal forest. In: *Fire, Climate Change and Carbon Cycling in the Boreal Forest,* Kasischke, E.S and Stokes, B.J. (eds), Ecological Studies Series, Springer, New York, pp. 274–288.

Nilsson, S., Shvidenko, A., Stolbovoi, V., Gluck, M., Jonas, M. and Obersteiner, M. (2000). Full carbon account for Russia. National Institute for applied systems analysis. Interim report 00-021.

Pattey, E., Desjardins, R.L. and St-Amor, G. (1997) Mass and energy exchanges over a black spruce forest during key periods of BOREAS 1994. *Journal of Geophysical Research* **102**: 28967–28975.

Pattey, E., Strachan, I.B., Desjardins, R.L. and Massheder, J. (2002) Measuring nighttime CO_2 flux over terrestrial ecosystems using eddy covariance and nocturnal boundary layer methods. *Agricultural and Forest Meteorology* **113**: 145–158.

Phillips, O.L., Malhi, Y., Higuchi, N., Laurance, W.F., Nunez, P.V., Vasquez, R.M. *et al.* (1998) Changes in the carbon balance of tropical forests: evidence from long-term plots. *Science* **282**: 439–442.

Pypker, T.G. and Fredeen, A.L. (2002) Ecosystem CO_2 flux over two growing seasons for a sub-boreal clearcut 5 and 6 years after harvest. *Agricultural and Forest Meteorology* **114**: 15–30.

Rannik, U., Altimir, N., Raittila, J., Suni, T., Gaman, A., Hussein, T. *et al.* (2002) Fluxes of carbon dioxide and water vapour over scots pine forest and clearing. *Agricultural and Forest Meteorology* **111**: 187–202.

Rayment, M.B. and Jarvis, P.G. (2000) Temporal and spatial variation of soil CO_2 efflux in a Canadian boreal forest. *Soil Biology and Biochemistry* **32**: 35–45.

Rayment, M.B., Loustau, D. and Jarvis, P.G. (2002) Photosynthesis and respiration of black spruce at three organizational scales: shoot, branch and canopy. *Tree Physiology* **22**: 219–229.

Running, S.W. and Coughlan, J.C. (1988). FOREST-BGC, a general model of forest ecosystem processes for regional application. 1. Hydrologic balance, canopy gas exchange and primary production processes. *Ecological Modelling* **42**: 125–154.

Running, S.W. and Gower, S.T. (1991). FOREST-BGC, a general-model of forest ecosystem processes for regional application. 2. Dynamic carbon allocation and nitrogen budgets. *Tree Physiology* **9**: 147–160.

Sarmiento, J.L. and Gruber, N. (2002) Sinks for anthropogenic carbon. *Physics Today* **55**: 30–36.

Schlesinger, W.H. (1991) *Biogeochemistry: An Analysis of Global Change.* Academic Press, San Diego, CA.

Schulze, E.D., Lloyd, J., Kelliher, F.M., Wirth, C., Rebbman, C., Luhker, B. *et al.* (1999) Productivity of forests in the Euro-Siberian boreal region and their potential to act as a carbon sink – a synthesis. *Global Change Biology* **5**: 703–722.

Schulze, E.D., Högberg, P., Van Oene, H., Persson, T., Harrison, A.F., Read, D., Kjøller, A. and Matteucci, G. (2000) Interactions between the carbon- and nitrogen cycle and the role of biodiversity: A synopsis of a study along a north–south transect through Europe. In: Schulze, E.-D. (ed.) *Carbon and Nitrogen Cycling in European Forest Ecosystems,* pp. 468–491. (*Ecological Studies* 142.) Springer Verlag, Heidelberg.

Sellers, P.J., Hall, P.G., Kelly, R.D., Black, T. A., Baldocchi, D., Berry, J. *et al.* (1997) BOREAS in 1997: experiment overview, scientific results, and future directions. *Journal of Geophysical Research* **102**: 28731–28769.

Stocks, B.J., Mason, J.A., Todd, J.B., Bosch, E.M., Wotton, B.M., Amiro, B.D., Flannigan, M.D., Hirsch, K.G., Logan, K.A., Martell, D.L. and Skinner, W.R. (2002) Large forest fires in Canada, 1959–1997. *Journal of Geophysical Research* **107 (D1)**: 4001, doi:10.1029/2001JD00084.

Suyker, A.E., Verma, S.B. and Arkebauer, T. (1997) Season-long measurement of carbon dioxide exchange in a boreal fen. *Journal of Geophysical Research* **102**: 29021–29028.

Swanson, R.B. and Flanagan, L.B. (2001) Environmental regulation of carbon dioxide exchange at the forest floor in a boreal black spruce ecosystem. *Agricultural and Forest Meteorology* **108**: 165–181.

Valentini, R., Matteucci, G., Doleman, A.J., Schulze, E.D., Rebmann, C., Moors, E.J. *et al.* (2000) Respiration as the main determinant of carbon balance in European forests. *Nature* **404**: 861–865.

Verseghy, D.L. (1991) CLASS- A Canadian land surface scheme for GCM. 1. Soil model. *International Journal of Climatology* **11**: 111–133.

Vesela, T. (2003) Hyytiälä. The Euroflux dataset 2000. In: Valentini, R. (ed.) Fluxes of Carbon, Water and Energy of European Forests. *Ecological Studies* **163** Springer Verlag, Heidelberg. 270pp.

Vesala, T., Haataja, J., Aalto, P., Altimir, N., Buzorius, G., Garam, E. *et al.* (1998) Long-term field measurements of atmosphere–surface interactions in boreal forest combining forest ecology, micrometeorology, aerosol physics and atmospheric chemistry. *Trends in Heat, Mass and Momentum Transfer* **4**: 17–35.

Volney, W.J.A. and Fleming, R.A. (2000) Climate change and impacts of boreal forest insects. *Agriculture Ecosystems and Environment* **82**: 283–294.

Wallin, G., Linder, S., Lindroth, A., Rantfors, M., Flemberg, S. and Grelle, A. (2001) Carbon dioxide exchange in Norway spruce at the shoot, tree and ecosystem scale. *Tree Physiology* **21**: 969–976.

Whittaker, R.H. and Likens, G.E. (1975) *Primary Production of the Biosphere.* Springer-Verlag, New York.

Widen, B. (2002) Seasonal variation in forest-floor CO_2 exchange in a Swedish coniferous forest. *Agricultural and Forest Meteorology* **111**: 283–297.

Widen, B. and Majdi, H. (2001) Soil CO_2 efflux and root respiration at three sites in a mixed pine and spruce forest: seasonal and diurnal variation. *Canadian Journal of Forest Research* **31**: 786–796.

Wofsy, S.C., Goulden, M.L., Munger, J.W., Fan, S., Bakwin, P.S., Daube, B.C., Bassow, S.L. and Bazzaz, F.A. (1993) Net exchange of CO_2 in a mid-latitude forest. *Science* **260**: 1314–1317.

9

Carbon exchange of deciduous broadleaved forests in temperate and Mediterranean regions

Dennis Baldocchi and Liukang Xu

'Form follows function', *Louis Henri Sullivan (1856–1924), architect*

'Form follows function – that has been misunderstood. Form and function should be one, joined in a spiritual union', *Frank Lloyd Wright (1869–1959), protégé of Louis Henri Sullivan and architect*

1. Introduction

Over the course of history, temperate and Mediterranean forests have been intertwined with mankind. First, they coexist with a large fraction of humanity. Second, they are major resources for fuel, timber, and derived wood products. And, third, they are a part of our psyche, as they have been the locale of many legends and much history. When walking through the woods, many key questions come to the mind of a forest ecologist, including what climate and soil factors influence the form of the forest? How does that forest form interact with climate to affect functioning? How will the form and function of forests change in the future as the environment changes?

One way to address these questions is through integrated field and modelling studies of carbon exchange (Canadell *et al.*, 2000; Running *et al.*, 199) and ecosystem dynamics (Bugmann, 2001; Foley *et al.*, 1998). The tools of this trade include long-term eddy covariance measurements of CO_2 exchange between the forest and the atmosphere, field campaigns on the physiological capacity and water relations of leaves and trees, studies on soil respiration, and periodic inventories of plant biomass and soil carbon (Gower, 2003; Reich and Bolstad, 2001).

The Carbon Balance of Forest Biomes, edited by H. Griffiths and P.G. Jarvis.
© 2005 Taylor & Francis Group

The objective of this chapter is to synthesize carbon flux field data that can be used to parameterize and validate models that couple biophysics, ecophysiology, biogeochemistry, and ecosystem dynamics; the intent of these models is to understand how the carbon balances of forests ecosystems respond to their environment and disturbances on multiple time-scales. Specifically, our goal is to describe the carbon balance of deciduous broadleaved forests in temperate zones and in the adjacent Mediterranean zones of the world. To accomplish this goal we adopt a biophysical and ecophysiological perspective. First, we survey information on the assimilation and respiration of leaves of trees in temperate forests and respiration from the underlying soil. Then we examine the processes controlling net and gross carbon exchange information at the scale of the canopy. Throughout the paper we attempt to examine and discuss the dynamics of carbon exchange components of temperate forests on hourly, daily, and annual time-scales.

2. Fundamental concepts

The net amount of carbon that a forest ecosystem is able to acquire over the course of a year is called net ecosystem productivity (N_e; g (C) m^{-2} per year). It is defined as the difference between gross primary productivity (G_p) and ecosystem respiration (R_{eco}):

$$N_e = G_p - R_{eco}. \tag{1}$$

Over the course of a year N_e is positive if the ecosystem is a sink for carbon. Alternatively, micrometeorologists assess net ecosystem carbon exchange, which is a metric opposite in sign from N_e ($N_e \simeq -N_{ec}$). Variations in G_p will occur temporally, on hourly, daily, and seasonal time-scales, as sunlight, temperature, and soil moisture, among other variables, alter photosynthesis. Spatial variation in G_p will arise because of differences in available nitrogen, soil type, climate/exposure, leaf area index, stand age, and species mix. Alternatively, one can define G_p as the sum of net primary productivity, N_p, and autotrophic respiration, R_{auto}. As a rule of thumb, autotrophic respiration constitutes about half of G_p (Gifford, 1994).

Ecosystem respiration, R_{eco}, consists of autotrophic respiration by the leaves, plant stems, and roots (R_{auto}) and heterotrophic respiration by the microbes and soil fauna (R_{hetero}):

$$R_{eco} = R_{plant} + R_{soil} = R_{auto} + R_{hetero} \tag{2}$$

If one studies an ecosystem for many years, then one must consider net biome productivity (N_B), which includes carbon losses resulting from disturbances such as fire, insect infestations, and pathogen damage (Schulze *et al.*, 2000).

3. Geographic distribution

The geographic distribution of deciduous broadleaved forests is wide. Temperate, deciduous broadleaved forests, also known as hardwoods, range between the 30 and 50 degree latitude bands of North America, Europe, and Asia (Barnes, 1991). In the Mediterranean climates of Europe, California, Chile, South Africa, and Australia one finds a mix of deciduous and evergreen broadleaved tree species. On an area basis, temperate broadleaved forests occupy about 1420 million ha (1 hectare (ha) = 10^4 m^2), a value that represents about 10% of the terrestrial biosphere (Melillo *et al.*, 1993).

The climate space inhabited by temperate deciduous broadleaved forests, by definition, is neither exceedingly wet or dry, nor exceedingly cold or hot (*Table 1*). In general, temperate deciduous forests occur in climates where annual precipitation ranges between 800 and 1400 mm and annual precipitation exceeds its annual potential evaporation (Holdridge, 1947). Mean minimum winter temperatures in this habitat range between 0 and –20 °C, with bud hardiness being a more important thermal factor for determining survival limits than is stem hardiness (Barnes, 1991; Woodward, 1987); hardening lowers the freezing point of the protoplasm by accumulating sugars and other osmotic reducing substances (Larcher, 1975).

By contrast, Mediterranean broadleaved forests (which are either deciduous or evergreen) experience warm and wet winters and hot and dry summers. Minimum winter temperatures for Mediterranean oak and walnut, for example, are above 5 °C, while mean maximum summer temperatures are higher than 33 °C (*Table 1*). With regard to annual precipitation, trees in Mediterranean climates have adapted to survive on less precipitation (500–1000 mm per year) and endure extended periods when potential evaporation exceeds precipitation. Mediterranean trees are able to survive on a limited supply of water by regulating their stomata and limiting transpiration during the dry season when potential evaporation is high (Eamus and Prior, 2001; Joffre *et al.*, 1999; Xu and Baldocchi, 2003).

4. Composition, form, and function

To many people, a general impression of the structure and beauty of temperate broadleaved forests comes from films such as 'The Adventures of Robin Hood', with widely spaced trees depicted in 'Sherwood Forest'. This idealized version is unrepresentative of temperate deciduous forests that are distributed across the globe. In this section we discuss many structural and functional attributes of temperate broadleaved forests.

4.1 *Composition and form*

Great diversity exists in composition, form, and function of temperate broadleaved forests (Barnes, 1991; Hicks and Chabot, 1985; Parker, 1995; Reich and Bolstad, 2001). Most significantly, temperate broadleaved forests come in two forms: deciduous and evergreen. The main genera of deciduous trees include oak (*Quercus*), beech (*Fagus*), poplar/aspen (*Populus*), maple (*Acer*), basswood (*Tilia*), hickory (*Carya*), and birch (*Betula*). The main genera of evergreen broadleaved forests include *Nothofagus*, *Eucalyptus*, and *Quercus*.

Structural diversity comes in the guise of differences in basal area, height, and leaf area index. One cause of structural diversity can be attributed to how climate and soil factors interact to affect functioning of forest. In principle, the function of an assemblage of trees in a forest is to intercept and absorb light and to use this light energy to assimilate carbon. The goal of this activity is to provide the substrate for structural components of the trees, and energy to build wood, leaves, and roots and to maintain their

Table 1. *Data on the relation between mean climate conditions and genera of hardwood trees growing in North America. Where denoted, the data are separated for the western and eastern portions of the continent. The data are derived from Thompson et al. (1999) (http://pubs.usgs.gov/pp/1999/p1650-a/datatables/hgtable.xls). Columns are included for the mean temperature (degrees Celsius) of the coldest month, the mean annual temperature, the mean maximum temperature during July, growing degree days above 5 °C, annual precipitation (millimeters), and the moisture index, which is defined as actual evaporation (E_a) divided by potential evaporation (E_p).*

Taxon	Coldest month mean temperature (°C)	Annual mean temperature (°C)	July mean maximum temperature (°C)	Above 5 °C growing degree days	Annual precipitation (mm)	Moisture index, E_a/E_p
Acer	−6.6	7.1	29.8	2 000	885	0.94
Acer Eastern NA	−5.2	9.2	28.8	2 500	1 010	0.97
Acer Western NA	−6.9	4.2	29.2	1 100	750	0.75
Alnus	−16.4	0.8	29.1	1 100	735	0.93
Alnus Eastern NA	−18.1	0.6	29.1	1 100	770	0.96
Alnus Western NA	−15.1	0.3	28.3	900	510	0.72
Betula	−19.8	−1.0	30.8	900	605	0.87
Carya	1.0	13.6	30.1	3 400	1 070	0.96
Castanea	3.5	14.9	28.6	3 700	1 200	0.97
Fraxinus	−4.8	9.5	33.1	2 500	875	0.94
Fraxinus Eastern NA	−6.1	8.6	29.5	2 400	910	0.95
Fraxinus Western NA	9.0	17.2	33.1	4 500	485	0.49
Juglans	0.1	12.9	31.3	3 300	995	0.95
Juglans Eastern NA	−1.0	12.4	29.4	3 100	1 030	0.96
Juglans Western NA	8.7	16.3	31.3	4 300	625	0.69
Ostrya/Carpinus	−2.1	11.4	29.7	2 900	1 045	0.96
Quercus	−0.9	11.7	33.8	2 900	905	0.94
Quercus Eastern NA	−2.8	11.2	31.8	2 900	960	0.95
Quercus Western NA	6.5	14.6	33.8	3 600	540	0.63
Tilia	−5.7	8.8	27.4	2 400	940	0.97
Ulmus	−4.9	9.7	31.9	2 600	920	0.96

metabolism. Of course there are costs towards achieving these goals. The opening of stomata to capture CO_2, for instance, allows water vapour to escape. Consequently, the trees need to grow where more water is available through precipitation, than is lost through evaporation and transpiration. Second, the respiratory costs of supporting the light-harvesting superstructure must be less than the ability of the system to harvest light energy, in the form of carbon. Over geological/evolutionary time-scales, inter- and intra-plant competition for light, water and nutrients, and reproductive success have conspired to affect the structure of trees in a forest (Bugmann, 2001).

Land-use history is another important factor for understanding the structure and function of contemporary temperate forests. Because temperate forests exist in highly populated regions, they have been greatly disturbed over the past few centuries. Consequently, few old-growth temperate broadleaved forests, like the archetypal Sherwood Forest, exist. Today, most temperate forests are aggrading (Nabuurs, 2004). This is because a large-scale change in land use occurred at the transition between the 19th and 20th Centuries. As significant fractions of rural populations moved to urban settings, they abandoned farmlands and changed their cooking and heating sources from wood to fossil fuels. The current age distribution of temperate forests has an important ramification on the net amount of carbon that these forests acquire over a year because the carbon balance of these stands is far from being in an equilibrium (Bugmann *et al.*, 2001; Friend *et al.*, 1997).

Temperate broadleaved forests generally form closed canopies, which intercept over 95% of incoming sunlight (Baldocchi and Collineau, 1994; Jarvis and Leverenz, 1983) and possess leaf area indices that typically range between 4 and 7 (Hutchison *et al.*, 1986; Parker *et al.*, 1989). A recent literature survey of leaf area index of temperate forests reports a centroid value of 5.1 ± 1.8 for temperate deciduous broadleaved forests (Asner *et al.*, 2003). The heights of many mature deciduous forest stands range between 20 and 30 m, but the vertical profile of leaf area depends on stand age (Aber, 1979; Harding *et al.*, 2001). The vertical distribution of leaf area in young stands (less than 20 years) is relatively even, as shade-tolerant and sun-tolerant trees compete against one another for light. As the stand closes at intermediate ages (about 40 years), a disproportionate amount of leaf area is positioned in the upper third of the canopy and the stem space is relatively open because efficient light capture by the canopy crown suppresses understorey plants (Harding *et al.*, 2001; Hutchison *et al.*, 1986; Parker, 1995). At maturity (older than 80 years), mortality starts to cause gaps, allowing understorey plants to flourish. This causes an appreciable amount of leaf area to be distributed throughout the forest. Finally, at old-age the profile of leaf area is bimodal, with a peak of leaf area density in the understorey and the tree crowns.

Mediterranean broadleaved forests, in contrast, form sparse, open canopies, with isolated trees in many circumstances. These forests cover between 20 and 80% of the land and are 5–10 m tall. Consequently, the leaf area is much less than in temperate forests; the centroid leaf area index for Mediterranean forests is 1.71 ± 0.76 (Asner *et al.*, 2003).

4.2 *Function*

Leaf photosynthesis and respiration are two measures of physiological functioning. In this section we survey measures of photosynthetic capacity and respiration, and examine their temporal dynamics and sources of spatial variation.

4.2.1 Leaf photosynthesis

Over the past decade, the process-based biochemical model of Farquhar *et al.* (1980) has emerged as the dominant paradigm for computing leaf photosynthesis (A). When light is ample and ribulose bisphoshate (RUBP) is CO_2 saturated, photosynthesis can be quantified as:

$$A = (1 - \frac{0.5O}{\tau C_i}) \frac{V_{c\,max}\, C_i}{C_i + K_c\,(1 + O/K_O)} - R_d \qquad (3)$$

where O and C_i are the concentrations of oxygen and CO_2 in the intercellular air space, τ is the RUBISCO-specific factor, and K_c and K_o are the Michaelis–Menten constants for CO_2 and O_2, respectively. R_d represents CO_2 evolution from mitochondria in the light, rather than that associated with photorespiratory carbon oxidation (Farquhar *et al.*, 1980). The maximum velocity of carboxylation, V_{cmax}, is one of the key model parameters and is derived from the initial slope of A–C_i curves at saturating light. The model parameter V_{cmax} is evaluated from the lower region of the A/C_i curve, when C_i is less than 150 µmol mol^{-1}.

Wullschleger (1993) has shown that other model parameters of the Farquhar model (dark respiration, R_d, and the maximum rate of electron transport, J_{max}) scale with V_{cmax}, and other researchers have shown that photosynthetic capacity (A_{max}) scales with leaf nitrogen (Reich *et al.*, 1998). So if one has one piece of information on photosynthesis, one can conceivably have all the model parameters needed to implement a leaf photosynthesis model. A caveat to this supposition is that most published values of V_{cmax} assume that mesophyll conductance is infinite. Their value will be error prone when mesophyll conductance is finite and the CO_2 concentration at the chloroplast does not equal that in the substomatal cavity (Ethier and Livingston, 2004).

In *Table 2* we present data from our survey of V_{cmax} measurements on temperate deciduous trees. A representative value for V_{cmax} of deciduous trees centres at about 50 µmol m^{-2} s^{-1}, but much lower and higher values have been observed. In an attempt to distill these data, we plotted the relationship between A_{max} and V_{cmax} from hundreds of leaf measurements of broadleaved deciduous trees in contrasting environments in *Figure 1*. V_{cmax} is clearly a function of A_{max}, but the relationship is not universal.

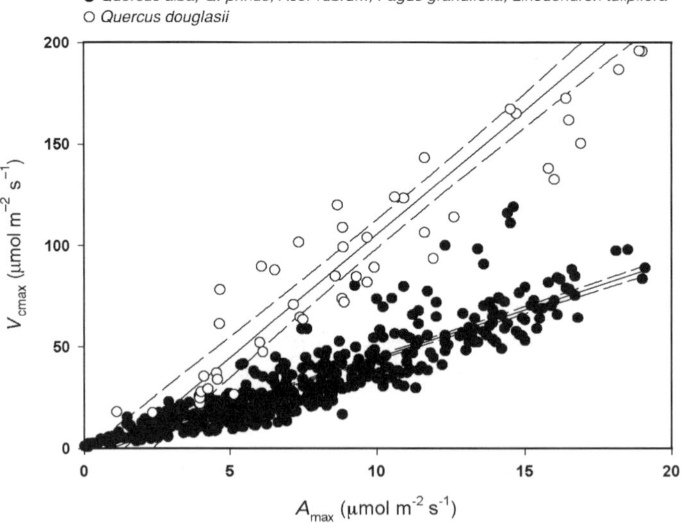

Figure 1. *Relationship between maximum photosynthetic rates at light saturation A_{max} ($Q_p = 1500$ µmol m^{-2} s^{-1}) and ambient CO_2 (350 parts per million (p.p.m.)) and V_{cmax}. Data of Xu and Baldocchi (2003) are for a xeric oak in a Mediterranean climate (Quercus douglasii) and data of Wilson et al. (2000) are for a mixture of deciduous tree species in an eastern deciduous forest biome.*

Table 2. *Survey of photosynthetic capacity (V_{cmax}) at 25 °C for tree species growing in the temperature deciduous forest zone.*

Species	Site	V_{cmax} (μmol m^{-2} s^{-1})	Reference	Comment
Deciduous trees		47 ± 33	Wullschleger, 1993	16 species
Acer pseudoplantus	Nancy, France	77.8	Dreyer *et al.*, 2001	seedlings
Betula pendula	Nancy, France	70.5	Dreyer *et al.*, 2001	
Fagus sylvatica	Nancy, France	66.3	Dreyer *et al.*, 2001	
Fagus excelsior	Nancy, France	84.6	Dreyer *et al.*, 2001	
Juglans regia	Nancy, France	63.6	Dreyer *et al.*, 2001	
Quercus petraea	Nancy, France	87.7	Dreyer *et al.*, 2001	
Quercus robur	Nancy, France	90.5	Dreyer *et al.*, 2001	
Quercus coccifera	Portugal	58.2	Tenhunen *et al.*, 1990	
Arbutus unedo	Portugal	46.6	Harley *et al.*, 1986	
Quercus alba	Oak Ridge, TN	73	Harley and Baldocchi, 1995	sun leaves
Quercus alba	Oak Ridge, TN	52	Harley and Baldocchi, 1995	shade leaves
Quercus alba	Oak Ridge, TN	28.9	Wilson *et al.*, 2000	spring D100–150
Acer rubrum	Oak Ridge, TN	30–50	Wilson *et al.*, 2000	spring D100–150
Quercus alba	Oak Ridge, TN	48.2	Wilson *et al.*, 2000	summer D150–250
Acer rubrum	Oak Ridge, TN	20–50	Wilson *et al.*, 2000	summer D150–250
Quercus alba	Oak Ridge, TN	25.7	Wilson *et al.*, 2000	late summer, D250–325
Acer rubrum	Oak Ridge, TN	~15	Wilson *et al.*, 2000	late summer
Quercus douglasii	Ione, CA	62.90	Xu and Baldocchi, 2003	April
Quercus douglasii	Ione, CA	103.06	Xu and Baldocchi, 2003	May
Quercus douglasii	Ione, CA	68.91	Xu and Baldocchi, 2003	June
Quercus douglasii	Ione, CA	57.88	Xu and Baldocchi, 2003	July
Quercus douglasii	Ione, CA	46.42	Xu and Baldocchi, 2003	August
Quercus douglasii	Ione, CA	33.64	Xu and Baldocchi, 2003	September
Quercus rubra	Blackrock, NY	55.5	Turnbull *et al.*, 2002	upper canopy
Quercus rubra	Blackrock, NY	57.6	Turnbull *et al.*, 2002	middle canopy
Quercus rubra	Blackrock, NY	48.9	Turnbull *et al.*, 2002	lower canopy
Quercus prinus	New York	49.3	Turnbull *et al.*, 2002	upper canopy
Quercus prinus	New York	57.3	Turnbull *et al.*, 2002	middle canopy
Quercus prinus	New York	44.4	Turnbull *et al.*, 2002	lower canopy
Acer rubrum	New York	53	Turnbull *et al.*, 2002	upper canopy
Acer rubrum	New York	43	Turnbull *et al.*, 2002	middle canopy
Acer rubrum	New York	32.4	Turnbull *et al.*, 2002	lower canopy
Plantanus orientalis	Kyoto, Japan	29.9	Kosugi *et al.*, 2003	expansion
Plantanus orientalis	Kyoto, Japan	47.8	Kosugi *et al.*, 2003	mid-summer
Plantanus orientalis	Kyoto, Japan	23.6	Kosugi *et al.*, 2003	fall
Plantanus orientalis	Kyoto, Japan	45	Kosugi *et al.*, 2003	June
Liriodendron tulipifera	Kyoto, Japan	34.9	Kosugi *et al.*, 2003	June
Prunus	Kyoto, Japan	51.9	Kosugi *et al.*, 2003	June
Cercidiphyllum	Kyoto, Japan	27.8	Kosugi *et al.*, 2003	June

From this growing body of photosynthesis data some general themes are emerging between photosynthesis and leaf structural and functional properties (Niinemets, 2001; Reich and Bolstad, 2001; Reich *et al.*, 1997). First, leaf photosynthesis on an area basis increases linearly with leaf nitrogen on a unit area basis (*Figure 2*). Second, leaf mass per unit area increases linearly with leaf nitrogen per unit area (*Figure 3*). Consequently, leaf photosynthesis per unit area increases linearly with leaf mass per unit area. The mechanism for this correlation stems from the large nitrogen content of RUBISCO, the CO_2-fixing enzyme.

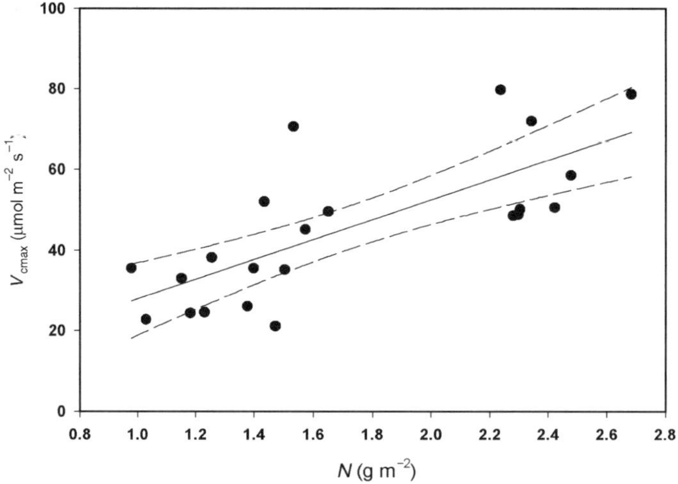

Figure 2. *Relationship between* V_{cmax} *and leaf nitrogen content N on an area basis. Data of D. Baldocchi and G. Geidt for Oak Ridge, Tennessee, USA, during summer 1996. Measurements were made on leaves of* Quercus alba.

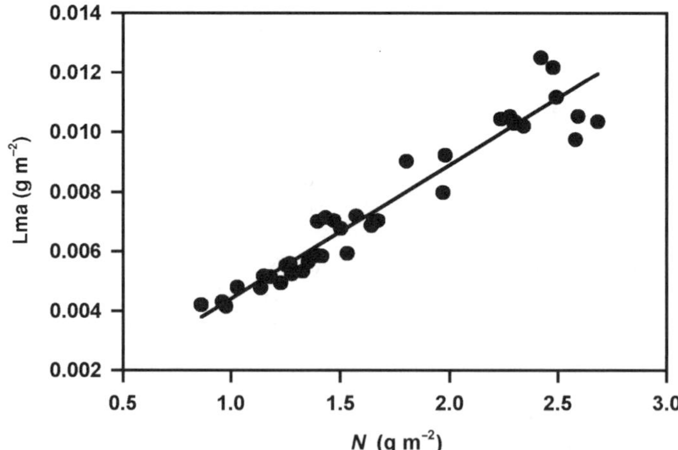

Figure 3. *Relationship between leaf mass per unit area (Lma) and nitrogen content N on an area basis for* Quercus alba *(data of D. Baldocchi and G. Geidt, Oak Ridge, Tennessee, USA, summer 1996).*

What are the consequences of these trends? Within a forest, upper sunlit leaves are thicker than those in the shaded understorey, thereby explaining the vertical canopy gradients of leaf nitrogen and photosynthetic capacity that also exist (Chen *et al.*, 1993). Another implication of these results is that canopies with small leaf area indices have more sunlit leaves (Niinemets, 2001; Reich and Bolstad, 2001), so the leaves of these canopies are thicker and have higher photosynthetic capacity (Xu and Baldocchi, 2003).

From our survey and our own data, we also conclude that the gradients of V_{cmax} and leaf nitrogen within a stand can be as large as the gradient of V_{cmax} and leaf nitrogen across the deciduous forest biome of sunlit leaves. Hence, one needs to consider sampling position when extracting such information from the literature and using it in a model.

Information is needed on the seasonality of V_{cmax} to apply photosynthesis models over the course of the growing season (Baldocchi and Wilson, 2001). Unfortunately, such data remain relatively rare. However, interesting patterns are emerging based on the limited data we have on-hand (*Figure 4*; Kosugi *et al.*, 2003; Wilson *et al.*, 2000; Xu and Baldocchi, 2003). For temperate forest trees in mesic climates, V_{cmax} experiences modest seasonality (Kosugi *et al.*, 2003; Wilson *et al.*, 2000). It attains a maximum value of about 50 µmol m^{-2} s^{-1} around midsummer and declines as fall approaches and eventual leaf senescence. In contrast, trees growing in a Mediterranean ecosystem achieve extremely high values of V_{cmax} ($\approx$110 µmol m^{-2} s^{-1}) during their short growing season, when moisture is ample (Xu and Baldocchi, 2003); these V_{cmax} values are comparable to values of fertilized crops (Wullschleger, 1993). Then V_{cmax} and leaf nitrogen drop dramatically as soil moisture is depleted. It is interesting to note that the areas under the V_{cmax} time curves in *Figure 4* are similar for the mesic and xeric trees. This result suggests that certain optimization strate-

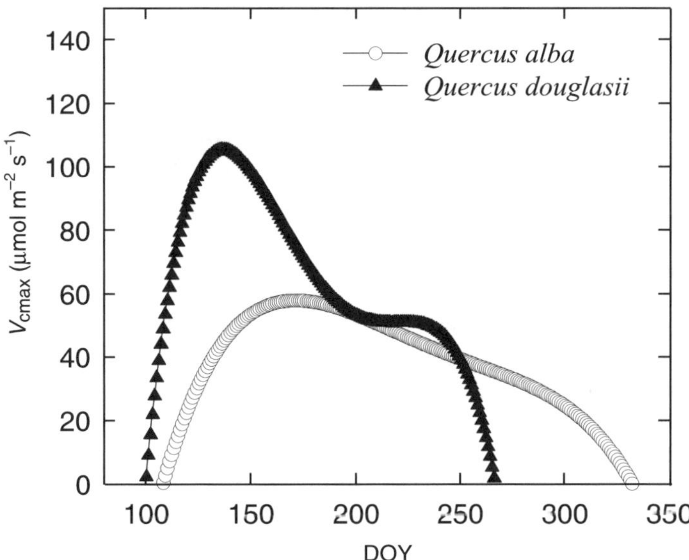

Figure 4. *Seasonal variation in V*cmax *for a xeric* (Q. douglasii) *and mesic* (Q. alba) *oak species. Data are from Xu and Baldocchi (2003) and Wilson et al. (2001).*

gies may be at play between the magnitude and the time course of V_{cmax} in xeric environments. If leaves on trees in short growing seasons are to acquire enough carbon to offset respiratory costs, they must develop light harvesting systems that have very high photosynthetic capacities during the short period when water is available. To achieve high rates of photosynthesis they need the mechanics to do so, and this comes at the expense of developing thick leaves. We find this line of logic interesting, as it is consistent with the findings of Niinemets (2001). There also has to be an additional constraint on the total leaf area that the ecosystem can support, as in dry environments a closed canopy with large leaf areas cannot be sustained because this behaviour would engender high evaporative costs, too.

4.2.2 Leaf respiration

The growing body of data on leaf respiration is delineating important relational patterns that can be exploited to parameterize models, too. First, a body of data is showing that R_d scales with leaf nitrogen, so consequently R_d scales with V_{cmax} (Reich *et al.*, 1998; Turnbull *et al.*, 2003). There are times when these relations break, however. During the early stages of leaf development, respiration rates can exceed mid-summer basal rates by a factor of five, as construction costs of the growing leaf are inordinately high (Xu and Baldocchi, 2003). Furthermore, dark respiration rates in sunlight are much reduced compared to nocturnal rates, owing to the 'Kok' effect (Turnbull *et al*, 2002; Xu and Baldocchi, 2003).

4.2.3 Stomatal conductance

Over the past decade the Ball–Berry–Collatz model (Collatz *et al.*, 1991) has emerged as a popular algorithm for computing stomatal conductance. Its strength rests on its coupling (though empirical) of stomatal conductance (g_s) to photosynthesis (A), relative humidity (rh), and leaf surface CO_2 concentration (C_s):

$$g_s \propto m \, \frac{A \cdot \mathrm{rh}}{C_s} \tag{4}$$

Drawing on the growing body of data in *Table 3* we conclude that the coefficient of proportionality (m) is fairly constant, with a mean value of 10.4 ± 0.56.

There is some controversy as to the behaviour of the coefficient, m, when trees are subject to drought. Sala and Tenhunen (1996) reported that m decreases as soil dries. Modellers using the simple biosphere model, SiB, assume that m is constant, and instead decrease V_{cmax} with drought (Colello *et al.*, 1998); members of this team were co-developers of equation (4). Most recently, Xu and Baldocchi (2003) conducted a seasonal study of V_{cmax} and m on a Mediterranean oak species and found that m remained relatively constant (8.8) as predawn water potential varied from –0.01 MPa to –4.0 MPa. Instead, leaf nitrogen content and, consequently, $V_{cmax,}$ decreased dramatically with progressive summer drought.

5. Stand-scale carbon fluxes

As we write this report, the FLUXNET community has over forty site-years of carbon flux data from more than ten deciduous broadleaved forest sites. In *Table 4* we survey published values of net ecosystem CO_2 exchange (N_{ec}) that have been reported over the

Table 3. *Literature survey of values for the coefficient, m, of the Ball–Berry–Collatz stomatal conductance equation.*

m	Taxon	Reference
8.1	*Populus tremuloides*	Berry, (pers. comm)
9.5	*Quercus alba*	Harley and Baldocchi, 1995
9.5	*Acer rubrum*	Harley and Baldocchi, 1995
7.2–11.1	*Eucalyptus grandis*	Leuning, 1990
13.5	*Populus tremuloides*	Nikolov *et al.*, 1995
16.0	*Quercus ilex* (well watered)	Sala and Tenhunen, 1996
5.4	*Quercus ilex* (drought)	Sala and Tenhunen, 1996
9.3–18.0	*Eucalyptus grandis*	Leuning, 1995
8.9	*Quercus douglasii*	Xu and Baldocchi, 2003
10.0	*Acer saccharum*	Ellsworth and Liu, 1994
13.4	*Acer saccharum*	Ellsworth and Reich, 1993
12.0	*Acer saccharum*	Tjoelker *et al.*, 1995
8.7	*Pinus flexilis*	Nikolov *et al.*, 1995
10.0	*Arbutus unedo*	Harley and Tenhunen, 1991
12.7	*Fagus sylvatica*	Medlyn *et al.*, 2001
10.1	*Phillyea augustifolia*	Medlyn *et al.*, 2001
8.2	*Pistacia lentiscus*	Medlyn *et al.*, 2001
6.2	*Quercus ilex*	Medlyn *et al.*, 2001
9.4	*Betula*	Medlyn *et al.*, 2001
2.9–6.4	*Picea abies*	Medlyn *et al.*, 2001
9.8	*Plantanus orientalis*	Kosugi *et al.*, 2003
9.3	*Liriodendron tulipifera*	Kosugi *et al.*, 2003
5.8	*Cercidiphyllum japonicum*	Kosugi *et al.*, 2003

past decade from long-term measurement studies over temperate forests. Values of N_{ec} range between –200 and –600 g (C) m^{-2} per year, and centre at –315 ± 200 g (C) m^{-2} per year. Some high values are the result of systematic bias errors at night, when respiration is underestimated (Baldocchi *et al.*, 2000). However, in general, high rates reflect the world-wide demographics of temperate forests, which tend to be relatively young (Nabuurs, 2004), with rates of net carbon exchange that have not reached equilibrium—a balance between ecosystem photosynthesis and respiration (see, for example, Friend *et al.*, 1997).

To arrive at these annual sums, a forest experiences gains and losses of carbon over the course of a day and season, as light, temperature, leaf area and soil moisture change. In the following sections we discuss the temporal dynamics of ecosystem carbon exchange of temperate and Mediterranean deciduous forests.

6. Temporal dynamics of net ecosystem CO$_2$ exchange

6.1 *Temperate deciduous forests*

The annual sums reflected in *Table 4* are the consequence of seasonal changes in the activities and magnitudes of photosynthesis and respiration. Deciduous broadleaved

Table 4. *Literature survey of published values of N_{ec}, G_p and R_{eco} using the eddy covariance method over broadleaved deciduous forests.*

Site	g(C) m^{-2} per year N_{ec}	Year	g(C) m^{-2} per year G_p	g(C) m^{-2} per year R_{eco}	Reference
Prince Albert, Saskatchwan, Canada	−160	1994			Black *et al.*, 1996
Prince Albert, Saskatchwan, Canada	−144	1994			Black *et al.*, 2000
Prince Albert, Saskatchwan, Canada	−80	1996			Black *et al.*, 2000
Prince Albert, Saskatchwan, Canada	−116	1997			Black *et al.*, 2000
Prince Albert, Saskatchwan, Canada	−290	1998			Black *et al.*, 2000
Prince Albert, Saskatchwan, Canada	−70	1996	1 120	1 050	Barr *et al.*, 2002
Prince Albert, Saskatchwan, Canada	−120	1997	1 170	1 050	Barr *et al.*, 2002
Prince Albert, Saskatchwan, Canada	−270	1998	1 350	1 090	Barr *et al.*, 2002
Harvard Forest, Petersham, MA	−220	1991			Wofsy *et al.*, 1993
Harvard Forest, Petersham, MA	−280	1991			Goulden, 1996
Harvard Forest, Petersham, MA	−220	1992			Goulden, 1996
Harvard Forest, Petersham, MA	−140	1993			Goulden, 1996
Harvard Forest, Petersham, MA	−210	1994			Goulden, 1996
Harvard Forest, Petersham, MA	−270	1995			Goulden, 1996
Harvard Forest, Petersham, MA	−200	1992			Barford *et al.*, 2001
Harvard Forest, Petersham, MA	−190	1993			Barford *et al.*, 2001
Harvard Forest, Petersham, MA	−200	1994			Barford *et al.*, 2001
Harvard Forest, Petersham, MA	−250	1995			Barford *et al.*, 2001
Harvard Forest, Petersham, MA	−200	1996			Barford *et al.*, 2001
Harvard Forest, Petersham, MA	−210	1997			Barford *et al.*, 2001
Harvard Forest, Petersham, MA	−120	1998			Barford *et al.*, 2001
Harvard Forest, Petersham, MA	−230	1999			Barford *et al.*, 2001
Harvard Forest, Petersham, MA	−210	2000			Barford *et al.*, 2001
Borden, Ontario	−130	1996			Lee *et al.*, 1999
Borden, Ontario	−160	1997			Lee *et al.*, 1999
Borden, Ontario	−310	1998			Lee *et al.*, 1999
Borden, Ontario	−60	1996	1 200	1 140	Barr *et al.*, 2002
Borden, Ontario	−240	1997	1 330	1 100	Barr *et al.*, 2002
Borden, Ontario	−170	1998	1 410	1 240	Barr *et al.*, 2002
Takayama, Japan	−120	1994			Yamamoto *et al.*, 1999
Takayama, Japan	−214	1999	1 146	984	Saigusa *et al.*, 2002
Italy	−470	1994			Valentini *et al.*, 1996
Italy	−660	1997	1 302	636	Valentini *et al.*, 2000
Hesse, France	−218	1996	1 011	793	Granier *et al.*, 2000
Hesse, France	−257	1997	1 245	988	Granier *et al.*, 2000
Hesse, France	−257	1997	1 245	988	Granier *et al.*, 2002
Hesse, France	−68	1998	1 314	1 235	Granier *et al.*, 2002
Hesse, France	−296	1999	1 335	1 036	Granier *et al.*, 2002
Belgium	−157	1997			Valentini *et al.*, 2000
Belgium	−430	1997			Valentini *et al.*, 2000
Belgium	−600	1996–1997			Aubinet *et al.*, 2001
Belgium	−519	1998	1 386	867	Granier *et al.*, 2003

Table 4. *continued*

Site	N_{ec} g(C) m^{-2} per year	Year	G_p g(C) m^{-2} per year	R_{eco} g(C) m^{-2} per year	Reference
Morgan-Monroe, Indiana	−240	2000			Schmid *et al.*, 2000
Morgan-Monroe, Indiana	−237	1998			Ehman *et al.*, 2002
Morgan-Monroe, Indiana	−287	1999			Ehman *et al.*, 2002
Oak Ridge, TN	−525	1994			Greco and Baldocchi 1996
Oak Ridge, TN	−660	1997			Baldocchi, *et al.*, 2000
Oak Ridge, TN	−470	1995			Wilson and Baldocchi, 2001
Oak Ridge, TN	−576	1996			Wilson *et al.*, 2001
Oak Ridge, TN	−618	1997			Wilson *et al.*, 2001
Oak Ridge, TN	−592	1998			Wilson *et al.*, 2001
Oak Ridge, TN	−629	1999			Wilson *et al.*, 2001
Denmark, Soroe	−169	1997	1 271	1 048	Pilegaard *et al.*, 2001
Denmark, Soroe	−124	1998	1 335	1 191	Pilegaard *et al.*, 2001
Denmark, Soroe	−122	1997	1 025	875	Granier *et al.*, 2002
Denmark, Soroe	−71	1998	1 255	1 249	Granier *et al.*, 2002
Denmark, Soroe	−227	1999	1 355	1 245	Granier *et al.*, 2002
Germany, Hainich	−494	2000	1 580	1 086	Knohl *et al.*, 2003
Germany, Hainich	−490	2001	1 540	1 050	Knohl *et al.*, 2003
Michigan, Douglas Lake	−210	1999	1 350	1 140	Schmid *et al.*, 2003
Michigan, Douglas Lake	−210	2000			Schmid *et al.*, 2003
Michigan, Douglas Lake	−170	2001			Schmid *et al.*, 2003

forests are dormant and respiring during the winter and assimilate carbon and grow during the spring and summer (*Figure 5*). During the winter, respiration is a strong function of soil temperature, as adequate soil moisture is generally available (Granier *et al.*, 2000; Greco and Baldocchi, 1996; Pilegaard *et al.*, 2001; Schmid *et al.*, 2000). The presence or absence of snow has a major impact on soil temperatures and the soil respiration from these forests. Snow insulates the soil surface, so higher soil temperatures occur when snow is present, causing higher rates of soil respiration (Goulden *et al.*, 1996).

During spring a pronounced peak in ecosystem respiration occurs because of stimulation of growth respiration as the soil warms and leaves emerge (Granier *et al.*, 2000; Greco and Baldocchi, 1996; Pilegaard *et al.*, 2001; Schmid *et al.*, 2000). Daily respiration rates between 5 and 7 g (C) m^{-2} d^{-1} can occur during this period (Granier *et al.*, 2000), thereby doubling or tripling respiration rates that occur earlier in the spring. As leaf-out occurs, the ecosystem experiences a pronounced switch from being a net source of carbon to being a net sink. This switch can represent a net change of carbon exchange on the order of 8 to 10 g (C) m^{-2} d^{-1}. Consequently, being able to predict or assess the date of leaf-out (by satellite observations or by phenological

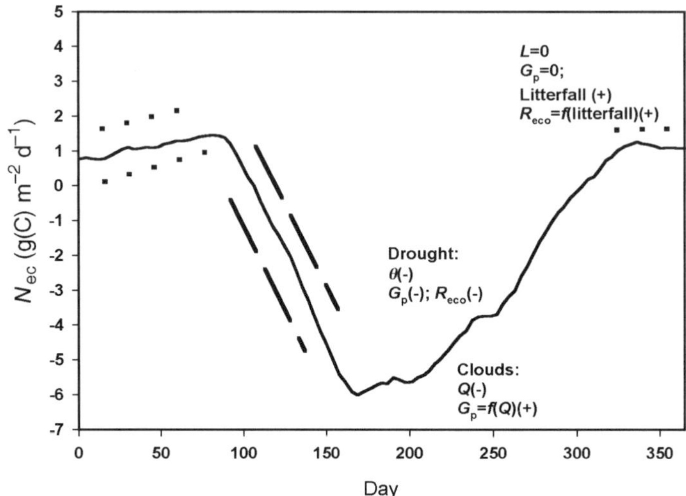

Figure 5. *Seasonal variation in daily sums of net CO_2 exchange of a temperate deciduous broad leaved forest. Also shown are the environmental and biological factors that affect the seasonal dynamics of CO_2 exchange. These variables include volumetric soil moisture (θ), photosynthetic photon flux density (Q), soil temperature (T_{soil}) and leaf area index (L) (adapted from Baldocchi and Valentini, 2004). Data sources: Baldocchi et al., 2000; Granier et al., 2000; Granier et al., 2003; Goulden et al., 1996; Greco and Baldocchi, 1996; Knohl et al., 2003; Pilegaard et al., 2001; Schmid et al., 2000; Valentini et al., 1996; Wofsy et al., 1993.*

modelling) will have a major impact on assessing carbon exchange of this biome correctly (Baldocchi *et al.*, 2001).

The date of leaf-out can vary by 30 days at a given site (Goulden *et al.*, 1996; Wilson and Baldocchi, 2001) or by over 100 days across the deciduous forest biome (*Figure 6*). Using simple regression analysis we have found that length of growing season predicts over 78% of the variance associated with N_{ec} across a global network of deciduous forest carbon flux measurement sites.

Once a forest has attained full-canopy closure, available sunlight (the photosynthetic photon flux density, Q) explains most of the variability in hourly rates of carbon exchange, but in a non-linear and saturating fashion (Baldocchi and Harley, 1995; Granier *et al.*, 2000, 2003; Goulden *et al.*, 1996; Pilegaard *et al.*, 2001; Schmid *et al.*, 2000; Valentini *et al.*, 1996). The transparency of the atmosphere, however, complicates this relation. A growing body of field studies is showing that the initial slope of the light response curve (also known as light use efficiency) increases as the fraction of diffuse radiation increases (Baldocchi, 1997; Gu *et al.*, 2002); in general short-term N_{ec} is twice as sensitive to changes in diffuse light as to changes in direct light. The presence of clouds produces a decrease in the absolute amount of sunlight, but the larger fraction of diffuse sunlight causes a shift in the N_{ec}-light response curve, so the reduction in N_{ec} is minimized. And in the presence of aerosols, an increase in the fraction of diffuse light can occur without decreasing Q and this will cause N_{ec} to increase.

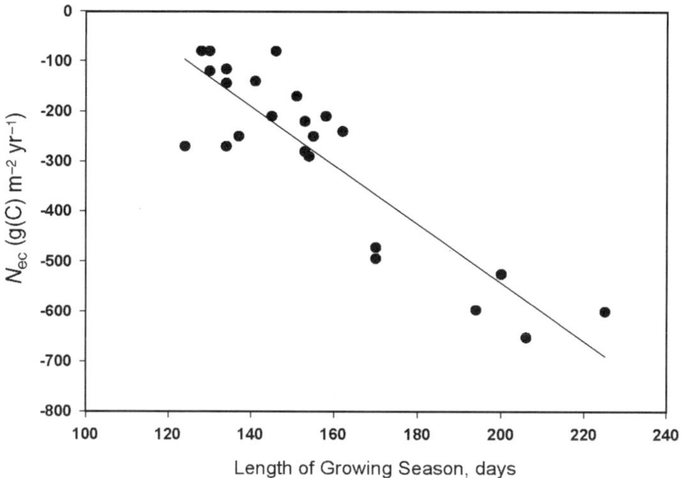

Figure 6. Impact of length of growing season on net ecosystem carbon exchange, N_{ec}, of temperate deciduous broadleaved forests. This plot updates the figure published by Baldocchi et al. (2001) by adding newer data from Schmid et al. (2003) and Knohl et al. (2003). The coefficient of determination is 0.78 and the regression slope is -5.87 g (C) m^{-2} d^{-1}.

Droughts tend to be episodic across the temperate deciduous forest biome. The occurrence of summer drought and large water vapour pressure deficits cause reductions in daytime photosynthesis (Baldocchi, 1997; Granier *et al.*, 2003; Goulden *et al.*, 1996). Drying of the soil will also force reductions in soil respiration (Hanson *et al.*, 1993). Together, these features reduce N_{ec} from values during normal summer conditions.

In the fall, leaves senesce, photosynthesis ceases and soil respiration experiences an enhancement resulting from the input of fresh litter that is readily decomposable. This leads to another respiratory pulse by the ecosystem as soils are still warm and fall rains stimulate respiration (Granier *et al.*, 2000; Goulden *et al.*, 1996).

With over 8000 hours of data available per year per site, alternative methods can be used to digest this information and quantify the temporal dynamics of the fluxes. Fast Fourier transforms are one method. In *Figure 7* we compare power spectra of CO_2 fluxes measured at North American and European flux sites. Dominant power is associated with the seasonal and diurnal forcings attributable to the Earth's daily revolution on its axis and its annual revolution around the sun. Of secondary note are spectral peaks at weekly time-scales. The passage of weather fronts occurs on this time-scale and they cause photosynthesis and respiration to vary by altering direct and diffuse radiation, light use efficiency and vapour pressure deficits.

6.2 *Deciduous Mediterranean forests*

Many deciduous forests in Mediterranean regions consist of two distinct layers, an annual grass understorey, with multiple species and functional groups (grasses, forbs

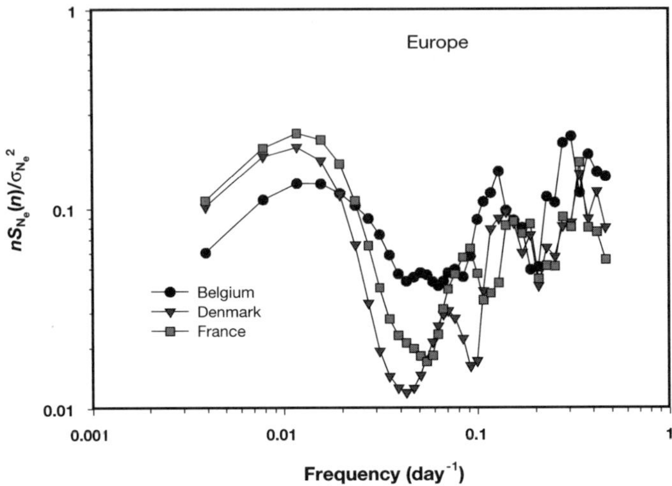

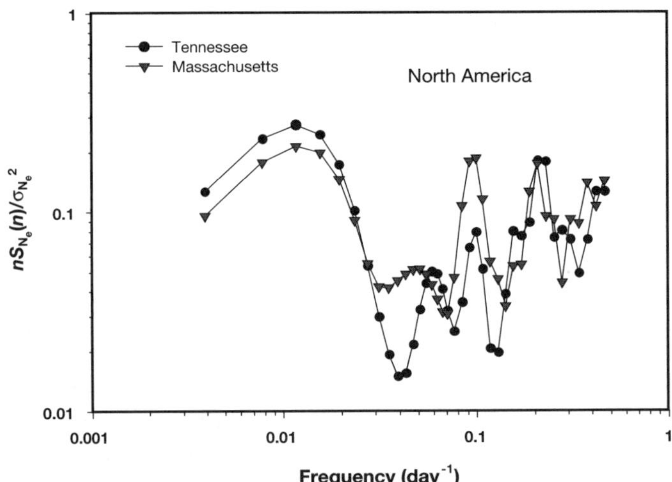

Figure 7. *Power spectra of CO_2 fluxes at European and North American temperate deciduous forest sites. The y-axis consists of the product of natural frequency (n) times the power spectra density (S(n)) for net ecosystem CO_2 exchange. This product is normalized by the variance (σ^2)of N_e. N.B. The data are from the FLUXNET database (adapted from Baldocchi et al., 2001).*

and nitrogen fixers), and tree overstorey. The grassland is green and physiologically functional during the late fall, winter and early spring. In contrast, the oak woodland is deciduous, dormant and respiring. During this period net rates of CO_2 uptake are symmetric about noon and the ecosystem is respiring at night (*Figure 8*). By spring, the tree leaves are expanding rapidly to reach full photosynthetic potential, causing a gradual increase in daytime CO_2 uptake rates. After the rains cease (around May), the grasses die and the trees gradually draw down the supply of moisture in the soil

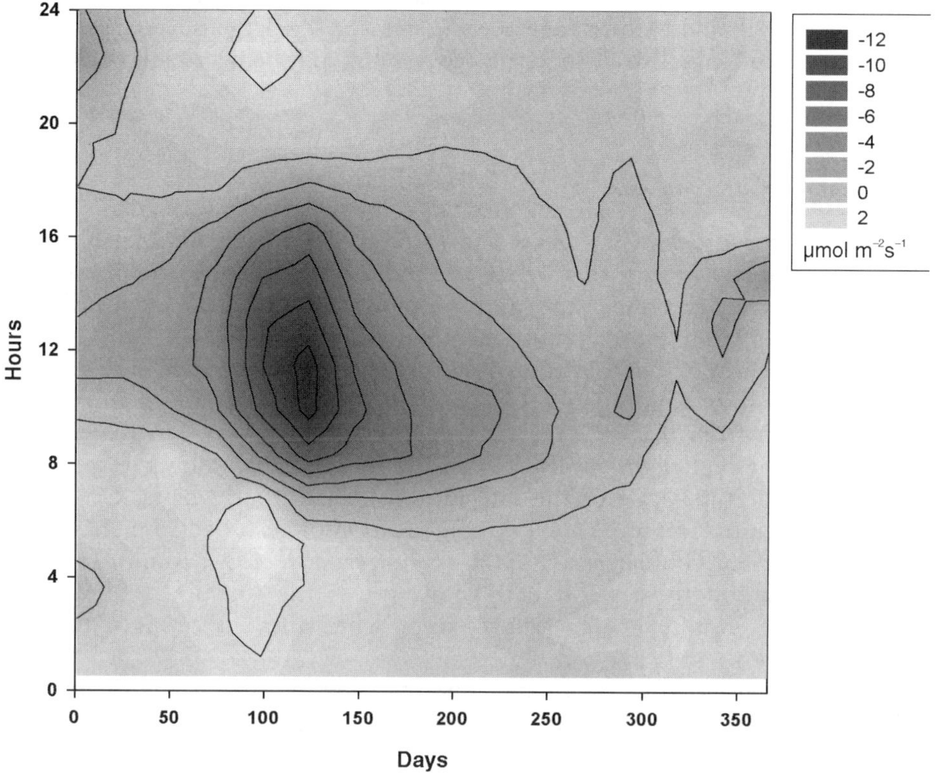

Figure 8. *Fingerprint plot of net carbon exchange N_{ec} of oak savannah growing in the Mediterranean climate of California during 2002. These data express how the diurnal curve of CO_2 exchange evolves over the course of a year. The annual integrated N_{ec} is $260g\,(C)\,m^{-2}$ (D. Baldocchi and L. Xu, unpublished data).*

as they transpire. During the hot dry and rainless summer, their stomata carefully regulate water loss to avoid lethal cavitation (Griffin, 1973; Kiang, 2002) while maintaining low rates of photosynthesis. As the intensity of the summer drought increases the time of peak CO_2 uptake shifts towards the morning, the magnitude of daytime photosynthesis decreases and the net CO_2 exchange approaches zero (*Figure 8*).

Episodic rain events during the summer period can cause huge pulses in soil respiration (Reichstein *et al.*, 2002b; Rey *et al*, 2002; Xu and Baldocchi, 2003). At least two mechanisms are possible for producing enhanced respiration rates after summer rainfall events. One is a physical displacement of soil air and CO_2 by the downward moving front of water in the soil. But this effect is short-lived and the volume of air in the soil profile is relatively small. Another effect is attributed to a rapid activation of heterotrophic respiration. A set of classic papers by Birch (1958) and Orchard and Cook (1983) show that dormant populations of microbes are able to start respiration, within hours, with the addition of water and microbial respiration ceases as soon as the soil layer dries. In ecosystems with low N_{ec}, such as savannah and Mediterranean ecosystems, a few large pulses of soil respiration have the potential to change the ecosystem from being a sink of carbon to a source.

6.3 *Canopy photosynthesis*

Canopy photosynthesis of a temperate forest is a function of available sunlight, whether it is direct or diffuse, the leaf area index of the canopy, the photosynthetic capacity of leaves, air temperature and soil moisture (Baldocchi and Amthor, 2001; Ruimy *et al.*, 1995). In most circumstances, variations in sunlight explain over 70% of the variance in canopy photosynthesis, so we focus on this variable in this chapter.

The response of canopy-scale CO_2 exchange rates to variations in sunlight are much different from that of leaves, which follow a Michaelis–Menten relationship and saturate at high light levels. The shape of the canopy light response curve depends on the time-scale of the integration. On an hourly basis, CO_2 exchange rates of temperate forests are a curvilinear function of absorbed sunlight (Baldocchi and Harley, 1995; Granier *et al.*, 2002; Hollinger *et al*, 1994; Ruimy *et al.*, 1995). Integrating carbon fluxes to a daily basis forces the functional relationship to become linear (Ruimy *et al.*, 1995), except when there is drought (Greco and Baldocchi, 1996).

The Michaelis–Menten function is a popular way of quantifying the response of hourly canopy photosynthesis measurements to changes in light (Falge *et al.*, 2001, 2002):

$$J_\mathrm{p} = \frac{\alpha \cdot Q \cdot J_{\mathrm{pmax}}}{J_{\mathrm{pmax}} + \alpha \cdot Q} \tag{5}$$

where α is quantum yield, Q is photon flux density of visible light and J_{pmax} is the maximum flux density of canopy photosynthesis. The quantum yield can be deduced by evaluating the derivative of this function with respect to Q when Q is zero.

In *Table 5* we present values of J_{pmax} and quantum yield that have been published for broadleaved forests. Maximum rates of canopy photosynthesis range between 20 and 30 μmol m^{-2} s^{-1} and quantum yields cluster around 0.04 mol CO_2 mol^{-1} photon for forests growing in North America and Europe. The lowest values of J_{pmax} and quantum yield are for the Mediterranean oak forest in Italy.

The reader, however, should be cognizant that the slope of the relationship between canopy CO_2 exchange rates and available sunlight (Q_a) is affected by whether the sky is clear or cloudy. For example, the initial slope of the canopy light response curve can double when sky conditions change from clear to cloudy (Baldocchi, 1997; Gu *et al.*, 2002; Hollinger *et al.*, 1994).

6.4 *Ecosystem respiration*

As noted in equation (2), ecosystem respiration consists of respiration by plants and soil, the latter comprising roots and soil heterotrophs. In general, soil respiration can be

Table 5. *Survey of model parameters for the relationship between canopy photosynthesis and photon flux density. The data are for deciduous forests and come from original sources and syntheses by Falge et al. (2002) and Ruimy et al. (1995).*

Forest	Location	Reference	J_{pmax} (μmol m^{-2} s^{-1})	Quantum yield (mol CO_2 mol^{-1} photon)
Betula/Quercus	Takayama, Japan	Saigusa *et al.*, 2002	n.a.	0.042–0.070
Fagus	Hesse, France	Granier *et al.*, 2002	24.8	0.0446
Fagus	Soroe, Denmark	Granier *et al.*, 2002	25.3	0.0398
Quercus	Morgan Monroe, IN	Schmid *et al.*, 2000	24.9	0.0597
Quercus/Acer	Walker Branch, TN	Baldocchi and Harley, 1995	29.9	0.040
Populus/Acer	Douglas Lake, MI	Schmid *et al.*, 2003	28–37.1	0.055–0.0695
Nothofagus	Maruia, New Zealand	Hollinger *et al.*, 1994	17.0	0.043
Quercus ilex	Castelporziano, Italy	Valentini *et al.*, 1991	17.5	0.015
Broadleaved forests	Average across the biome	Ruimy *et al.*, 1995	24.6	0.037

two-thirds to three-fourths of ecosystem respiration and one-third to one-half is root respiration (Raich and Tufekcioglu, 2000; Hanson *et al.*, 2000). On an annual time-scale, soil respiration correlates with litterfall (Raich and Tufekcioglu, 2000). In contrast, ecosystem respiration correlates strongly with soil temperature on hourly, daily, and monthly time-scales. However, recent studies show that ecosystem respiration is also a function of photosynthesis (Högberg *et al.*, 2001) and soil moisture (Reichstein *et al.*, 2002b).

An example of the seasonal variation in soil respiration of a temperate forest is shown in *Figure 9*. Modulations in carbon effluxes are caused by seasonal changes in temperature, photosynthetic activity, rain events, drought and pulsed inputs of litter.

The Q_{10} function is a popular algorithm for expressing how ecosystem respiration varies as temperature (T) deviates from a base temperature, T_b:

$$R_{eco}(T) = R_{eco}(T_b)Q_{10}^{\frac{(T-T_b)}{10}} \tag{6}$$

The Q_{10} factor represents the proportionality factor by which ecosystem respiration increases with a 10 °C increase in temperature. On the basis of the temperature dependency of the kinetic rates of basic enzyme reactions one would expect a Q_{10} of 2. In *Table 6* we survey reference respiration rates and Q_{10} coefficients for temperate and Mediterranean broadleaved forests, and observe that Q_{10} values range between 1 and 5. In many circumstances, these anomalously high values are an artifact of how these quotients are assessed. First, respiration data are noisy and measurements may be biased low during cool periods with stable thermal stratification. Second, the range of temperature can be narrow during certain phenological periods. To circumvent this problem, investigators often compute Q_{10} values by using data from across the growing season. This procedure, on the other hand, can force Q_{10} values to be artificially

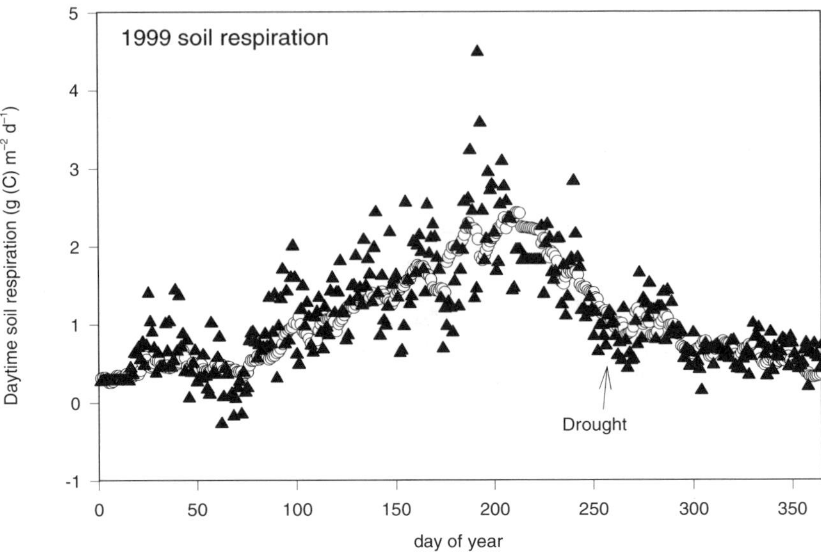

Figure 9. *Seasonal variation of CO_2 exchange measured (▲) with an understorey eddy covariance system and modelled (○). Data from Wilson and Baldocchi (2001).*

Table 6. *Survey of the basal (b) or maximum (m) rates of ecosystem respiration and the Q_{10} respiration factor. Data are from original sources or the synthesis of Falge et al. (2002).*

Forest	Location	Reference	R_{eco}	Q_{10}
Oak–maple	Oak Ridge, TN	Greco and Baldocchi, 1996	3.9, b	1.62 (a)*
Maple–tulip poplar	Morgan Monroe, IN	Schmid *et al.*, 2000	1.08, b	1.89 (s)
Oak maple	Petersham, MA	Goulden *et al.*, 1996	4.7, m	2.1 (s)*
Beech	Central Italy	Valentini *et al.*, 1996		2.2 (a)
Boreal aspen	Prince Albert, Sask	Black *et al.*, 1996		5.4 (s)
Beech	Denmark	Granier *et al.*, 2002	6.8, m	2.53 (s)*
Beech	Nancy, France	Granier *et al.*, 2002	7.4, m	3.53 (s)*
Quercus ilex/pubescens	France	Reichstein *et al.*, 2002a	3.91, b	2.10
Quercus ilex/suber	Italy	Reichstein *et al.*, 2002b	4.54, b	1.65
Betula/Quercus	Japan	Saigusa *et al.*, 2002	1.96, b	2.67
	Willow Creek, WI		5.7, m	
	Park Falls, WI		7.9, m	

*(a) air temperature.
*(s) soil temperature.

high. This is because they reflect the combined effects of growth, maintenance and reproductive respiration of both roots and micro organisms during the warmest months and only maintenance respiration during the coolest months.

Based on equation (6), one could expect annual ecosystem respiration to increase as one moves south in the northern hemisphere, to warmer climates. But one has to be careful about drawing this conclusion, because as soils get warmer they also get drier, and this inhibits ecosystem respiration and forces Q_{10} to decrease, as found by Reichstein *et al.* (2002b) for an evergreen Mediterranean forest.

7. Ecosystem carbon balances

At annual and inter-annual time-scales ecologists study net primary productivity of temperate forests by measuring changes in the above and below-ground carbon pools. The classic definition of N_p, as assessed by biometry, is:

$$N_p = \Delta L + \Delta D + \Delta H \tag{7}$$

where ΔL is the live increment, ΔD is the increment in detritus, and ΔH the increment lost to herbivory. In practice, equation (7) can be evaluated by examining the carbon increments in wood, soil , leaves, woody and fine roots (Curtis *et al.*, 2002; Gower, 2003). In *Table 7* we report data from recent carbon balance studies at deciduous forest field sites.

We note that N_e produced by biometric measurements is smaller than values produced by eddy covariance measurements in many cases (Curtis *et al.*, 2002). These differences occur for a variety of reasons. For biomass measurements there are errors associated with the allometric scaling functions, the time duration between samples and biased sampling of below ground carbon pools. Eddy covariance measurements

Table 7. Published carbon budgets of deciduous, broadleaved forests in North America and Europe. These data were produced by stand and soil carbon inventories. Data sources are Barford et al., 2001; Granier et al., 2003; Ehman et al., 2002; Curtis et al., 2002.

	Oak Ridge, TN	Nancy, France	Harvard Forest, MA	Morgan Monroe, IN	Willow Creek, WI	Colleglongo, Italy
	Curtis *et al.*, 2002	Granier *et al.*, 2003	Barford *et al.*, 2001	Ehman *et al.*, 2002	Curtis *et al.*, 2002	Granier *et al.*, 2003
Dominant genus	*Quercus/Acer*	*Fagus*	*Quercus/Acer*	*Quercus*	*Populus*	*Quercus ilex*
C Flux g (C) m^{-2}yr^{-1}						
NEP	252	218	160	73	106	557
NPP	727	448	565	1049	511	659
R_{soil}	950	575	800	1207	810	566
R_{hetero}	475	230		604		
R_{auto}						
Biomass Increment	264	456	175	311	186	

suffer from systematic bias errors that occur at night when the atmosphere is stable, causing this method to overestimate N_e in many circumstances.

8. Future issues

The study of temperate forest carbon balances remains an active and challenging problem. Five years ago the priorities for future work would have been along the lines of conducting eddy flux studies on annual time-scales. Viewed from today, we have crossed this barrier and we need to extend the length of the flux time records to at least a decade to study the inter-annual scales of variance that occur in the temperate zone (Chapter 8, this volume; Baldocchi and Wilson, 2001). Secondly, we need to assess the seasonality of photosynthesis and respiration model parameters, so we can apply biophysical models on these time-scales with confidence. One approach to flux partitioning involves using a combination of soil respiration chambers, understorey eddy flux measurements and carbon isotopes (Baldochi and Vogel, 1996; Bowling *et al.*, 1999). New soil CO_2 sensors provide an alternative means of measuring soil respiration continuously and without artifacts by coupling them with flux-gradient theory (Tang *et al.*, 2003). Thirdly, we need to assess carbon fluxes in complex terrain with confidence and we need to understand mechanistically how to correct fluxes at night when the atmosphere is stable and drainage flows transfer CO_2 at preferred locations. For up-scaling fluxes to the landscape and regional scales, we need to relate tower-based eddy fluxes with remote sensing signals and biomass contained within flux footprints. This work will involve partitioning measurements of N_{ec} into its assimilatory and respiratory components and the production of better footprint models. Finally, we need to conduct biometric inventory studies over longer time-scales to understand lag effects and inter-annual variability on the production of biomass increments.

Acknowledgements

This work was supported by the US Department of Energy's Terrestrial Carbon Program (DE-FG03-00ER63013), WESTGEC and the University of California Agricultural Experiment Station. We thank Yadvinder Malhi for providing an internal review, and Howard Griffiths and Paul Jarvis for organizing the workshop and commissioning and editing this paper. We also thank and acknowledge the numerous FLUXNET scientists whose data were used to derive relationships presented in this report. Those meriting special notice include Steve Wofsy, Mike Goulden, Riccardo Valentini, Kell Wilson, Eva Falge, Kim Pilegaard, Andre Granier, HaPe Schmid, and Gretchen Geidt.

References

Aber, J.D. (1979) Foliage height profiles and succession in northern hardwood forests. *Ecology,* **60:** 18–23.

Asner, G.P., Scurlock, J.M.O. and Hicke, J.A. (2003) Global synthesis of leaf area index observations: implications for ecological and remote sensing studies. *Global Ecology and Biogeography* **12:** 191–205.

Aubinet, M., Chermanne, B., Vandenhaute, M., Longdoz, B., Yernaux, M. and Laitat, E. (2001) Long term carbon dioxide exchange above a mixed forest in the Belgian Ardennes. *Agricultural and Forest Meteorology* **108**: 293–315.

Baldocchi, D.D. (1997) Measuring and modeling carbon dioxide and water vapor exchange over a temperate broad-leaved forest during the 1995 summer drought. *Plant, Cell and Environment* **20**: 1108–1122.

Baldocchi, D.D. and Amthor, J.S. (2001) Canopy photosynthesis: history, measurements and models. In: *Terrestrial Global Productivity* (eds J. Roy, B. Saugier, and H.A. Mooney), pp. 9–31. Academic Press, San Diego, CA.

Baldocchi, D.D. and Collineau, S. (1994) The physical nature of light in heterogeneous canopies: spatial and temporal attributes. In: Caldwell, M.M and Pearcy, R.W. (eds) *Exploitation of Environmental Heterogeneity by Plants: Ecophysiological Processes above and below Ground*, pp. 21–72. Academic Press, San Diego, CA.

Baldocchi, D.D. and Harley, P.C. (1995) Scaling carbon dioxide and water vapor exchange from leaf to canopy in a deciduous forest: model testing and application. *Plant Cell and Environment* **8**: 1157–1173.

Baldocchi, D. and Valentini, R. (2004) Geographic and temporal variation of the mechanisms controlling carbon exchange by ecosystems and their sensitivity to environmental perturbations. In: Field, C. and Raupach, M. (eds) *Towards CO_2 Stabilization: Issues, Strategies and Consequences*. A SCOPE/GCP Rapid Assessment Project. Island Press pp 295–315.

Baldocchi, D.D. and Vogel, C. (1996) A comparative study of water vapor, energy and CO_2 flux densities above and below a temperate broadleaf and a boreal pine forest. *Tree Physiology* **16**: 5–16.

Baldocchi, D.D. and Wilson, K.B. (2001) Modeling CO_2 and water vapor exchange of a temperate broadleaved forest across hourly to decadal time scales. *Ecological Modelling* **142**: 155–184.

Baldocchi, D.D., Finnigan, J.J., Wilson, K.W., Paw U, K.T. and Falge, E. (2000) On measuring net ecosystem carbon exchange in complex terrain over tall vegetation. *Boundary Layer Meteorology* **96**: 257–291.

Baldocchi, D., Falge, E., Gu, L.H., Olson, R., Hollinger, D., Running, S. *et al.* (2001) FLUXNET: A new tool to study the temporal and spatial variability of ecosystem-scale carbon dioxide, water vapor, and energy flux densities. *Bulletin of the American Meteorological Society* **82**: 2415–2434.

Barford, C.C., Wofsy, S.C., Goulden, M.L., Munger, J.W., Pyle, E.H., Urbanski, S.P., Hutyra, L., Saleska, S.R., Fitzjarrald, D. and Moore, K. (2001) Factors controlling long- and short-term sequestration of atmospheric CO_2 in a mid-latitude forest. *Science* **294**: 1688–1691.

Barnes, B.V. (1991) Deciduous forests of North America. In: (eds E. Rohrig and U. Ulrich), *Temperate Deciduous Forests Ecosystems of the World*, pp. 219–344. Elsevier, Amsterdam.

Barr, A.G., Griffis, T.J., Black, T.A., Lee, X., Staebler, R.M., Fuentes, J.D., Chen, Z. and Morgenstern, K. (2002) Comparing the carbon budgets of boreal and temperate deciduous forest stands. *Canadian Journal of Forest Research* **32**: 813–822.

Birch, H.F. (1958) The effect of soil drying on humus decomposition and nitrogen availability. *Plant and Soil* **10**: 9–31.

Black, T.A., den Hartog, G., Neumann, H., Blanken, P., Yang, P., Nesic, Z., Chen, S., Russel, C., Voroney, P. and Staebler, R. (1996) Annual cycles of CO_2 and water vapor fluxes above and within a Boreal aspen stand. *Global Change Biology* 2: 219–230.

Black, T., Chen, W., Barr, A., Arain, M., Chen, Z., Nesic, Z., Hogg, E., Neumann, H. and Yang, P. (2000) Increased carbon sequestration by a boreal deciduous forest in years with a warm spring. *Geophysical Research Letters*, 27: 1271–1274.

Bowling, D.R., Baldocchi, D.D. and Monson, R.K. (1999) Dynamics of isotope exchange of carbon dioxide in a Tennessee deciduous forest. *Global Biogeochemical Cycles* 13: 903–921.

Bugmann, H. (2001) A review of forest gap models. *Climatic Change* 51: 259–305.

Bugmann, H.K.M., Wullschleger, S.D., Price, D.T., Ogle, K., Clark, D.F. and Solomon, A.M. (2001) Comparing the performance of forest gap models in North American. *Climatic Change* 51: 349–388.

Canadell, J., Mooney, H., Baldocchi, D., Berry, J., Ehleringer, J., Field, C.B. *et al.* (2000) Carbon metabolism of the terrestrial biosphere. *Ecosystems* 3: 115–130.

Chen J., Reynolds, J., Harley, P. and Tenhunen, J. (1993) Coordination theory of leaf nitrogen distribution in a canopy. *Oecologia* 93: 63–69.

Colello, G.D., Grivet, C., Sellers, P.J. and Berry, J.A. (1998) Modeling of energy, and CO_2 flux in a temperate grassland ecosystem with SiB2: May–October 1987. *Journal of Atmospheric Science* 55: 1141–1169.

Collatz, G.J., Ball, J.T., Grivet, C. and Berry, J.A. (1991) Physiological and environmental regulation of stomatal conductance, photosysnthesis and transpiration: a model that includes a laminar boundary layer. *Agricultural and Forest Meteorology* 54: 107–136.

Curtis, P.S., Hanson, P.J., Bolstad, P., Barford, C., Randolph, J.C., Schmid, H.P. and Wilson, K.B. (2002) Biometric and eddy-covariance based estimates of annual carbon storage in five eastern North American deciduous forests. *Agricultural and Forest Meteorology* 113: 3–19.

Desai, A.R., Bolstad, P.V., Cook, B.D., Davis, K.J. and Carey, E.V. (2005) Comparing net ecosystem exchange of carbon dioxide between an old-growth and mature forest in the upper midwest, USA. *Agricultural and Forest Meteorology* 128: 33–55.

Dreyer, E., Le Roux, X., Montried, P., Daude, F.A. and Masson, F. (2001) Temperature response of leaf photosynthetic capacity in seedlings from seven temperate tree species. *Tree Physiology* 21: 223–232.

Eamus, D. and Prior, L. (2001) Ecophysiology of trees of seasonally dry tropics: Comparisons among phenologies. *Advances in Ecological Research* 32: 113–197.

Ehman, J.L., Schmid, H.P., Grimmond, C.S.B., Randolph, J.C., Hanson, P.J., Wayson, C.A. and Cropley, F.D. (2002) An initial intercomparison of micrometeorological and ecological inventory estimates of carbon exchange in a mid-latitude deciduous forest. *Global Change Biology* 8: 575.

Ellsworth, D.S. and Reich, P.B. (1993) Canopy structure and vertical patterns of photosynthesis and related leaf traits in a deciduous forest. *Oecologia* 96: 169–178.

Ellsworth, D. and Liu, X. (1994) Photosynthesis and canopy nutrition of 4 sugar maple forests on acid soils in northern Vermont. *Canadian Journal of Forest Research* 24: 2118–2127.

Ethier, G.J. and Livingston, N.J. (2004) On the need to incorporate sensitivity to CO_2 transfer conductance into the Farquhar-von Caemmerer-Berry leaf photosynthesis model. *Plant, Cell and Environment* **27**: 137–153.

Falge, E., Baldocchi, D., Olson, R., Anthoni, P., Aubinet, M., Bernhofer, Ch. *et al.* (2001) Gap filling strategies for defensible annual sums of net ecosystem exchange. *Agricultural and Forest Meteorology* **107**: 43–69.

Falge, E., Baldocchi, D., Tenhunen, J., Aubinet, M., Bakwin, P., Berbigier, P., Bernhofer, C., Burba, G., Clement, R. and Davis, K.J. (2002) Seasonality of ecosystem respiration and gross primary production as derived from FLUXNET measurements. *Agricultural and Forest Meteorology* **113**: 53–74.

Farquhar, G. D., von Caemmerer, S. and Berry, J.A. (1980) A biochemical model of photosynthetic CO_2 assimilation in leaves of C_3 species. *Planta* **149**: 78–90.

Foley, J.A., Levis, S., Prentice, I.C., Pollard, D. and Thompson, S.L. (1998) Coupling dynamic models of climate and vegetation. *Global Change Biology* **4**: 561–579.

Friend, A.D., Stevens, A.K., Knox, R.G. and Cannell, M.G.R. (1997) A process-based, terrestrial biosphere model of ecosystem dynamics (Hybrid v3.0). *Ecological Modelling* **95**: 249–287.

Gifford, R.M. (1994) The global carbon cycle–a viewpoint on the missing sink. *Australian Journal of Plant Physiology* **21**: 1–15.

Goulden, M.L. (1996) Carbon assimilation and water use efficiency by neighboring Mediterranean-climate oaks that differ in water access. *Tree Physiology* **16**: 417–424.

Goulden, M.L, Munger, J.W., Fan, S.-M., Daube, B.C. and Wofsy, S.C. (1996) Exchange of Carbon Dioxide by a Deciduous Forest: Response to Interannual Climate Variability. *Science* **271**: 1576–1578.

Gower, S.T. (2003) Patterns and mechanisms of the forest carbon cycle. *Annual Review of Environment and Resources* **28**: 169–204

Granier, A., Ceschia, E., Damesin, C., Dufrene, E., Epron, D., Gross, P. *et al.* (2000) The carbon balance of a young Beech forest. *Funct Ecology,* **14**: 312–325.

Granier, A., Pilegaard, K. and Jensen, N.O. (2002) Similar net ecosystem exchange of beech stands located in France and Denmark. *Agricultural and Forest Meteorology* **114**: 75–82.

Granier, A., Aubinet, M., Epron, D., Falge, E., Gudmundsson, J., Jenson, N.O., Kostner, B., Matteucci, G., Pilegaard, K., Schmidt, M. and Tenhunen, J. (2003) Deciduous forests: carbon and water fluxes, balances and ecophysiological determinants. In: Valentini, R. (ed) *Fluxes of Carbon, Water and Energy of European Forests,* pp. 55–70. Ecological Series 163. Springer Verlag, Berlin.

Greco, S. and Baldocchi, D.D. (1996) Seasonal variations of CO_2 and water vapor exchange rates over a temperate deciduous forest. *Global Change Biology* **2**: 183–198.

Griffin, J.R. (1973) Xylem sap tension in three woodland oaks of central California. *Ecology* **54**: 152–159.

Gu, L., Baldocchi, D.D., Verma, S.B., Black, T.A., Vesala, T., Falge, E.M. and Dowty, P.R. (2002) Superiority of diffuse radiation for terrestrial ecosystem productivity. *Journal of Geophysical Research* **107**: 107(DE):4050, doi:10.1029/2001JD001242.

Hanson, P.J., Edwards, N.T., Garten, C.T. and Andrews, J.A. (2000) Separating root and soil microbial contributions to soil respiration: A review of methods and observations. *Biogeochemistry* **48**: 115–146.

Hanson, P.J., Wullschlegar, S.D., Bohlman, S. and Todd, D. (1993) Seasonal and topographic patterns of forest floor CO^2 efflux from an upland oak forest. *Tree Physiology* **13**: 1–15.

Harding, D.J., Lefsky, M.A., Parker, G.G. and Blair, J.B. (2001) Laser altimeter canopy height profiles: methods and validation for closed-canopy, broadleaf forests. *Remote Sensing of Environment* **76**: 283–297.

Harley, P.C. and Baldocchi, D.D. (1995) Scaling carbon dioxide and water vapor exchange from leaf to canopy in a deciduous forest .1. Leaf model parameterization. *Plant Cell Environment* **18**: 1146–1156.

Harley, P. and Tenhunen, J. (1991) Modeling the photosynthetic response of C3 leaves to environmental factors. In: *Modeling Crop Photosynthesis from Biochemistry to Canopy* (eds K.J. Boote and R.S. Loomis), pp. 17–39. Crop Science Society of America, Madison, WI.

Harley, P.C., Tenhunen, J.D. and Lange, O.L. (1986) Use of an Analytical Model to Study Limitations on Net Photosynthesis in Arbutus-Unedo under Field Conditions. *Oecologia* **70**: 393–401.

Hicks, D.J. and Chabot, B.F. (1985) The deciduous forest. In: Chabot, B.F. and Mooney, H.A. (eds), *Physiological Ecology of North American Vegetation*, pp. 257–277. Chapman Hall, San Diego, CA.

Högberg, P., Nordgren, A., Buchmann, N., Taylor, A., Ekblad, A., Högberg, M., Nyberg, G., Ottosson-Lofvenius, M. and Read, D. (2001) Large-scale forest girdling shows that current photosynthesis drives soil respiration. *Nature* **411**: 789–792.

Holdridge, L.R. (1947) Determination of world plant formations from simple climatic data. *Science* **105**: 367–368

Hollinger, D.Y., Kelliher, F.M., Byers, J.N., Hunt, J.E., McSeveny, T.M. and Weir, P.L. (1994) Carbon-dioxide exchange between an undisturbed old-growth temperate forest and the atmosphere. *Ecology* **75**: 134–150.

Hutchison, B.A., Matt, D.R., McMillen, R.T., Gross, L.J., Tajchman, S.J. and Norman, J.M. (1986) The architecture of a deciduous forest canopy in eastern Tennessee, USA. *Journal of Ecology* **74**: 635–646.

Jarvis, P.G. and Leverenz, J.W. (1983) Productivity of temperate, deciduous and evergreen forests. In: *Encyclopedia of Plant Physiology* (ed O.L. Lange *et al.*). Springer-Verlag, Berlin.

Joffre, R., Rambal, S. and Damesin, C. (1999) Functional attributes in Mediterranean-type ecosystems. In: Pugnaire, F.I. and F. Valladares, F. (eds) *Handbook of Functional Plant Ecology*, pp. 347–380. Marcel Dekker, Inc. New York.

Kiang, N. (2002) *Savannas and seasonal drought: The landscape-leaf connection through optimal stomatal control.* University of California, Berkeley.

Knohl, A., Schulze, E.-D., Kolle, O. and Buchmann, N. (2003) Large carbon uptake by an unmanaged 250-year-old deciduous forest in Central Germany. *Agricultural and Forest Meteorology* **118**: 151–167.

Kosugi, Y., Shibata, S. and Kabashi, S. (2003) Parameterization of the CO_2 and H_2O gas exchange of several temperate deciduous broad-leaved trees at the leaf scale considering seasonal changes. *Plant Cell and Environment*, **26**: 285.

Larcher, W. (1975) *Physiological Plant Ecology*. Springer-Verlag, Berlin. 252 pp.

Lee, X., Fuentes, J.D., Staebler, R.M. and Neumann, H.H. (1999) Long-term observation of the atmospheric exchange of CO_2 with a temperate deciduous forest in southern Ontario, Canada. *Journal of Geophysical Research* **104**: 15975–15984.

Leuning, R. (1990) Modeling Stomatal Behavior and Photosynthesis of Eucalyptus-Grandis. *Australian Journal of Plant Physiology* **17**: 159–175.

Leuning, R. (1995) A Critical-Appraisal of a Combined Stomatal-Photosynthesis Model for C-3 Plants. *Plant Cell and Environment* **18**: 339–355.

Medlyn, B.E., Bartib, C.V.M., Broadmeadow, M.S.J., Ceulemans, R., Angelis, P.D., Forstreuter, M. *et al.* (2001) Stomatal conductance of forest species after long-term exposure to elevated CO_2 concentration: a synthesis. *New Phytologist* **149**: 247–264.

Melillo, J.M., McGuire, A.D., Kicklighter, D.W., Moore, B., Vorosmarty, C.J. and Schloss, A.L (1993) Global climate change and terrestrial net primary production. *Nature* **22**: 234–240.

Nabuurs, G.-J. (2004) Current consequences of past actions. In: Field, C. and Raupach, M. (eds) *Towards CO_2 stabilization: Issues, Strategies and Consequences*, A SCOPE/GCP Rapid Assessment Project. Island Press.

Niinemets, U. (2001) Global-scale climatic controls of leaf dry mass per area, density, and thickness in trees and shrubs. *Ecology* **82**: 453–469.

Nikolov, N.T., Massman, W.J. and Schoettle, A.W. (1995) Coupling biochemical and biophysical processes at the leaf level: an equilibrium photosynthesis model for leaves of C3 plants. *Ecological Modelling* **80**: 205–235.

Orchard, V.A. and Cook, F.J. (1983) Relationship between soil respiration and soil moisture. *Soil Biology & Biochemistry* **15**: 447–453.

Parker, G.G. (1995) Structure and microclimate of forest canopies. In: *Forest Canopies*, pp 73–106. Academic Press.

Parker, G.G., Oneill, J.P. and Higman, D. (1989) Vertical profile and canopy organization in a mixed deciduous forest. *Vegetatio* **85**: 1–11.

Pilegaard, K., Hummelshoj, P., Jensen, N.O. and Chen, Z. (2001) Two years of continuous CO_2 eddy-flux measurements over a Danish beech forest. *Agricultural and Forest Meteorology* **107**: 29–41.

Raich, J.W. and Tufekcioglu, A. (2000) Vegetation and soil respiration: correlations and controls. *Biogeochemistry* **48**: 71–90.

Reich, P.B. and Bolstad, P. (2001) Productivity of evergreen and deciduous temperate forests. In: *Terrestrial Global Productivity* (eds. J. Roy, B. Saugier. and H. Mooney), pp. 245–283. Academic Press, San Diego, CA.

Reich, P.B., Walters, M.B. and Ellsworth, D.S. (1997) From tropics to tundra: Global convergence in plant functioning. *Proceedings of the National Academy of Sciences of the USA* **94**: 13730–13734.

Reich, P.B., Walters, M.B., Ellsworth, D.S., Vose, J.M., Volin, J.C., Gresham, C. and Bowman, W.D. (1998) Relationships of leaf dark respiration to leaf nitrogen, specific leaf area and leaf life-span: a test across biomes and functional groups. *Oecologia* **114**: 471–482.

Reichstein, M., Tenhunen, J.D., Roupsard, O., Ourcival, J.-M., Rambal, S., Miglietta, F., Peressotti, A., Pecchiari, M., Tirone, G. and Valentini, R. (2002a) Severe drought effects on ecosystem CO_2 and H_2O fluxes at three Mediterranean evergreen sites: revision of current hypotheses? *Global Change Biology* **8**: 999.

Reichstein, M., Tenhunen, J.D., Roupsard, O., Ourcival, J.M., Rambal, S., Dore, S. and Valentini, R. (2002b) Ecosystem respiration in two Mediterranean evergreen Holm Oak forests: drought effects and decomposition dynamics. *Functional Ecology* **16**: 27–39.

Rey, A., Pegoraro, E., Tedeschi, V., De Parri, I., Jarvis, P.G. and Valentini, R. (2002) Annual variation in soil respiration and its components in a coppice oak forest in central Italy. *Global Change Biology* **8**: 851–866.

Ruimy, A., Jarvis, P.G., Baldocchi, D.D. and Saugier, B. (1995) CO_2 fluxes over plant canopies and solar radiation: a review. *Advances in Ecological Research* **26**: 1–53.

Running, S.W., Baldocchi, D.D., Turner, D., Gower, S.T., Bakwin, P. and Hibbard, K. (1999) A global terrestrial monitoring network, scaling tower fluxes with ecosystem modeling and EOS satellite data. *Remote Sensing of the Environment* **70**: 108–127.

Saigusa, N., Yamamoto, S., Murayama, S., Kondo, H. and Nishimura, N. (2002) Gross primary production and net ecosystem exchange of a cool–temperate deciduous forest estimated by the eddy covariance method. *Agricultural and Forest Meteorology* **112**: 203–215.

Sala, A. and Tenhunen, J.D. (1996) Simulations of canopy net photosynthesis and transpiration in *Quercus ilex* L. under the influence of seasonal drought. *Agricultural and Forest Meteorology* **78**: 203–222.

Schmid, H.P., Grimmond, C.S., Cropley, F., Offerle, B. and Su, H.-B. (2000) Measurements of CO_2 and energy fluxes over a mixed hardwood forest in the midwestern United States. *Agricultural and Forest Meteorology* **103**: 357–374.

Schmid, H.P., Su, H.-B., Vogel, C.S. and Curtis, P. S. (2003) Ecosystem–atmosphere exchange of carbon dioxide over a mixed hardwood forest in northern lower Michigan. *Journal of Geophysical Research* **108**: doi:10.1029/2002JD003011.

Schulze, E.-D., Wirth, C. and Heimann, M. (2000) CLIMATE CHANGE: Managing Forests After Kyoto. *Science* **289**: 2058–2059.

Tang, J., Baldocchi, D.D., Qi, Y. and Xu, L. (2003) Assessing soil CO_2 efflux using continuous measurements of CO_2 profiles in soils with small solid-state sensors. *Agricultural and Forest Meteorology* **118**: 207–220.

Tenhunen, J.D., Serra, A.S., Harley, P.C., Dougherty, R.L. and Reynolds, J.F. (1990) Factors influencing carbon fixation and water use by Mediterranen sclerophyll shurbs during summer drought. *Oecologia* **82**: 381–393.

Thompson, R. S., Anderson, K.H. and Bartlei, P.J. (1999) Atlas of relations between climatic parameters and distributions of important trees and shrubs in North America. U.S. Geological Survey Professional Paper 1650 A and B, http://pubs.usgs.gov/pp/1999/p1650-a/.

Tjoelker, M.G., Volin, J.C., Oleksyn, J. and Reich, P.B. (1995) Interaction of ozone pollution and light effects on photosynthesis in a forest canopy experiment. *Plant Cell and Environment* **18**: 895–905.

Turnbull, M.H., Whitehead, D., Tissue, D.T., Schuster, W.S.F., Brown, K.J., Engel, V.C. and Griffin, K.L. (2002) Photosynthetic characteristics in canopies of *Quercus rubra*, *Quercus prinus* and *Acer rubrum* differ in response to soil water availability. *Oecologia* **130**: 515–524.

Turnbull, M.H., Whitehead, D., Tissue, D.T., Schuster, W.S.F., Brown, K.J. and Griffin, K.L. (2003) Scaling foliar respiration in two contrasting forest canopies. *Functional Ecology* **17**: 101–114.

Valentini, R., DeAngelis, P., Matteucci, G., Monaco, R., Dore, S., and Scarascia-Mugnozza, G.E. (1996) Seasonal net carbon dioxide exchange of a beech forest with the atmosphere. *Global Change Biology* **2**: 199–207.

Valentini, R., Mugnozza, G.E.S., Deangelis, P. and Bimbi, R. (1991) An experimental test of the eddy correlation technique over a Mediterranean macchia canopy. *Plant Cell and Environment* **14**: 987–994.

Valentini, R., Matteucci, G., Dolman, A.J., Schulze, E.D., Rebmann, C., Moors, E.J. *et al.* (2000) Respiration as the main determinant of carbon balance in European forests. *Nature* **404**: 861–865.

Wilson, K.B., Baldocchi, D.D. and Hanson, P.J. (2000) Spatial and seasonal variability of photosynthesis parameters and their relationship to leaf nitrogen in a deciduous forest. *Tree Physiology* **20**: 565–587.

Wilson, K.B. and Baldocchi, D.D. (2001) Comparing independent estimates of carbon dioxide exchange over five years at a deciduous forest in the southern United States. *Journal of Geophysical Research* **106**: 34167–34178.

Wofsy, S., Goulden, M., Munger, J., Fan, S., Bakwin, P., Daube, B., Bassow, S. and Bazzaz, F. (1990) Net exchange of CO_2 in a midlatitude forest. *Science* **260**: 1314–1317.

Wofsy, S.C., Goulden, M.L., Munger, J.W., Fan, S.M., Bakwin, P.S., Daube, B.C., Bassow, S.L. and Bazzaz, F.A. (1993) Net Exchange of CO_2 in a Midlatitude Forest. *Science* **260**: 1314–1317.

Woodward, F.I. (1987) *Climate and Plant Distribution*. Cambridge University Press.

Wullschleger, S. (1993) Biochemical limitations to carbon assimilation in C3 plants–a retrospective analysis of the A/Ci curves from 109 species. *Journal of Experimental Botany* **44**: 907–920.

Yamamoto, S., Murayama, S., Saigusa, N. and Kondo, H. (1999) Seasonal and interannual variation of CO_2 flux between a temperate forest and the atmosphere in Japan. *Tellus Series B – Chemical and Physical Meteorology* **51**: 402–413.

Yi, C., Davis, K.J., Bakwin, P.S., Denning, A.S., Zhang, N., Desai, A. *et al.* (2004) Observed covariance between ecosystem carbon exchange and atmospheric boundary layer dynamics at a site in northern Wisconsin. *Journal of Geophysical Research* **109**: D08302, doi10.1029/2003JD004164.

Xu, L. and Baldocchi, D.D. (2003) Seasonal trend of photosynthetic parameters and stomatal conductance of blue oak (Quercus douglasii) under prolonged summer drought and high temperature. *Tree Physiology* **23**: 865–877.

Xu, L. and Baldocchi, D. (2004) Seasonal variation in carbon dioxide exchange over a Mediterranean grassland in California. *Agricultural and Forest Meteorology*. Submitted.

10

The carbon balance of the tropical forest biome

Yadvinder Malhi

1. Tropical forests in the global carbon cycle

Tropical forests are nature's great green engines, a lush green girdle across the earth's moist equatorial regions, occupying approximately $17\,000 \times 10^3$ km^2 in 1990 (Food and Agriculture Organization of the United Nations (FAO), 2000). These include lowland evergreen rainforests, moist deciduous and dry deciduous forests, and montane forests. With tropical savannahs, they account for up to 60% of terrestrial photosynthesis, harbour 50% of global diversity, and hold a large fraction of the carbon held in live vegetation. Every year they cycle about 12% of the CO_2 held in the atmosphere through photosynthesis, respiration and microbial decay. Hence small shifts in the carbon cycling of tropical forests can have profound consequences for the global carbon cycle. Moreover, the turnover and response times of tropical forests are short, and hence the response to global perturbations (and adaptation to new environmental states) may be rapid.

The principal components of the global atmospheric carbon CO_2 cycle in the 1990s are summarized in *Figure 1* (Royal Society, 2001), derived from the Intergovernmental Panel on Climate Change (IPCC, 2001). Fossil fuel combustion, the reversal of millions of years of organic carbon accumulation in sediments, has resulted in the release of almost 6.4 ± 0.4 Pg (C) per year in the 1990s (cement production is a minor additional anthropogenic term). These emissions can be estimated with reasonable accuracy from economic data. The other major anthropogenic source of CO_2, land-use change in the tropics, is known with much less certainty but was estimated to contribute 1.7 Pg (C) ha^{-1} per year. We will revisit this estimate later in this chapter.

As CO_2 is fairly well mixed in the global atmosphere, the build up of CO_2 in the atmosphere is accurately quantified at $3.2 + 0.1$ Pg (C) per year in the 1990s. This is only 40% of estimated emissions, and necessitates the existence of 'sinks' of carbon in the land or oceans. The carbon sink in the oceans in fairly well quantified at 1.7 ± 0.5 Pg (C) per year through a combination of direct measurements, models, and measurements of oxygen shifts in the atmosphere; the remaining carbon must be

The Carbon Balance of Forest Biomes, edited by H. Griffiths and P.G. Jarvis.
© 2005 Taylor & Francis Group

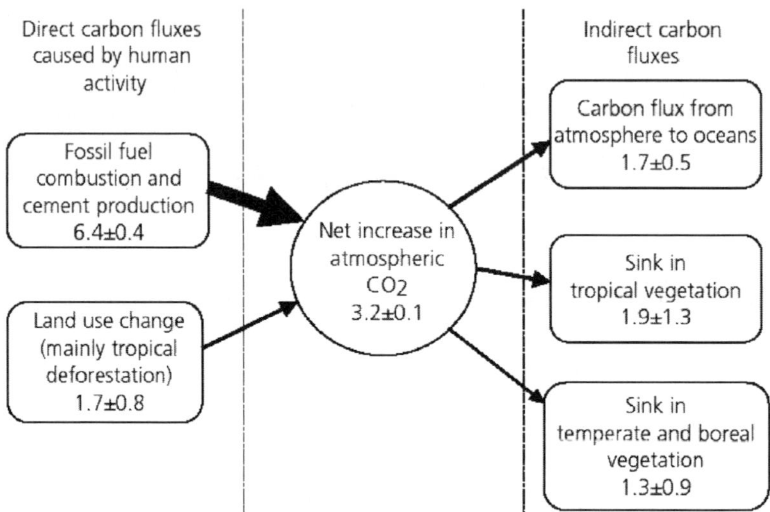

Figure 1. *The major components of the anthropogenic global carbon cycle in the 1990s, as reported by The Royal Society (2001). All units are in petagrams (10^{15}g) of carbon per year. Royal Society, "The role of land ...," The Royal Society: 27, Copyright (2001). Used with permissions.*

disappearing into some form of terrestrial carbon sink. The two prime candidates are tropical and temperate/boreal vegetation; analyses of CO_2 concentration gradients in the atmosphere (see later) suggest some form of equipartitioning (with large uncertainty bars) between these possibilities.

This terrestrial carbon 'sink' is notoriously difficult to measure directly because of the high spatial heterogeneity in terrestrial ecosystem processes. A great deal of research has gone into understanding and quantifying these terms more effectively; here, we will focus on recent thinking about the magnitude of deforestation emissions and biomass carbon 'sink' in tropical forest regions.

2. Deforestation and land-use change

Since the discovery of fire management, most human societies have relied on modifications of natural landscapes with consequent changes in the carbon storage densities of forests, savannahs, and grasslands (Perlin, 1999). In particular, most of the temperate forests of Europe and China, and the monsoon and dry forests of India and the Middle East, have been progressively cleared with the spread of pasture and cropland since 7000 before present (BP) (Williams, 1990) and only a fraction of the original forest area survived into the industrial era. Long-term clearance of tropical rainforests was much less, with notable localized exceptions such as the areas occupied by Mayan civilization in the Americas (Whitmore *et al.*, 1990) and the Khmer civilization in Cambodia. However, some recent evidence argues that permanent clearance of tropical forests, or at least seasonally dry tropical forests, was much greater than previously estimated (Willis *et al.*, 2004).

Figure 2 shows the net carbon emissions since 1850 caused by land-use change (Houghton, 1999; updated by R.A. Houghton, personal communication). Because deforestation, logging, and regrowth are more difficult to monitor than industrial activity, and the net carbon emissions from deforestation more difficult to quantify, there is still substantial uncertainty in the absolute magnitude of these figures. Houghton estimates that a total of 124 Gt (C) were emitted between 1850 and 1990, mostly from the tropics. There are noteworthy variations over time. In particular, temperate deforestation rates have slowed greatly and there has even been a net expansion of forest cover in North America and Europe, whereas deforestation in the tropics has surged since the 1950s to account for almost all current emissions. Different regions have surged at different times in response to political priorities: for example, temperate Latin America in the 1900s, the Soviet Union in the 1950s and 1960s, the tropical Americas in the 1960s and 1980s, and tropical Asia in the 1980s and 1990s.

The major types of land-use change that affect carbon storage are: (i) the permanent clearance of forest for pastures and arable crops; (ii) shifting cultivation, which may vary in extent and intensity as populations increase or decline; (iii) logging with subsequent forest regeneration or replanting; and (iv) abandonment of agriculture and replacement by regrowth or planting of secondary forest (i.e., deforestation, afforestation, and reforestation). Many of these processes (shifting cultivation, logging, clearing for pasture, and abandonment) involve dynamics between forest destruction and subsequent recovery, although the net effect has been a loss of carbon from forests. A review of these processes is provided by Houghton (1995, 1996).

Of the various processes, the expansion of croplands at the expense of natural ecosystems has dominated and continues to dominate the net efflux of carbon from the terrestrial biosphere, with the regions of highest activity being the tropical forests of southeast Asia (0.76 Pg (C) per year in 1990) and Africa (0.34 Gt (C) per year in 1990). The conversion of forest into cattle pasture has gradually increased in

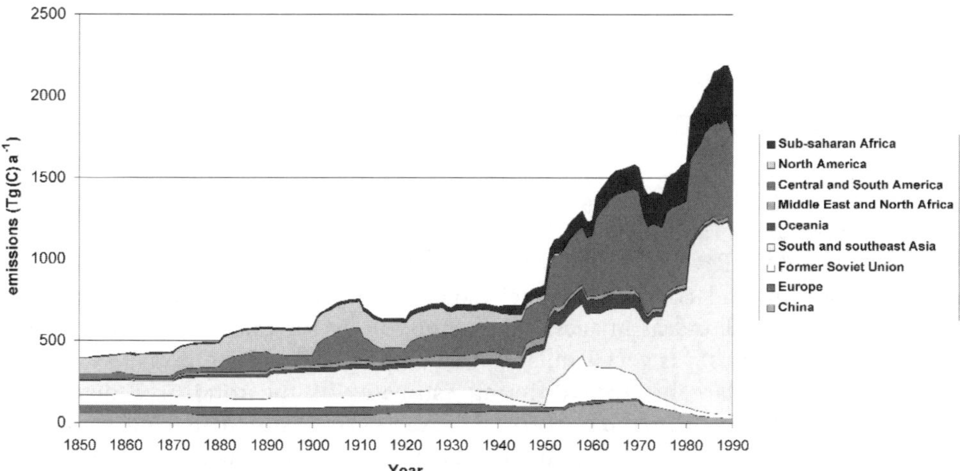

Figure 2. The net cumulative carbon emissions in different regions since 1850 caused by land-use change (Houghton, 1999; updated by Houghton, personal communication). The vertical order of the key reflects the vertical order of the data series. Houghton, R.A., "The Annual net flux . . .", Tellus Series B-Chemical and Physical Meteorology *51.* Copyright (1999), used with permissions from Blackwell Publishing.

importance and is concentrated almost entirely in Latin America (0.34 Pg (C) in 1990 compared with 0.16 Pg (C) from cropland expansion in the same region). Net emissions from wood harvesting are dominated by logging in southeast Asia (0.19 Pg (C) in 1990; 66% of total emissions from harvesting), and China (0.07 Pg (C) in 1990; 23% of total emissions). In the 19th century, shifting cultivation was in decline and thus generated a net sink of carbon, because of the abandonment of agricultural lands (often under tragic conditions) by indigenous peoples in the Americas and southeast Asia. Since the mid-20th Century, however, increasing population pressures have led to an increase of cultivation area and a decrease in rotation times, and shifting cultivation has become an increasing source of carbon. In North America and Europe, there has been a gradual abandonment of agricultural lands and regrowth of forests, which has resulted in a carbon sink.

The results presented above (Houghton, 1999) are based on estimates of land-cover change derived from the FAO. However, recent studies in several countries have suggested that the FAO may be overestimating land-cover change. Results of these studies are compared with the Houghton/FAO estimates in *Table 1*.

Table 1. *Estimated carbon emissions from deforestation from three sources: Houghton (2003), DeFries et al. (2002), and Achard et al. (2004). All three analyses are broadly similar in the model calculation of carbon dynamics, and the major source of differences between the estimates comes from the estimated rate of deforestation. Houghton (2003) used FAO deforestation data; the analyses by DeFries et al. (2002) and Achard et al. (2004) are based on independent satellite data.*

Reference	Tropical Americas Pg (C) per year	Tropical Africa Pg (C) per year	Tropical Asia Pg (C) per year	Pan-tropical Pg (C) per year
Houghton, 2003	0.75 ± 0.3	0.35 + 0.2	1.09 ± 0.5	2.20 ± 0.6
DeFries *et al.*, 2002	0.43 (0.21–0.62)	0.12 (0.08–0.14)	0.35 (0.19–0.59)	0.91 (0.50–1.36)
Achard *et al.*, 2004	0.44 (0.35–0.52)	0.16 (0.13–0.19)	0.39 (0.32–0.56)	0.98 (0.81–0.12)

DeFries *et al.* (2002) used a continuous global time-series of satellite advanced very high resolution radiometer (AVHRR) data to estimate pan-tropical deforestation rates. The AVHRR data are only at a resolution of 8 km: to relate fractional (sub-pixel) changes in land cover with pixel radiometric properties, they calibrated off numerous available Landsat images. The methodology for calculating carbon emissions from forest areas was very similar to that employed by Houghton (1999): hence discrepancies arise largely from differing estimates of the total area deforested. DeFries *et al.* (2002) estimated pan-tropical net carbon emission from land-use change to be 0.91 (0.50–1.36) Gt (C) per year, about half of the rate estimated by the FAO.

Similarly, Achard and colleagues presented a new analysis of deforestation rates from satellite data (Achard *et al.*, 2002) and presented new estimates of net carbon emissions (Achard *et al.*, 2004) that covered the humid and dry tropics. They concluded that deforestation had resulted in net carbon emissions of 0.98 ± 30 Pg (C) per year in the 1990s (with forest degradation contributing 4% of this, and the regrowth sink –3.3%), a value very close to the independent estimate of DeFries *et al.* (2002) and again lower than the FAO. The estimates from Achard *et al.* (2004) and DeFries *et al.* (2002) are also very consistent at continental scale (*Table 1*), despite being based on different sensors. The Americas, Asia, and Africa contribute about 45%, 40%, and 15% of total land-use change emissions, respectively. As tropical Asia holds less than half the forest area of the tropical Americas, the *fractional* rates of deforestation are much higher in Asia. Achard *et al.* (2004) also added an estimate of 0.88 ± 0.07 Pg (C) emissions from the 'exceptional' burning of 2.4 million ha (1 hectare (ha) $= 10^4$ m^2) of peatland in the Indonesian forest fires in 1997–98 (Page *et al.*, 2002), to arrive at a mean C emission of the terrestrial tropical biosphere of 1.1 ± 0.3 Pg (C) per year for the 1990s.

Why would the FAO have overestimated deforestation rates? FAO estimates for the 1980s are largely derived from country reports, and it seems that there is a tendency for these reports to overestimate forest clearance. Detailed comparisons between satellite analyses and FAO estimates in specific regions suggest that the country reports overestimated deforestation rates in Brazil (Houghton *et al.*, 2001) and Bolivia (Steininger *et al.*, 2001). In the 1990s, FAO estimates also used satellite images for 10% of the forest area: the persistence of the discrepancy with estimates derived from pan-tropical satellite coverage suggests that this 10% area may have been biased to high-disturbance areas.

Although the estimates for the area of forest cleared have declined, there are arguments that the amount of carbon emitted per unit area of forest may be higher. There are several processes that may be contributing to carbon emissions but which are not included in these calculations. These include forest degradation without loss of forest cover, illegal, unmonitored logging, and hidden ground fires (Nepstad *et al.*, 1999), and may add about 0.4 Gt (C) per year to the estimate of net carbon emissions (Fearnside *et al.*, 1999). Another important uncertainty is the amount of biomass held in intact forests. Fearnside *et al.* (1999) suggested that biomass may be higher than usually estimated; on the other hand, recent evidence suggests that the estimates of Fearnside are too high (Wood *et al.*, submitted).

If we take the DeFries/Achard estimates of carbon emissions from tropical land-use change (ca. 0.9 ± 0.4 Pg (C) per year) as our revised best estimate, and add 0.8 Pg (C) per year as a result of the Indonesian peatland fires of 1997–98 (Page *et al.*, 2002), the net biosphere carbon sink is lower than previously suggested (*Figure 1*). As an upper limit, if we leave the latitudinal gradient of CO_2 and the inferred net balance of the tropics unaffected, the tropical carbon sink is reduced from 1.9 ± 1.3 to 1.0 ± 1.2 Pg (C) per year.

Hence, consideration of the global carbon cycle alone suggests that the net carbon balance of the tropics is -0.2 ± 1.1 Pg (C) per year, i.e., the tropics could plausibly be a net source of 1 Pg (C) (if there is no carbon sink in intact forests), or a net sink of 1 Pg (C) (if the carbon sink is very strong). Can consideration of the atmospheric distribution of CO_2 help us clarify this picture? What does the atmosphere 'see' as coming out of the tropics?

3. The view from the atmosphere

Although CO_2 is well mixed in the global atmosphere, it is not perfectly mixed, and there are small horizontal gradients in CO_2 concentration that are driven by the spatial pattern of sources and sinks for CO_2. For example, CO_2 concentrations are higher in the Northern Hemisphere (the source of most fossil fuel combustion) than in the Southern Hemisphere. In principle, if the transport of CO_2 in the atmosphere can be accurately represented by meteorological models, it should be possible to use observations of atmospheric CO_2 concentrations to derive the spatial pattern of CO_2 sources and sinks. As the spatial pattern of CO_2 sources from fossil fuel combustion can be accurately predicted from economic data, the remainder will be the spatial pattern of biosphere CO_2 sources and sinks (including the effects of land-use change). Further information can be derived by also including observations of the stable isotopes of CO_2 (biosphere photosynthesis has a different isotopic signature than fossil fuel combustion or oceanic dissolution), and observations of O_2 concentrations.

The primary requirements for these 'inverse modelling' studies are atmospheric transport models, which developed in the 1980s with the expansion of computing power, and a global network of observation stations. However, most of these observation stations arc in North America and Europe, with very little coverage over the oceans and no coverage over tropical land masses. This dearth of observations is the main problem that currently constrains this approach (see Chapter 7, this volume).

Several research groups around the world are applying this atmospheric inversion technique. Recently, the TRANSCOM3 experiment compared the results from applying 16 different transport models to the same dataset (Gurney *et al.*, 2002). Gurney *et al.* found that results were relatively robust for the northern and southern extratropics, but very poorly constrained in the tropics. There appeared to be a significant carbon sink uniformly distributed across northern land regions, with a total northern land sink of 2.3 ± 0.7 Pg (C) per year. The study suggests that the tropical lands are probably a net source of carbon, of magnitude 1.2 ± 1.4 Pg (C) per year, implying that the source from tropical deforestation more than compensates any carbon sink in intact forests. The uncertainties are so large, and the observations so few, however, that Gurney *et al.* emphasized that inversion techniques cannot yet provide major constraints on the tropical carbon balance.

What causes this variation between studies? One major constraint is limited data over much of the globe. This will be gradually ameliorated as the observation network expands, and will be revolutionized with the arrival of satellite-borne CO_2 sensors (see below). There are also still problems with the atmospheric transport models. Different models differ in their description of the boundary layer between the atmosphere and the land surface, and these differences explain much of the variation in estimated high-latitude sinks. In tropical latitudes, the problem is even more severe. Air is mixed rapidly in the tropics by vigorous convective storms and rapidly transported to higher latitudes. Therefore small spatial gradients in CO_2 are rapidly smoothed out, and high-accuracy observations and transport models would be required to interpret these regions. As a result, tropical regions are almost entirely unconstrained in present-day inversion studies, and the carbon balance of tropical regions simply calculated as a residue of the calculations for high-latitude regions.

The TRANSCOM synthesis is far from the end of the story, however. An important feature in the inversion is how monthly variations of both CO_2 and atmospheric transport are incorporated. The TRANSCOM synthesis used long time-series data of CO_2 concentrations, but used a year-averaged meteorological model that varied between months, but not between years. More recent studies have explicitly used reconstructed atmospheric transport for different months and years. Perhaps the most state-of-the-art analysis so far has been presented by Rödenbeck *et al.* (2003), who also included a very careful investigation of the importance of the duration of particular datasets. This approach seems to provide values of tropical carbon balance of –0.8 ± 1.3 Pg (C) per year, outside the range of the TRANSCOM values (+1.0 ± 1.3 Pg (C) per year), implying that the tropics are a net carbon sink, and that the intact tropics must, therefore, be a very strong carbon sink of about 2 Pg (C) per year. This is far from being the final word on the topic, but it is clear that inversion studies do not yet yield enough accuracy and confidence to quantify the tropical carbon sink to the precision required.

Although they have difficulty is defining mean values for the tropics, inversion studies are able to describe seasonal and inter-annual variations about the mean with much greater confidence. All inversion studies agree that the terrestrial tropics are the major source of inter-annual variability in the global carbon budget. The primary source of variability is in the tropical land flux, which was a source of 50 g (C) m^2 per year during the strong El Niño years of 1982–83 and 1997–98 (Rödenbeck *et al.*, 2003). Variation in the ocean flux, and in the temperate land flux, is an order of magnitude smaller. The analysis of Rödenbeck *et al.* (2003) suggests that, for the El Niño period June 1997–1998, the primary CO_2 source regions appear to be Amazonia and Indonesia (the longitudinal distribution of these sources is determined by the tropical mid-ocean CO_2 observations in the mid-Pacific, Atlantic, and Indian oceans, and hence are highly sensitive to the reliability of data from single stations). Both Amazonia and Indonesia were strongly affected by drought in this period (reducing photosynthesis and perhaps increasing respiration), which also led to high fire incidence in forests and peatlands. Both these factors led to a strong tropical carbon source that doubled the rate of CO_2 increase in the atmosphere.

Another interesting feature is the carbon flux anomaly during the period June 1991 to May 1993. During this period the rate of CO_2 increase in the atmosphere slowed considerably, despite the existence and persistence of a moderate El Niño. An unusual feature of this period is that it lies in the wake of the Pinatubo volcanic eruption, which injected enormous volumes of aerosol into the stratosphere. The analysis of Rödenbeck *et al.* (2003) suggests that much of the extra carbon absorbed in this period was in Amazonia (and eastern North America). One possible explanation is that the aerosol haze increased the amount of diffuse radiation hitting the forest canopy. Diffuse radiation is more effective at penetrating the forest canopy and hence moderate amounts of haze may actually increase net photosynthesis (Gu *et al.*, 2002). An alternative explanation is that the haze lowered temperatures slightly, and reduced the rates of tropical respiration. There is, however, little evidence of a cooling in tropical surface temperatures in the immediate post-Pinatubo period (Malhi and Wright, 2004).

Returning to the issue of the mean tropical carbon balance, a combination of inversion studies and deforestation estimates suggest that the intact tropics (away from the deforestation zones) are anything between carbon-neutral or a strong carbon sink, favouring the latter if the recent analyses by Rödenbeck *et al.* (2003) are accurate. A carbon sink of 1 Pg (C) year, if it were uniformly distributed across wet and moist tropical forests, would correspond to a per unit area sink of 0.55 Mg (C) ha^{-1} per year. If this carbon is being absorbed by the forest, it must be going either into biomass or soil. In recent years, an increasing amount of field research has attempted to directly measure this carbon sequestration.

4. A carbon sink away from the degradation?

Despite the gnawing threats of deforestation and degradation, there are still large areas of tropical forests that are among the world's great wilderness areas, with fairly light human impact. A prime example is much of northern and western Amazonia. What is the carbon balance of these regions?

The net carbon balance of an area of forest can be considered as the difference between net primary production of plant organic matter, N_p, and the breakdown and heterotrophic respiration, R_h, of this matter through herbivory or microbial decay, plus any export of matter through extraction or transport of carbon in water runoff. If export of carbon can be neglected, these two terms must be in approximate balance because any change in N_p eventually produces a corresponding change in heterotrophic respiration, with a lag time equal to the mean residence time of carbon in biomass and soil. For example, if there is a short-term increase in soil respiration driven by increasing soil temperatures, the labile soil carbon stocks will eventually decrease to bring heterotrophic respiration back to a level with net primary production. However, variations in N_p and R_h on a time-scale shorter than the carbon residence times can result in a net flux of carbon to or from the forest system. Thus, inter-annual variation in cloudiness, precipitation, and temperature might result in significant inter-annual variations in forest carbon balance.

However, there may also be longer-term shifts superimposed on this inter-annual variation. The ongoing global rise in mean atmospheric CO_2 concentration from a pre-industrial value of 280 p.p.m. to more than 375 p.p.m. by 2004 may be stimulating plant photosynthesis, with a lagged response in respiration leading to a net carbon sink. Similarly, the pan-tropical rise in air temperatures by 0.26 °C per decade in recent decades (Malhi and Wright, 2004) may be enhancing nutrient mineralization and stimulating forest growth, or alternatively enhancing water stress and increasing respiration rates. Regional changes in precipitation, such as the drying of the northern African tropics (Malhi and Wright, 2004) may have strong local effects on carbon balance. Changes in disturbance regime (fires, storm frequency, etc.) may also affect the equilibrium biomass content of forest regions.

Tropical forests are a prime candidate for the CO_2 fertilization hypothesis, because of their intrinsic high productivity. A crucial question has been the extent to which such a response is limited by other factors, such as low nutrient availability (Chambers and Silver, 2004; Körner, 2004; Lewis *et al.*, 2004). Several studies have attempted to model the effect of rising CO_2 on tropical forest productivity, with the resulting increase being largely dependent on how the nutrient cycle is modelled. It has been argued that forests might simply increase their nutrient acquisition processes by investing in mycorrhizal colonization, and by mineralizing nutrient reserves in the soil through the production of surface enzyme systems and organic acid exudates.

A small, steady increase in forest productivity can produce a large net carbon sink. For a linear increase in N_p, followed by a linear increase in respiration with lag time τ, the rate of carbon sequestration is $\tau dN_p/dt$ (Malhi and Grace, 2000; Taylor and Lloyd, 1992). For a forest in central Amazonia, Malhi and Grace (2000) estimated a mean biomass residence time of 16 years, and a net primary productivity of

15.6 Mg (C) per year. Hence an increase in tropical forest productivity at a rate of 0.1% per year would result in a net biomass carbon sink of 0.25 Mg (C) ha^{-1} per year, and a rate of 0.75% per year would result in a sink of 0.75 Mg (C) ha^{-1} per year. Soil carbon and coarse woody debris may also be an additional sink of similar magnitude, but there is strong evidence that many tropical soils are saturated in their potential to bind fine soil carbon, and hence the soil may not be a significant carbon sink (Telles *et al.*, 2003).

A tropical forest typically holds 150–250 Mg (C) ha^{-1} in live biomass. Hence a biomass carbon sink of 0.25–0.75 Mg (C) ha^{-1} per year should become directly measurable after about a decade. In recent years, we have attempted to detect directly any net increase in the biomass of old-growth Amazonian forests. There are complications, however. Most mature forests have their own dynamic of local tree death (through storms, disease, fires, etc.) and subsequent regrowth, and it is necessary to measure over a sufficiently wide area and/or sufficient temporal duration to ensure adequate sampling of these events. Secondly, how do we actually measure plant biomass? For simple forests comprising a few tree species, it is usually sufficient to harvest a few trees, determine the relationship between tree diameter, tree height, wood density, and wood volume for each species, and then rely on tree diameter and height measurements to estimate the above ground biomass. If a few trees are completely excavated out of the soil, the relationship between tree diameter and root biomass can also be used to estimate below ground biomass. In tropical forests, which typically host 200–300 different tree species in each hectare, the situation becomes more complicated, and usually an average (species-independent) relationship between tree diameter and tree biomass is used. Different relationships can be used for different functional groups (e.g., palms, gap colonizers, understorey species, canopy species).

Tracking changes in soil carbon content can be more difficult, because the distribution of carbon within the soil depends on previous positions of trees, and is therefore extremely patchy. This spatial heterogeneity makes it difficult to monitor long-term trends without a very intensive sampling network. In tropical regions the data do not yet exist to compile a systematic inventory across the whole landscape. Instead, our work has focused on the above-ground biomass of mature, undisturbed tropical forests. These are ecosystems that are expected, on average, to be in carbon balance in a stable atmosphere, and therefore provide a test bed for looking for the effects of global atmospheric change. Phillips *et al.* (1998) initially compiled a database of mature tropical forest sites, mainly in Amazonia, that had been inventoried at least twice, and looked for evidence of changes in biomass. They concluded that there was large variability between plots, because of the natural dynamics of tree death and regrowth, but that in sum, the forest plots did appear to be accumulating carbon. This was tantalizing evidence that these pristine natural ecosystems were responding to global climate change. Phillips *et al.* (1998) estimated that South American tropical forests were accumulating carbon at a mean rate of 0.7 ± 0.2 Mg (C) ha^{-1} per year, implying a total Neotropical live biomass carbon sink of 0.6 ± 0.3 Pg (C) per year (Phillips *et al.*, 1998). Subsequently, there has been some debate over the methodological uncertainties of measuring biomass change in tropical forests, with particular focus on the bias introduced by incorrect measurement of buttressed trees (Clark, 2002; Phillips *et al.*, 2002a).

Since this initial study, we have been working on building a more extensive tropical forest dataset, and on improving and standardizing data collection and statistical analysis. We have developed a project RAINFOR (the Amazon Forest Inventory Network: www.geog.leeds.ac.uk/projects/rainfor/), which involves revisiting forest plots in key locations across Amazonia, remeasuring all trees to a standard methodology, and analysing soil and leaf nutrient and carbon content to build up a standardized database (Malhi *et al.*, 2002b). Our hope is that these study sites will become reference sites that will be visited regularly in the future for monitoring how 21st Century atmosphere change is affecting tropical ecosystems.

In the latest RAINFOR analysis, Baker *et al.* (2004) carefully re-examined methodological issues in an Amazon forest plot dataset, eliminating any sites over which there was methodological uncertainty. Their results confirmed the findings of Phillips *et al.* (1998): out of 59 plots included in the analysis, the mean biomass increment was 0.61 ± 0.22 Mg (C) ha^{-1} per year. Interestingly, the biomass carbon sink was highest in western Amazonia (mean 0.51 ± 0.25 Mg (C) ha^{-1} per year), and only marginally significant in eastern Amazonia (mean 0.35 ± 0.32 Mg (C) ha^{-1} per year). One of the most fascinating discoveries has been that forests in western Amazonia are two to three times more productive than forests in the east (Malhi *et al.*, 2004). Both productivity and sink strength appear related to soil fertility rather than rainfall: the seasonally dry but relatively fertile sites of southern Peru seem to be increasing in biomass at a rate similar to the aseasonal forests of northern Peru and Ecuador.

The latest RAINFOR results confirm the apparent increase in tree biomass in old-growth forests. Baker *et al.* (2004) estimate a mean above-ground carbon sink of 0.61 ± 0.22 Pg (C) ha^{-1} per year. Following Phillips *et al.* (1998), we can add a 10% correction for small trees and lianas, and a 33% correction for below-ground biomass, leading to an estimated total biomass carbon sink of 0.89 ± 0.32 Mg (C) ha^{-1} per year. Multiplying this by the FAO estimate of Neotropical moist forest area (5 987 000 km^2), we can estimate a total Neotropical moist forest biomass sink of 0.54 ± 0.19 Pg (C) per year (a scaling-up taking into account soil type would increase this value to about 0.62 ± 0.18 Pg (C) per year (Malhi and Phillips, 2004)). We make no assumptions about biomass sinks in soil or coarse woody debris, although these pools would be expected to increase slightly at least. Finally, if the as yet uncompiled data from the African and Asia tropics (which account for 50% of global tropical moist forest area) were to show a similar trend to Amazonia, the global moist tropical live biomass sink would be about 1.2 ± 0.4 Pg (C) per year.

There are still some methodological issues to be resolved, however. Is there a bias in where tropical foresters choose to locate their forest plots? How large a sample is required to ensure that natural forest dynamics are adequately covered? Can changes in soil carbon be measured? Will the mean wood density of these changing tropical forests remain the same, or will it decrease? These are issues that we plan to tackle over the coming years.

5. Pulling it all together

We can return to *Figure 1* and see where these new data lead us. The new, reduced estimates of carbon emissions from tropical deforestation (DeFries *et al.*, 2002; Achard *et al.*, 2004: 1.0 ± 0.4 Pg (C) per year) imply, as a lower limit, a reduced tropical carbon sink of 1.0 ± 1.2 Pg (C) per year (if the net balance of the tropics were unaltered). We looked to forest biomass data to place further constraints on this figure, and arrive at a very similar (but more tightly constrained) value of 1.2 ± 0.4 Pg (C) per year. This value makes untested guesses about the Asian and African tropics, but at the same time does not include other tropical biomes such as seasonally dry forests, savannahs, montane forests, and mangroves, and also does not include any possible sinks in dead wood or soil carbon. There are clearly further aspects to be explored, but for the moment it appears that the biomass changes we are observing on the ground:

(i)   are consistent with what would be expected from consideration of the global carbon cycle;
(ii) suggest a net tropical carbon sink of -0.3 ± 0.6 Pg (C) per year (i.e., the tropics are close to carbon balance or a modest sink);
(iii) suggest a net rate of stimulation of productivity of the order of 0.3% per year, consistent with CO_2 fertilization results from laboratory experiments, but at the upper end of what is expected from mature natural ecosystems.

6. Implications of a biosphere carbon sink for biodiversity

As outlined above, there is now clear evidence that terrestrial ecosystems are accumulating carbon from the atmosphere, and thereby are performing a global service by slowing the projected rate of climate change. However, it is unclear whether this is necessarily beneficial for the ecosystems themselves. Superficially, it may seem that a larger supply of a limiting factor (CO_2) may be a good thing, but in fact many of the richest and most diverse ecosystems, such as coral reefs or tropical forests, thrive in nutrient-poor habitats. A paucity of nutrients can encourage evolutionary innovation and diversity in predation strategies, and a variety of strategies to access sufficient nutrients. As nutrient supply increases, it is possible that a few species will be poised to exploit this new abundance, increasingly dominating over other species. It is interesting to speculate whether rising atmospheric CO_2 may lead to such a phenomenon. In tropical forests, for example, it is possible that fast-growing plant types, such as lianas, or trees that fix their own atmospheric nitrogen, or trees that exploit gaps in the forest canopy, may be better positioned to exploit rising CO_2. One of the aims of our RAINFOR project is to look for evidence that the composition of undisturbed tropical forests is shifting. This is a field that has barely been explored, but the early results are tantalizing. For example, our RAINFOR results appear to show that in Amazonia, lianas may have doubled in abundance in the past ten years (Phillips *et al.*, 2002b). This result is backed up with a recent study demonstrating an increase in liana leaf production at the 50 ha forest plot at Barro Colorado Island in Panama (Wright *et al.*, 2004), and by laboratory studies suggesting that lianas may benefit disproportionately from CO_2 fertilization (Granados and Körner, 2002). As the weight of lianas has a strong influence on the likelihood of tree death, this may have significant effects on forest dynamics and forest structure, possibly even leading to an eventual decline in forest biomass. There are likely to be many other shifts in forest composition that have simply not yet been looked for. The first detailed study of trends in forest composition in mature tropical forests (Laurance *et al.*, 2004) revealed that canopy trees appeared to be increasing in abundance and growth rate at the expense of slow-growing understorey trees, but that there was no increase in abundance of pioneer species. Further signals of global change are probably sitting there in the Amazonian forests. We simply have to look for them and know how to interpret them.

7. Future prospects

The evidence outlined above suggests that terrestrial biomes, and tropical forests in particular, are already playing a clear role in slowing the rate of atmospheric and climatic change. This has led to the debate whether management of the terrestrial biosphere could be employed as a tool to further slow atmospheric change. This management could take the form of enhanced reforestation, reduced forest degradation through logging, slowed tropical deforestation, and sequestration of carbon in soils through 'no-till' agricultural practices (Royal Society, 2001). These options would also have several beneficial environmental side-effects, such as protection of biodiversity, watershed protection, and reduced soil erosion.

One of the most immediate options is that of locking up carbon in tropical forests. Biosphere carbon management options could include: (i) the prevention of deforestation; (ii) the reduction of carbon loss from forests by changing harvesting regimes, converting from conventional to reduced impact logging, and controlling other anthropogenic disturbances such as fire and pest outbreaks; (iii) reforestation/ afforestation of abandoned or degraded lands; (iv) sequestration in agricultural soils through change is tilling practices; and (v) the increased use of biofuels to replace fossil fuel combustion.

How much potential does the carbon biosphere management option have? As an upper limit, we note that across all biomes worldwide, by the end of 2000 about 190 Pg (C) had been released from the biosphere by human activity (Malhi *et al.*, 2002a). Thus, if *all* agricultural and degraded lands worldwide were reverted back to original vegetation cover (an extremely unlikely scenario), a similar amount of carbon would be sequestered.

Slightly more realistically, (Kauppi *et al.*, 2001) confirm previous estimates (Brown *et al.*, 1996) that the potential avoidance and removal of carbon emissions that could be achieved through the implementation of an aggressive programme of changing forestry practices over the next 50 years is about 60–87 Pg (C). About 80% of this amount would be achieved in the tropics. Changes in agricultural practices could result in a further carbon sink of 20–30 Pg over the same period (Cole *et al.*, 1996), resulting in a maximum land management carbon sink of 80–120 Pg, and a mean annual sink of about 2 Pg (C) per year.

It is important to emphasize that here we are discussing an *additional, deliberately planned land carbon sink* that would complement the 'natural' sink mentioned above. The two sinks are sometimes confused in discussion. For example, although the natural carbon sink may reduce with climatic warming and possibly become a net source, intact, well-managed forests will almost always contain more carbon than degraded forests or agricultural lands. Therefore improved, carbon-focused, forest management will almost always result in net carbon sequestration.

How does this land management sink potential compare with expected carbon emissions over the 21st Century? The IPCC 'business as usual' scenario suggests that about 1400 Pg (C) would be emitted by fossil fuel combustion and land-use change over the 21st Century. The more detailed Special Report on Emissions Scenarios (SRES) (IPCC 2000) suggests that, without conscious environment-based decision-making, emissions will total between 1800 Pg (C) (scenario A2, a regionalized world) and 2100 Pg (C) (scenario A1F, a fossil-fuel-intensive globalized world) over the 21st Century. With more environmentally focused policies, total emissions are expected to vary between 800 and 1100 Pg (C). Whatever the details of the scenario, it is clear that even an extensive land carbon sink program could only offset a fraction of likely anthropogenic carbon emissions over the coming century. Fossil fuel emissions alone (not considering further emissions from ongoing tropical deforestation) over the 21st Century may exceed by five to ten times even the maximum possible human-induced forest carbon sink. Using an approximate conversion factor (3 Pg (C) emissions = 1 p.p.m. atmospheric CO_2 = 0.01 °C temperature rise (Malhi *et al.*, 2002a)) *a managed land-use sink of 100 Pg (C) over the 21st Century would reduce projected CO_2 concentrations in 2100 by about 33 p.p.m., and reduce the projected global mean temperature increase by 0.3 °C*: a modest but significant effect.

Figure 3 illustrates the impact of the future of tropical forests on scenarios of temperature increase by the end of the 21st Century. The mid-range of IPCC scenarios suggests an anthropogenic global warming of 3 °C by the end of the century (range 1.5–5.5 °C). This assumes that the terrestrial and oceanic carbon sinks continue to operate as they do today (i.e., the proportion of anthropogenic CO_2 that remains in the atmosphere is fairly fixed). However, a sink of 1 Pg (C) per year over a century would mean a substantial (larger than 50%) increase in tropical forest biomass by the end of the century, and it is very likely that other structural considerations would limit this sink. If we now assume that a current tropical carbon sink is immediately switched off, and that some of this 'extra' CO_2 would be absorbed by the oceans, temperatures at the end of this century would be 0.5 °C higher. Hence the presence of the tropical carbon sink has a significant but not overwhelming influence on global climate. Furthermore, if we assume that 75% of present-day global tropical forest biomass is released to the atmosphere, through either drought induced by climate change or direct deforestation, global temperature rises by a further 0.5 °C. This calculation only deals with carbon, and does not take into account the influence of tropical forests on the global hydrological and energy transfer cycles, which also substantially affect climate. The tropical forests may be slightly mitigating the effects of climate change, but they are far from 'saving us' from the consequences of anthropogenic climate change.

Absorbing carbon in trees clearly cannot 'solve' the global warming problem on its own. Where forest carbon absorption can be effective, however, is in being a significant component in a package of CO_2 mitigation strategies, and providing an immediate carbon sink while other mitigation technologies are developed. Carbon absorbed early in the century has a larger effect on reducing end-of-century temperatures than carbon absorbed late in the century. The immediate potential of forestry is illustrated in *Figure 4*. Suppose that a global carbon emissions goal for the 21st Century is to move our emissions pathway from the IPCC 'business-as-usual' scenario (IS92A) to a low-emissions scenario (SRES scenario B1), as illustrated in *Figure 4a*. The required carbon offset is the difference between these two emission curves. Now let us suppose that an ambitious forest carbon sink programme can be implemented immediately, aiming to absorb 75 Pg (C) by 2050, at a uniform rate of 1.5 Pg (C) per year. *Figure 4b* illustrates the proportional contribution that such a carbon

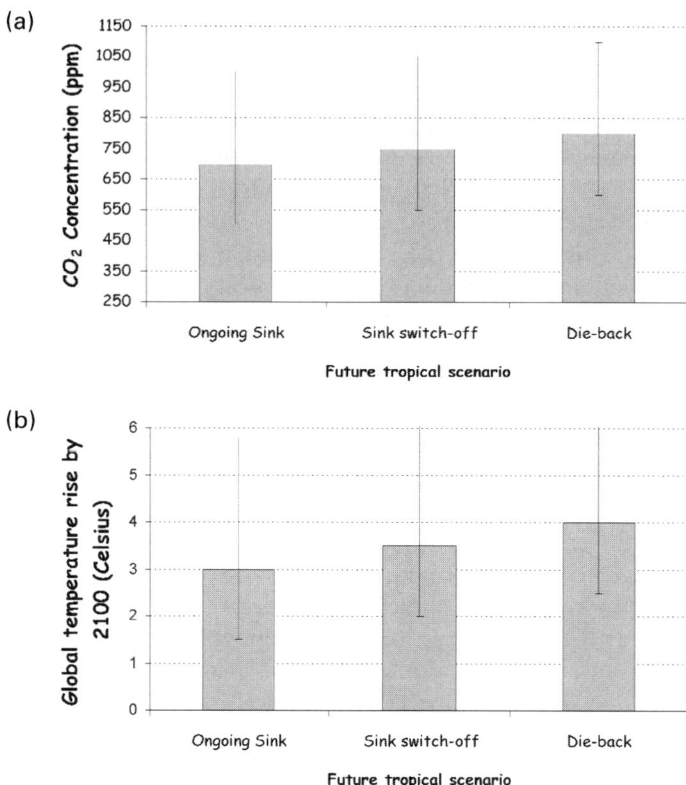

Figure 3. *An estimate of the impact of the future of tropical forests on (a) CO_2 concentrations, and (b) mean global surface temperature rise by the end of the 21st Century. Error bars indicate the range between IPCC scenarios.*

sink could make to the required total carbon offset. *For the next decade, such a land carbon sink would on its own be sufficient to move us onto the low emissions pathway, and for the subsequent two decades it could provide about half of the required carbon offsets.* Thus even a less ambitious carbon offset program could play a significant role over the next few decades. As the Century progresses and the magnitude of the required carbon reductions increases, the relative effect of forest carbon sinks declines. A forest carbon offset program implemented in 2050 would only be able to produce between 10 and 30% of the required offsets. Thus, to be relevant, a forest carbon sequestration program has to absorb most of its carbon within the next few decades. Tropical ecosystems have the highest productivities, and are therefore likely to be the most effective sinks on this short timescale. In addition, improved protection and management of tropical forests can bring numerous other benefits, including protection of the harbours of half of terrestrial biodiversity, watershed protection, and reduction in soil erosion.

In conclusion, the managed absorption of carbon in forests has the potential to play a significant role in any carbon emissions reduction strategy over the next few decades. Such a strategy can be viewed as partly undoing the negative effects of previous centuries of forest clearance, both in climatic and biological terms. The relative potential contribution of a forest carbon sink declines later in the century, and therefore forest carbon absorption cannot be viewed as a long-term solution to the global warming problem. It can only be a useful stopgap. As a stopgap, however, it is essential that carbon sinks are not allowed to divert resources and attention from required developments and changes in technology, energy use, and energy supply, the only developments that can provide a long-term solution to the great carbon disruption.

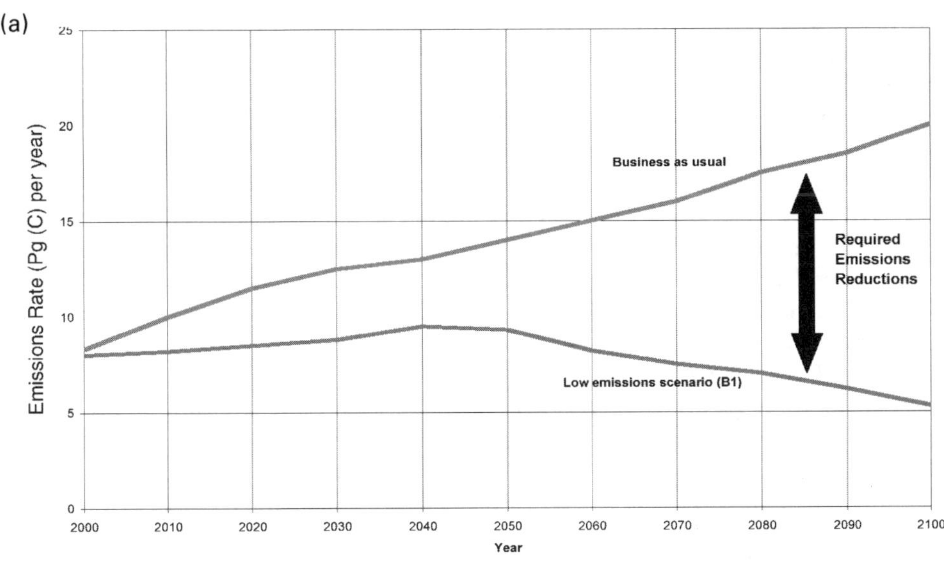

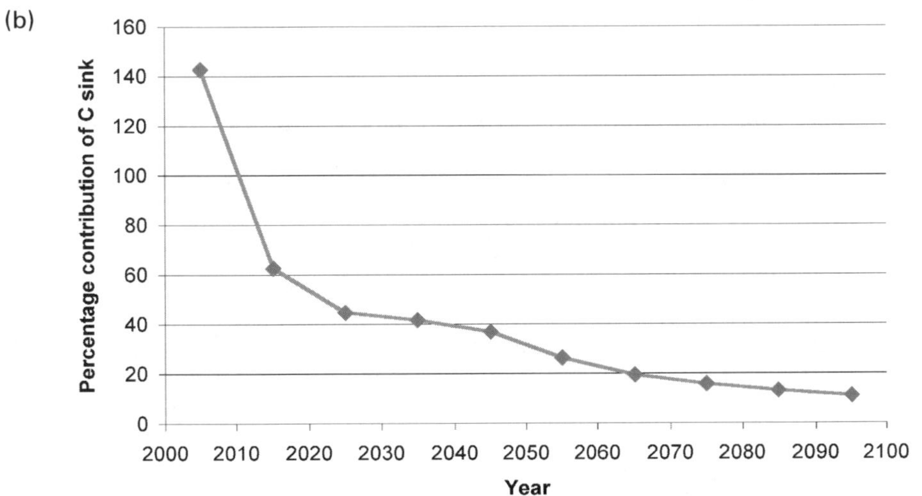

Figure 4. (a) Projected anthropogenic global carbon emissions goal for the 21st Century for a shift from the IPCC 'business-as-usual' scenario (IS92A) to a low-emissions scenario (SRES scenario B1). (b) Proportional contribution that a land-use carbon sink at a uniform rate of 1.5 Pg (C) per year could make to the required total carbon offset.

Acknowledgements

I thank Oliver Phillips, Tim Baker, Simon Lewis, Jon Lloyd, and other members of the RAINFOR consortium for their continued collaboration in determining the carbon sink of old-growth tropical forests, and acknowledge the generous support of a Royal Society University Research Fellowship.

References

Achard, F., Eva, H.D., Stibig, H.J., Mayaux, P., Gallego, J., Richards, T. and Malingreau, J.P. (2002) Determination of deforestation rates of the world's humid tropical forests. *Science* **297**: 999–1002.

Achard, F., Eva, H.D., Mayaux, P., Stibig, H.J. and Belward, A. (2004) Improved estimates of net carbon emissions from land cover change in the tropics for the 1990s. *Global Biogeochemical Cycles* **18**: article number GB2008.

Baker, T.R., Phillips, O.L., Malhi, Y., Almeida, S., Arroyo, L., DiFiore, A. *et al.* (2004) Increasing biomass in Amazonian forest plots. *Philosophical Transactions of the Royal Society of London Series B – Biological Sciences* **359**: 353–365.

Brown, S., Sathaye, J., Cannell, M. and Kauppi, P. (1996) Management of forests for mitigation of greenhouse gas emissions. In: Watson, R.T. Zinyowera, M.C. and Moss, R.H. (eds) *IPCC Climate Change 1995 – Impacts, Adaptations and Mitigation of Climate Change*, pp. 773–797. Cambridge University Press, Cambridge, UK.

Chambers, J.Q. and Silver, W.L. (2004) Some aspects of ecophysiological and biogeochemical responses of tropical forests to atmospheric change. *Philosophical Transactions of the Royal Society of London Series B – Biological Sciences* **359**: 463–476.

Clark, D.A. (2002) Are tropical forests an important carbon sink? Reanalysis of the long-term plot data. *Ecological Applications* **12**: 3–7.

Cole, C.V., Cerri, C., Minami, K., Mosier, N., Rosenberg, N. *et al.* (1996) Agricultural options for mitigation of greenhouse gas emissions. In: Watson, R.T., Zinyowera, M.C. and Moss, R.H. (eds) *IPCC Climate Change 1995 – Impacts, Adaptations and Mitigation of Climate Change*, pp. 745–771. Cambridge University Press, Cambridge, UK.

DeFries, R.S., Houghton, R.A., Hansen, M.C., Field, C.B., Skole, D. and Townshend, J. (2002) Carbon emissions from tropical deforestation and regrowth based on satellite observations for the 1980s and 1990s. *Proceedings of the National Academy of Sciences of the United States of America* **99**: 14256–14261.

FAO (2000) *Forest resources assessment 2000*. FAO Forestry papers 140. Rome, FAO.

Fearnside, P.M., Graca, P., Leal, N., Rodrigues, F.J.A. and Robinson, J.M. (1999) Tropical forest burning in Brazilian Amazonia: measurement of biomass loading, burning efficiency and charcoal formation at Altamira, Para. *Forest Ecology and Management* **123**: 65–79.

Granados, J. and Körner, C. (2002) In deep shade, elevated CO_2 increases the vigor of tropical climbing plants. *Global Change Biology* **8**: 1109–1117.

Gu, L.H., Baldocchi, D., Verma, S.B., Black, T.A., Vesala, T., Falge, E.M. and Dowty, P.R. (2002) Advantages of diffuse radiation for terrestrial ecosystem productivity. *Journal of Geophysical Research – Atmospheres* **107(D5–D6)**: doi: 10.1029/2001 JD001242.

Gurney, K.R., Law, R.M., Denning, A.S., Rayner, P.J., Baker, D., Bousquet, L. *et al.* (2002) Towards robust regional estimates of CO_2 sources and sinks using atmospheric transport models. *Nature* **415**: 626–630.

Houghton, R.A. (1995) Land-use change and the carbon-cycle. *Global Change Biology* **1**: 275–287.

Houghton, R. A. (1996) Terrestrial sources and sinks of carbon inferred from terrestrial data. *Tellus Series B – Chemical and Physical Meteorology* **48**: 420–432.

Houghton, R.A. (1999) The annual net flux of carbon to the atmosphere from changes in land use 1850–1990. *Tellus Series B – Chemical and Physical Meteorology* **51**: 298–313.

Houghton, R.A. (2003) Revised estimates of the annual net flux of carbon to the atmosphere from changes in land use and land management 1850–2000. *Tellus Series B – Chemical and Physical Meteorology* **55**: 378–390.

Houghton, R.A., Lawrence, K.T., Hackler, J.L. and Brown, S. (2001) The spatial distribution of forest biomass in the Brazilian Amazon: a comparison of estimates. *Global Change Biology* **7**(7): 731–746.

IPCC (2001) *Climate Change 2001: The Scientific Basis.* Cambridge University Press, Cambridge, UK.

Kauppi, P.S., Sedjo, R., Apps, M., Cerri, C., Fujimori, T., Janzen, H. *et al.* (2001) Technological and economic potential of options to enhance, maintain, and manage biological carbon reservoirs and geo-engineering. In: *Climate Change 2001: Mitigation*, pp. 302–343. IPCC. Cambridge University Press, Cambridge, UK.

Körner, C. (2004) Through enhanced tree dynamics carbon dioxide enrichment may cause tropical forests to lose carbon. *Philosophical Transactions of the Royal Society of London Series B – Biological Sciences* **359**: 493–498.

Laurance, W.F., Oliveira, A.A., Laurance, S.G., Condit, R., Nascimento, H.E.M., Sanchez-Thorin, A.C. *et al.* (2004) Pervasive alteration of tree communities in undisturbed Amazonian forests. *Nature* **428**: 171–175.

Lewis, S.L., Malhi, Y. and Phillips, O.L. (2004) Fingerprinting the impacts of global change on tropical forests. *Philosophical Transactions of the Royal Society of London Series B – Biological Sciences* **359**: 437–462.

Malhi, Y. and Grace, J. (2000) Tropical forests and atmospheric carbon dioxide. *Trends in Ecology and Evolution* **15**: 332–337.

Malhi, Y. and Wright, J. (2004) Spatial patterns and recent trends in the climate of tropical rainforest regions. *Philosophical Transactions of the Royal Society of London Series B – Biological Sciences* **359**: 311–329.

Malhi, Y., Meir, P. and Brown, S. (2002a) Forests, carbon and global climate. *Philosophical Transactions of the Royal Society of London Series A – Mathematical, Physical and Engineering Sciences* **360**: 1567–1591.

Malhi, Y., Phillips, O.L., Lloyd, J., Baker, T., Wright, J., Almeida, S. *et al.* (2002b) An international network to monitor the structure, composition and dynamics of Amazonian forests (RAINFOR). *Journal of Vegetation Science* **13**: 439–450.

Malhi, Y., Baker, T.R., Phillips, D.L., Almeida, S., Alvareg, E., Arroyo, J. *et al.* (2004) The above-ground coarse wood productivity of 104 Neotropical forest plots. *Global Change Biology* **10**: 563–591.

Malhi, Y., Phillips, O.L. (2004) Tropical forests and global atmospheric change: a synthesis. *Philosophical Transactions of The Royal Society of London Series B-Biological Sciences* **359**: 549–555.

Nepstad, D.C., Verissimo, A., Alencar, A., Nobre, C., Lima, E., Lefebvre, P. *et al.* (1999) Large-scale impoverishment of Amazonian forests by logging and fire. *Nature* **398**: 505–508.

Page, S.E., Siegert, F., Rieley, J.O., Boehm, H.D.V., Jaya, A. and Limin, S. (2002) The amount of carbon released from peat and forest fires in Indonesia during 1997. *Nature* **420**: 61–65.

Perlin, J. (1999) *A forest journey: the role of wood in the development of civilization.* Harvard University Press, Cambridge, MA.

Phillips, O.L., Malhi, Y., Higuchi, N., Laurance, W.F., Nunez, P.V., Vasquez, R.M. *et al.* (1998) Changes in the carbon balance of tropical forests: Evidence from long-term plots. *Science* **282(5388)**: 439–442.

Phillips, O.L., Malhi, Y., Vinceti, B., Baker, T., Lewis, S.L., Higuchi, N. *et al.* (2002a) Changes in growth of tropical forests: Evaluating potential biases. *Ecological Applications* **12**: 576–587.

Phillips, O.L., Martinez, R.V., Arroyo, L., Baker, T.R., Killeen, T., Lewis, S.L. *et al.* (2002b) Increasing dominance of large lianas in Amazonian forests. *Nature* **418**: 770–774.

Rödenbeck, C., Houweling, S., Gloor, M. and Heimann, M. (2003) CO_2 flux history 1982–2001 inferred from atmospheric data using a global inversion of atmospheric transport. *Atmospheric Chemistry and Physics* **3**: 1919–1964.

Royal Society (2001) The role of land carbon sinks in mitigating global climate change. The Royal Society, London: 27 pages.

Steininger, M.K., Tucker, C.J., Townshend, J.R.G., Killeen, T.J., Desch, A., Bell, V. and Ersts, P. (2001) Tropical deforestation in the Bolivian Amazon. *Environmental Conservation* **28**: 127–134.

Taylor, J.A. and Lloyd. J. (1992) Sources and sinks of atmospheric CO_2. *Australian Journal of Botany* **40**: 407–418.

Telles, E.D.C., de Camargo, P.B., Martinelli, L.A., Trumbore, S.E., da Costa, E.S., Santos, J., Higuchi, N. and Oliveira, R.C. (2003) Influence of soil texture on carbon dynamics and storage potential in tropical forest soils of Amazonia. *Global Biogeochemical Cycles* **17**: article number 1040.

Whitmore T.M., T.I.B.L., Johnson, D.L., Kates, R.W. and Gottschang, T.R. (1990) Long-term population change. In: Turner II, W.C.C.B.L., Kates, R.W., Richards, J.F., Mathews, J.T. and Meyer, W.B. (eds) *The Earth as Transformed by Human Action*, pp. 25–39. Cambridge University Press, Cambridge, UK.

Williams, M. (1990) Forests. In: Turner II, W.C.C.B.L., Kates, R.W., Richards, J.F., Mathews, J.T. and Meyer, W.B. (eds) *The Earth as Transformed by Human Action*, pp. 179–201. Cambridge University Press, Cambridge, UK.

Willis, K.J., Gillson, L. and Brncic, T.M. (2004) How virgin is virgin rainforest? *Science* **304**: 402–403.

Wood, D.M., Malhi, Y., Baker, T.A. and Phillips, O.L. (2004) Regional variation in the biomass of Amazonian forests, *in review.*

Wright, S.J., Calderon, O., Hernandez, A. and Paton, S. (2004) Are lianas increasing in importance in tropical forests? A 17–year record from Pannma. *Ecology* **85**: 484–489.

11

The carbon balance of forest soils: detectability of changes in soil carbon stocks in temperate and Boreal forests

Frauz Conen, Argyro Zerva, Dominique Arrouays, Claude Jolivet, Paul G. Jarvis, John Grace and Maurizio Mencuccini

1. Introduction to soil carbon stocks

Vegetation sits at the interface between the global atmosphere and the soil, where it effectively transfers carbon as organic detritus from the stock in the atmosphere to the stock in the soil. Forests in particular bring about this transfer in a very effective manner, with the result that large amounts of carbon have accumulated in the soil within many forests over thousands of years. Globally, there is about three times as much carbon in soils as in vegetation, with the largest proportions in the northern temperate and Boreal forests (Schlesinger, 1997; Watson *et al.*, 2000). Thus, changes in the stocks of soil carbon, as a result of climate, land-use changes and management practices, in particular, may be very important components of local and regional carbon budgets.

The increase in atmospheric CO_2, and the likely increase in global temperatures that will result, is expected to alter the distribution of carbon between atmosphere, vegetation, and soils (Watson *et al.*, 2000). However, both the direction and the magnitude of the possible changes in soil carbon stocks are still

The Carbon Balance of Forest Biomes, edited by H. Griffiths and P.G. Jarvis.
© 2005 Taylor & Francis Group

being discussed, because of their dependence on several inputs and outputs. For example, rising atmospheric CO_2 concentration and increased nitrogen availability are expected to enhance CO_2 uptake by trees from the atmosphere and lead to enhanced growth and detritus production. By contrast, it is commonly supposed that an increase in temperature will lead to an increase in rates of respiratory oxidation of soil organic matter and a reduction of the soil carbon stock. In neither case is the long-term acclimation of the processes involved sufficiently well understood or the feedbacks adequately appreciated for reliable projections to be made.

The responses of soil carbon pools to climate change can be viewed as the results of an experiment lacking a control treatment. Consequently it is important to establish a firm baseline against which later observations can be compared. Because climate change is expected to occur on a time-scale of several decades, repeated direct measurements of soil carbon stocks at intervals of several years seem a good option. The choice of methods for determining changes in soil carbon stocks is large (Post *et al.*, 2001) but repeated direct measurements are relatively cheap and applicable in all parts of the world where soil carbon stocks are currently in a quasi-steady-state. Where soils are not in a quasi-steady-state as a result of land-use change, the same method could be used to assess the combined effects of land-use and climate change. Well-established methods for measuring changes in soil carbon pools will be one of several tools used to monitor and verify changes in soil carbon stocks as part of the Kyoto Agreement and resulting initiatives such as carbon trading.

The aim of this paper is to assess the site-to-site variability in the detectability of changes in soil carbon stocks. We do this by looking at 24 datasets of soil carbon in temperate and Boreal forests, from either published studies or our own recent measurements. It is not our aim to derive optimal sampling strategies that minimize the sample size to determine either area-averaged stocks or stock changes over time. Rather, we want to summarize the available literature and attempt to derive general principles that could be of use elsewhere. To do this, we explore the form of the relationship between mean and variance across the studies published thus far and we discuss its implication for future studies. Also, we propose a generalized scaling relationship between reductions in area sampled and proportion of variance remaining to identify the degree of spatial dependency of forest soil carbon densities. Finally, to illustrate the consequences of our findings, we present the case of the simplest sampling strategy (i.e., simple random sampling), and we calculate the minimum detectable differences in soil carbon stocks across this sample of forest sites.

First, the study sites are described and the methodology presented, then the concept of minimum detectable difference is briefly introduced and explained. Finally, the major results are presented and discussed.

2. Background to sites compared in this study

2.1 *Perthshire, UK*

Samples were collected in Griffin Forest, on the north-facing slope of the Tay Valley near Aberfeldy in Perthshire, Scotland, at an elevation of about 340 m (56°36′ N, 3°48′ E). Annual precipitation averages 1200 mm and mean annual temperature is 8.2 °C. The formerly grazed heathland on a podsolic soil of a sandy loam texture was ploughed and planted with Sitka spruce (*Picea sitchensis* (Bong.) Carr.) in 1981. The ploughing regime used to establish the forest inverted the material from furrows onto adjacent strips of land to form elevated ridges, resulting in three distinct surface zones: furrows, ridges, and undisturbed land. Ridges occupied about 50% of the total area, furrows and undisturbed ground about 25% each. There was no understorey and only little vegetation on the forest floor of occasional patches of mosses and grasses. Soil cores of 5.7 cm diameter were taken in a stratified random design from a 0.85 ha (1 hectare (ha) = 10^4 m^2) plot. Sampling depth was determined by the lower boundary of the original A horizon. This boundary was identified by the sudden change from dark brown to bright pale colour. In the deepest parts of the furrows, where the A horizon had been completely removed by the plough, sampling depth was limited to the thickness of the L and O horizons formed since ploughing.

Samples were dried to constant mass at 60 °C and crushed in a mortar. Roots (larger than 1 mm) and the few stones (larger than 4 mm) were removed and the remaining sample was weighed, milled, and mixed. A sub-sample (approximately 1 g) was further ground and mixed in an agate mortar before being sub-sampled again (approximately 10 mg) for analysis in an elemental analyser (Carlo-Erba 1106).

2.2 *Northumberland, UK*

Samples were collected at several localities within Harwood Forest (55°10′ N, 2°3′ W), in Northumberland, England. Harwood forest mostly consists of even-aged stands of pure Sitka spruce (*Picea sitchensis* (Bong.) Carr.), although in poorly drained sites lodgepole pine (*Pinus contorta* Dougl. ex Loud.) is also present. The forest was created over several decades during the period between the two world wars, by ploughing and subsequent planting of trees on ericaceous moorland and upland pasture. The dominant soil type is peaty gley. The area rises from 200 m in the southeast to 400 m in the northwest. Average annual precipitation is 950 mm, mean annual temperature is 7.6 °C. Understorey and forest floor vegetation were generally sparse or absent in the closed-canopy stands; however, a dense cover of heather, rushes, and graminoids was present in the clearfelled and replanted areas.

In several of the age classes, surface zones associated with the original ploughing could no longer be identified (either because of time or because of subsequent preparation practices), so that random samples were taken within plots nested within stands (Anderson and McLean, 1974). Soil samples were taken from eleven stands of pure Sitka spruce of varying age: 40 years (first rotation) and 12, 20, and 30-years-old (second rotation), all located on peaty gley soils. Samples were also collected from one recent (3-year-old) clearfell and from one unplanted moorland plot outside the forest.

The entire area of the forest from which samples were taken was about 578 ha. In each stand (area varying between 40 and 60 ha) four or five plots were randomly selected and in each plot samples were taken from eight to ten points at randomly selected distances from the centre. The randomly selected distance was chosen such that the points would fall within a circle of 10 m radius around the center of the plot. The samples were taken with a soil auger of 5.7 cm diameter to a depth of 45 cm. Each sample was divided into L, O, and A layers for later analysis.

The samples were oven-dried at 105 °C for 24 h. Coarse fragments were removed by hand, and the soil was ground to pass a 0.5 mm mesh. Thirty per cent of all samples were analysed both by C/N analyser and by loss on ignition and three separate regressions were calculated for the litter, organic, and mineral layers (all $r^2 > 0.98$, $p < 0.001$). The carbon content of the other samples was measured by loss on ignition only, and the results converted to total carbon by using the previously calculated regression coefficients.

2.3 *Les Landes, France*

Soil samples were collected from a 9 ha mature forest (44°38′ N, 1°14′ W) of maritime pine (*Pinus pinaster* Ait.). The forest understorey vegetation is mainly composed of *Molinia caerulea* (L.) Moench., *Pteridium aquilinum* (L.) Kuhn, *Erica cinerea* L. and *Calluna vulgaris* (L.) Hull. This area receives about 900 mm per year of rain, and has a mean annual temperature of 12.7 °C. The land surface consists of a succession of dunes/interdunes, with the top of the dunes ranging from about 0.30–1.50 m in height and 10–50 m in width. Soils are hydromorphic sandy podsols, developed from Quaternary coarse sand, aeolian deposits. The vegetation, soil profile development, and organic matter storage are directly related to the microrelief and to seasonal variation of the superficial water-table level, with reference to the soil surface. Podsols with iron pans or rather cemented, enriched B-horizons are found in the well-drained upper parts of the ridges. More humus-rich podsols, with friable enriched B-horizons, sometimes overlaying a hydromorphic Cg-horizon, occur in the lower areas and poorly drained situations.

A stratified sampling design was applied. The forest was divided into 49 square subplots (strata). Half of the samples were collected along the lines of trees, the other half between the lines. The location of the sampling point was randomly selected, on the condition that at least one sample was located in each subplot. Samples were collected from 60 randomly selected sampling points. The O horizon samples were collected separately. Then, samples of the organo-mineral topsoil layer were collected down to depths of 0–0.4 m, 0.4–0.6 m, 0.6–0.8 m, and 0.8–1 m using a core sampler of 0.2 m inner diameter, to give samples of known volume for bulk density determination.

Bulk samples were oven dried to constant mass at 105 °C and a sub-sample was finely crushed (less than 50 µm), to obtain reliable homogeneity before organic carbon and nitrogen microanalysis. Organic carbon and nitrogen contents were determined by dry combustion using an automated carbon-nitrogen-sulphur elemental analyser (Fisons). Bulk densities were determined after oven-drying to a constant mass at 105 °C.

To estimate scales of spatial dependence, a classical estimator of the semivariogram, i.e., the spatial semivariance with lag h, was calculated:

$$\hat{\gamma}(h) = 1/2N(h) \sum_{i=1}^{N(h)} [z(x_i) - z(x_i + h)]^2 \qquad (1)$$

where z is a regionalized variable, $z(x_i)$ and $z(x_i + h)$ are measured samples at points x_i and $x_i + h$, and $N(h)$ is several pairs separated by distance or lag h.

A spherical model describing the experimental semivariograms was fitted through a weighted least squares procedure. As recommended by Webster and Oliver (1990), the weighting used was the number of couples, $N(h)$.

3. Previous studies on soil carbon content

Growing interest in soil carbon as a likely variable in climate change has resulted in several published studies on the estimation of soil organic carbon content and its variability (e.g., Conant and Paustian, 2002; Conant et al., 2003; Fernandez et al., 1993; Homann et al., 2001; Huntington et al., 1988; Liski, 1995) or detailed descriptions of soil carbon stocks (Garten et al., 1999; Weber, 1999). Depending on the aim of the study, sampled areas have ranged from 10 m² (Liski, 1995) to at least 126 ha, with sampling points as far as 8 km apart (Homann et al., 2001). Only one study (Palmer et al., 2002) appears to have reported data on the variability in carbon stocks for several plots across an entire state (30 plots across Georgia, USA). In most studies the forest floor material (litter layer and O horizon) and several underlying layers within the upper 30 cm of mineral soil have been analysed separately. Liski (1995) sampled to 40 cm, Huntington et al. (1988) to 54 cm, and Fernandez et al. (1993) to about 70 cm depth. Samples were taken with corers of cross-sections varying from 4.5 cm² (Garten et al., 1999) to 150 cm² (Homann et al., 2001) or excavated from 0.5 m² pits (Fernandez et al., 1993; Huntington et al., 1988). In the study using the smallest diameter corer, composite samples were made from two to four cores (Garten et al., 1999). Sample numbers in each study ranged from 18 (Garten et al., 1999; Weber, 1999) to 271 (Homann et al., 2001) (*Table 3*). All these sites were in old-growth forests except for one site studied by Conant et al. (2003), which was in a 40-year-old second rotation forest, planted after harvesting of the previous old-growth stand.

4. Estimation of the mean and variance in soil carbon stocks across studies

As explained above, available data covered both our own field sites in Europe (two in the UK and one in France) and data obtained from the literature. No two examples are directly comparable to each other as they all employed different methods. However, all the examples had in common that soil carbon concentration (per cent carbon) and soil mass per unit area (kilograms per square metre) had been determined on the same samples used for calculating soil carbon content per unit area (kilograms of carbon per square metre) for each sampling point. The more common practice of determining the two variables separately was deemed unsuitable for this type of investigation because the product of two averages is not necessarily the same as the average of the individual products, as will be shown later on.

The most important differences across studies were in the size of the sampled area, sampling depth, area covered by each sampling point, sampling design, and sample size. The magnitude of the area sampled was potentially the most important difference, as estimated variance should increase when larger areas are sampled. To correct for the differences across studies in the area sampled, we derived an empirical relationship between these two variables to enable us to scale the estimated variances to a common sampled area of one hectare.

To build this relationship, we identified those studies where estimates of variance were available at least at two different scales, i.e., where studies had deliberately been done with a 'nested' design (e.g., Conant and Paustian, 2002; Conant *et al.*, 2003; Homann *et al.*, 2001; Palmer *et al.*, 2002; Northumberland, this study) or studies where an explicit geostatistical approach had been followed and estimates of variance derived in relation to distance (e.g., Liski, 1995; Les Landes, this study). For each of these studies, we calculated the ratios of plot sizes and variances of the smaller plots relative to the largest areas given (e.g., stands, forests, region), whereby a variance of 1 and a plot size of 1 were attributed to the largest spatial scale sampled and proportionally lower values of variance and area were attributed to the lower spatial scales. For Palmer *et al.* (2002), we calculated plot areas based on the given distances for which variances were indicated (7.85×10^{-5}, 1, and 1 200 000 ha). By taking relative measures of size and variance, we postulated that a general relationship could be found across all studies, indicating the generality of the scaling of variance with distance across several orders of magnitude.

Variance estimates may also be affected by choice of the sampling design (see, for example, Papritz and Webster, 1995a, b). This is far more difficult to assess and account for. When we reviewed the available literature data, we assumed that the estimates of population variance were directly comparable. Finally, sample size can also affect estimates of variance. For instance, Conant and Paustian (2002) showed that a different allocation of a fixed sample size between plots and sub-plots resulted in different estimates of within-plot *versus* among-plots variance. We assessed the impact of initial sample size by regressing it, alone or in combination with other variables, against the corresponding estimates of variance. As mentioned already, our intention was not to determine the absolute sample sizes required at specific sites to achieve a specified aim, but rather to investigate the nature and magnitude of the variability reported for several studies across temperate and Boreal forests. The design of optimal sampling schemes, and the consequences of this choice for sample size, was outside the scope of this study.

5. Estimation of minimum detectable difference

We define the minimum detectable difference, δ, as the statistically significant difference between two estimates of mean soil carbon content at the same site on two different occasions (μ_{t1} and μ_{t2}). As significance limits, we set the probability for falsely rejecting the null hypothesis at 5% ($\alpha = 0.05$). The probability for falsely accepting the null hypothesis, we set at 10% ($\beta = 0.10$, i.e., the statistical power = 0.9). Assuming simple random sampling, δ can then be estimated in a one-sample t-test (Zar, 1999) from the following:

$$\delta = \sqrt{\frac{s^2}{n}}\,(t_{\alpha,\upsilon} + t_{\beta,\upsilon}) \tag{2}$$

where s^2 is an estimate of the population variance (σ^2), n is the sample size ($n_{t1} = n_{t2} = n$), $\sqrt{s^2/n}$ is the estimated standard error of the mean, and $t_{\alpha,v}$ and $t_{\beta,v}$ are the critical t-values for the specified values of α and β with $n - 1$ degrees of freedom (v). The estimate of δ assumes a normal distribution, that repeated sampling is performed by the same method (simple random sampling), and that the variance is equal on successive occasions.

For practical purposes, we may want to test the hypotheses that soil carbon either increases over time ($\mu_{t1} < \mu_{t2}$, for example soil carbon gain in recently afforested sites) or that it decreases over time ($\mu_{t1} > \mu_{t2}$, for example soil carbon loss from old growth forests). In both cases we should apply a one-tailed t-test. However, in the case of the effects of climate change on soil carbon stocks, the direction of the prediction is more uncertain and both increases and decreases have been suggested as possible trends. In this case, a two-tailed t-test would be more appropriate. Our calculated sample sizes are reported based on a one-tailed t-test. Estimates for two-tailed t-tests will obviously be larger and can easily be obtained from them.

If we have an estimate of the population variance, we can estimate δ as a function of sample size or, alternatively, estimate the sample size required to achieve a desired δ. In both cases, the estimate is subject to error in the variance estimate. If sample sizes are required within certain prescribed confidence limits, they could be substantially larger, depending on the error estimate of the population variance (Johnson *et al.*, 1990).

6. Soil carbon stocks in temperate biomes

6.1 *Sampling procedure*

In this study, we have only included estimates of soil carbon content and its variance that were produced from direct measurements of soil carbon content on several individual samples (cf., Schwager and Mikhailova, 2002). The results for the site in Perthshire are used to demonstrate the bias introduced when carbon content is estimated as the product of average carbon concentration (grams of carbon per gram of soil) and average bulk density (or soil mass per unit area) and depth. The area-weighted mean soil mass of the three sampled strata was 119 kg m^{-2} and the mean carbon concentration within these layers was 9.3 g (C) g^{-1} soil. The area-weighted mean carbon content of the individual sampling points was 9.7 kg (C) m^{-2}. Had carbon content been calculated as the product of mean soil mass and mean carbon concentration from the same samples, the area-weighted carbon content would have been overestimated by 13% (*Table 1*). Overestimation is a general characteristic arising from the negative correlation between carbon concentration (grams of carbon per gram of soil) and soil bulk density (grams of soil per cubic centimeter), which again is a result of different densities (grams per cubic centimeter) of organic and mineral soil particles.

6.2 *Scaling of variance with plot size*

Because of the complexity of the processes leading to carbon accumulation, soil carbon stocks vary from place to place and over many spatial scales, from the very local

Table 1. *Mean depth, soil mass, carbon concentration and content of the organically rich layers (L, O, A) in different strata in a ploughed and afforested podsol in Perthshire, UK. Carbon content was calculated in two different ways. Numbers in brackets indicate one standard error.*

Stratum	Fraction of total area	Sample size, n	Sample depth (cm)	Soil mass (kg m^{-2})	Carbon concentration (g (C) g^{-1} soil)	Carbon content (kg (C) m^{-2})	
						Mean of individual samples	Mean soil mass × mean of carbon concentration
Furrows	0.25	20	10.7 (1.3)	65 (10)	6.5 (0.7)	3.8 (0.65)	4.2 (0.79)
Undisturbed	0.25	20	19.8 (1.1)	87 (9)	13.1 (1.3)	9.8 (0.65)	11.4 (1.63)
Ridges	0.5	40	30.1 (1.5)	162 (10)	8.7 (0.8)	12.6 (0.90)	14.2 (0.17)
Total area	1	80	22.7 (0.9)	119 (6)	9.3 (0.5)	9.7 (0.51)	11.0 (0.46)

micropatch to the landscape and the regional scales. Geostatistical tools have often been applied to determine the spatial dependency of soil physical and chemical properties, but most studies have focused on changes occurring over only two or three orders of magnitude of changes in distance (e.g., from metres to hundreds of metres). In fewer cases, attention has been focused on the variability occurring at the regional scale (e.g., from kilometres to hundreds of kilometres). For logistical reasons, a sampling scheme investigating changes in soil carbon stocks over several orders of magnitude of distance has not, to our knowledge, been implemented yet.

The study made in the forest at Les Landes provides an example of the spatial dependency of soil carbon stocks in the range from tens of metres to hundreds of metres. In the upper 40 cm, where most of the carbon was located, 72% of the total variance of the 9 ha plot was nugget variance (i.e., the estimated vertical intercept of the regression model of semi-variance against distance) and the remaining 28% of the variance occurred over a range of 35.7 m (*Table 2*). Thus, compared with the size of the sampled plot (9 ha), subplots two orders of magnitude smaller (0.1 ha, i.e., a circle with diameter of up to 36.9 m) yielded lower estimates of variance. However, because of the nugget effect, even the smallest subplots still contain more than two-thirds of the total variance estimated at the plot scale. Although the nugget variance at greater depths was smaller, the range was similar. Large nugget values can be reduced by bulk sampling and by enlarging the size of the sampling point. However, this example is indicative of the low spatial dependency of variance on plot size. Decreasing plot size does not necessarily result in large decreases in variance.

The ranges of spatial dependence at Les Landes were larger than those found at a site in Finland (Liski, 1995). Liski (1995) identified scales of spatial dependence between 1.1 m for the 0–10 cm layer and 8.4 m for the 0–40 cm layer. On a larger scale, Homann *et al.* (2001) found that 81% (0–15 cm depth) and 83% (15–30 cm depth) of the total variance in a forest of over 126 ha in size was already contained inside 2 ha plots. Although the plots in this study were larger in absolute size, the relative change in size from larger to smaller plots (from 126 to 2 ha, i.e., a factor of about 60) was more modest than in the example from Les Landes. In the forest floor, only 36% of

Table 2. *Geostatistical parameters for soil carbon stocks at different depth intervals at a forest site in Les Landes, France.*

Depth interval (cm)	Mean (kg (C) m^{-2})	Coefficient of variation (%)	Nugget variance (fraction of sill)	Range (m)	Sill ((kg (C) m^{-2})^2)
0–40	6.91	69.8	0.72	36.9	22.8
40–60	1.63	67.5	0.24	37.71	1.13
60–80	1.22	62.4	0.03	40.05	1.02
80–100	1.66	132.5	0.52	40.05	1.00

the total variance was found at the 2 ha scale. However, this was of minor importance as the forest floor only contributed about 13% of the total carbon measured.

Over a wider range of plot sizes, there can be larger changes in variance. At our site in Northumberland, for example, 0.03 ha plots had only 54% of the variance found within an area of 578 ha (four orders of magnitude change in plot size). Some of this difference might result from the difference in age of the trees among the plots in different forest compartments. Over an even wider range, Palmer *et al.* (2002) studied variability in forest soil carbon across Georgia (USA), and found on average 21% of the total variance over distances of '*generally less than one meter*', another 38% over 1 ha plots, and the remaining 41% over an area of more than one million hectares (i.e., 11 and 7 orders of magnitude change in plot size, respectively). Clearly, variability of soil carbon stocks occurs over several orders of magnitude (cf., Webster and Oliver, 2001).

When relative variances are plotted against relative plot sizes for those studies where multiple estimates were available, a strong logarithmic relationship is indeed found between these two variables (*Figure 1*; $r^2 = 0.60$, $n = 10$, non-linear regression with the intercept for the logarithmic relationship forced through the point (1;1)). One point (an old-growth Douglas fir (*Psuedotsuga menziesii* (Mirb.) Franco) forest where buried logs could occasionally be found in the sample micro-plots (Conant *et al.*, 2003)), was noticeably outside the scatter of the regression line. For this example, reducing the area sampled from around 300 ha to a micro-plot of 2 m × 5 m size only reduced the variance by about 11% (Conant *et al.*, 2003)! The regression without this point is shown in *Figure 1* ($y = 0.035 \ln(x + 1)$, $r^2 = 0.84$, $n = 9$). This relationship indicated an approximate halving of the variance only for every six orders of magnitude decrease in plot size. In other words, reducing the plot size from 100 ha to 0.0001 ha (i.e., 1 m^2) (or, equivalently, from 10^6 ha to 1 ha) would only decrease the variance by an estimated 50%. We used this relationship to correct the estimated variance (hence, the estimated δ as well) to account for differences in sampling area across studies (δ_{corr}, *Table 3*). This correction changed the estimated δ only marginally, confirming that across-study changes in plot size did not significantly affect our estimates of variance.

6.3 *Factors affecting variance*

Considering all the sites presented in *Table 3*, there was a sevenfold difference between the smallest and the largest mean carbon content (3.0–21.3 kg (C) m^{-2}) and

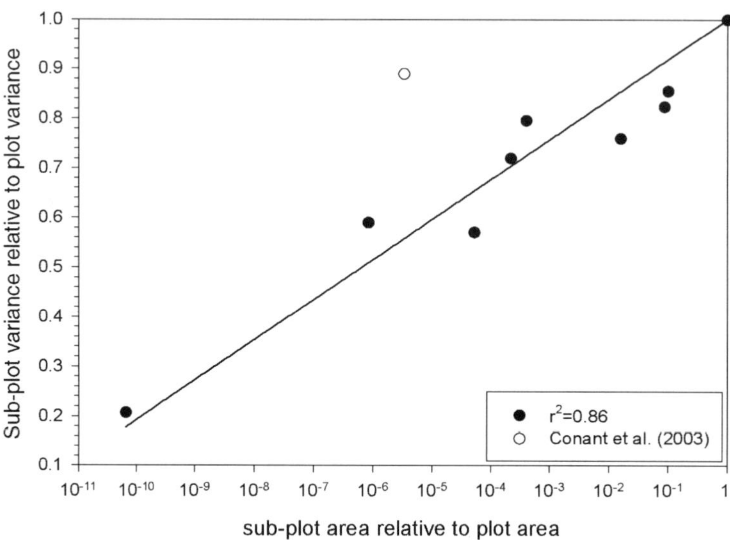

Figure 1. *Relationship between relative plot size and relative plot variance in soil carbon stocks across studies for which estimates were available at different spatial scales (Les Landes and Northumberland, this study; Homann et al., 2001; Palmer et al., 2002; Conant et al., 2003). Relative plot size was calculated as the ratio of the size of the sub-plots to the size of the plots. Relative variance was calculated as the ratio of the variance of the sub-plots to the total variance of the plots. The logarithmic regression line was forced to pass through the (1:1) point. The point marked by the white circle was excluded from the regression.*

coefficients of variation ranged from 10 to 112%. A strong curvilinear relationship was found between mean carbon content and variance corrected to 1 ha across all the sites reported in *Table 3* ($y = 0.030\,x^{2.446}$, $r^2 = 0.50$, $p < 0.001$). The relationship became even stronger ($y = 0.020\,x^{2.508}$, $r^2 = 0.63$, $p < 0.001$) when one site with a particular disturbance history was discarded from the analysis. One of the two Washington studies was heavily affected by the presence of buried wooden logs (old growth) that, as mentioned already, created high levels of very local variability. Some other 'disturbed' sites also showed a tendency towards larger scatter. For instance, the burning of the residues of the previous old-growth forest (second growth stand) (Conant *et al.*, 2003) appeared to increase variance. Estimates of variance at the plot scale for both the Perthshire and Northumberland sites may have been affected by the ploughing before planting. At the site in Perthshire, ploughing increased the total variance fourfold (*Table 1*), assuming initial variance was the same as we found 20 years later in the still-undisturbed bands between the ridges. The site at Les Landes was characterized by uneven topography of sandy, C-poor dunes with low water holding capacity, and shallow depressions where organic matter accumulated and water was more freely available. Such conditions create stark contrasts in conditions for plant growth and carbon mineralization rates. As mentioned in the site description, this leads to vegetation, soil profile and carbon storage being directly related to the micro-relief.

Table 3. *Estimated mean carbon content and its coefficient of variation (CV) for various forest sites. For each site, the minimum detectable change (δ) for a sample size of 100 was estimated, as well as the sample size required to obtain a δ of 0.5 kg (C) m^{-2}. These calculations assume simple random sampling.*

Location	+Area (ha)	Sample size (n)	Carbon content (kg (C) m^{-2})	CV (%)	δ for $n = 100$ (kg (C) m^{-2})	δ_{corr} for $n = 100$ (kg (C) m^{-2})	Sample size for $\delta = 0.5$ kg (C) m^{-2} (n)	Reference
Tennessee, USA (235 m altitude)	0.02	18	4.0	10	0.11	0.12	8	Garten *et al.*, 1999
Tennessee (plot)	0.001	12	3.0	16	0.15	0.16	12	Conant *et al.*, 2003
Tennessee (stand)	0.1	3	3.0	18	0.16	0.17	13	Conant *et al.*, 2003
Tennessee, USA (335 m altitude)	0.02	18	3.8	15	0.16	0.17	13	Garten *et al.*, 1999
Helsinki area, Finland	0.005	126	4.5	15	0.20	0.21	20	Liski, 1995
Tennessee, USA (1000 m altitude)	0.02	18	7.4	13	0.29	0.31	40	Garten *et al.*, 1999
Tennessee, USA (940 m altitude)	0.02	18	10.7	10	0.32	0.34	48	Garten *et al.*, 1999
Tennessee, USA (1670 m altitude)	0.02	18	9.6	12	0.34	0.36	53	Garten *et al.*, 1999
Oregon, USA	126	271	8.6	15	0.40	0.36	53	Homann *et al.*, 2001
Tierra del Fuego, Argentina	48.6	18	6.6	32	0.43	0.40	64	Weber, 1999
Tennessee, USA (1650 m altitude)	0.02	18	8.9	18	0.47	0.50	100	Garten *et al.*, 1999
Perthshire, UK (undisturbed)	0.85	20	9.8	30	0.86	0.86	292	This study
Maine, USA	0.4	24	11.1	26	0.85	0.87	295	Fernandez *et al.*, 1993
Washington (second growth, stand)	2.50	3	4.8	67	0.94	0.93	341	Conant *et al.*, 2003
Washington (second growth, plot)	0.001	12	4.8	59	0.84	0.94	347	Conant *et al.*, 2003
Les Landes, France (0–40 cm)	9	60	6.9	70	1.43	1.37	739	This study
Perthshire, UK (ploughed)	0.85	80	9.7	49	1.45	1.46	837	This study
New Hampshire, USA	23	55	16.0	38	1.86	1.76	1218	Huntington *et al.*, 1988
Washington (old growth, stand)	300.8	3	7.1	112	2.36	2.11	1752	Conant *et al.*, 2003
Northumberland, UK (stand)	50	5	21.3	37	2.31	2.15	1811	This study
Northumberland, UK (forest)	578	6	21.3	40	2.53	2.23	1951	This study
Northumberland, UK (plot)	0.03	8	21.3	35	2.26	2.40	2260	This study
Washington (old growth, plot)	0.001	12	7.1	106	2.22	2.48	2415	Conant *et al.*, 2003

Equation 1 can be re-written as follows:

$$n = \frac{s^2}{\delta^2}\left(t_{\alpha,\upsilon} + t_{\beta,\upsilon}\right)^2 \tag{3}$$

It is now apparent that the number of samples required, for a fixed minimum detectable difference, is a function of the population variance, not of its coefficient of variation or of its standard deviation. Because sample variance was strongly related to mean carbon content across sites, the number of samples required to estimate a δ of 5 Mg (C) ha^{-1} (assuming simple random sampling, cf., equation 1) was positively correlated to mean carbon content ($y = 0.8967\,x^{2.4232}$, $r^2 = 0.63$, $p < 0.001$), i.e., carbon-rich sites require a far larger sampling effort than carbon-poor sites (*Figure 2*).

The estimated values of variance were finally regressed against the initial sample size, either alone or in a multiple regression including site carbon stocks. In neither case was a significant relationship found, suggesting that the available estimates of variance were reasonably robust.

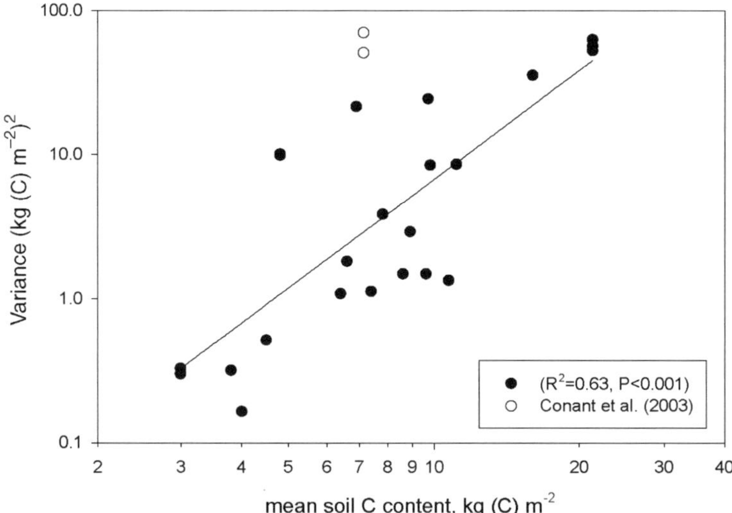

Figure 2. *Relationship between mean soil carbon stocks (kilograms of carbon per square metre) and estimated variance in soil carbon stocks across temperate and Boreal forests. For a change in soil carbon stocks between 3.0 and 21.3 kg (C) m^{-2}, the estimated population variance changed over three orders of magnitude (closed symbols). One site (Conant et al., 2003) fell outside the range of the regression between these two variables (labelled as outliers in the graph, open symbols; the two estimates are referred to plot-level and stand-level variance, respectively). At this old-growth Douglas fir site, spatial variability was large because of the presence of buried logs.*

6.4 *Sampling intensity and related minimum detectable changes*

Based on a sample size of 100, the minimum change that could be detected with a statistical significance ($\alpha = 0.05$, $\beta = 0.1$) ranged from 0.12 to 2.48 kg (C) m^{-2} and was log-normally distributed among the studied sites. For sites disturbed by wind-throw (Huntington *et al.*, 1988; Weber, 1999) or ploughing (Perthshire and Northumberland, plot scale), we found the geometric mean δ for a sample size of 100 to be 1.37 kg (C) m^{-2} (median 1.60). At undisturbed sites, the geometric mean δ based on 100 samples would be 0.39 kg (C) m^{-2} (median 0.34).

To attain a δ of 0.5 kg (C) m^{-2}, the sample sizes that would be required are estimated to lie between 8 and 2415 depending on the particular site carbon content (*Table 3*). The geometric mean sample size for a δ of 0.5 kg (C) m^{-2} at the disturbed sites is 777 (median 1069). At the Perthshire site, the ploughing disturbance calls for a nearly threefold increase in sample size, despite the stratified sampling regime (*Table 3*). For all the other 'undisturbed' sites, the geometric mean sample size for achieving a Δ of 0.5 kg (C) m^{-2} is 63 samples (median 46).

7. Conclusions and summary

Estimating soil carbon content as the product of mean carbon concentration and bulk density can result in considerable overestimation. Carbon concentration and soil mass need to be measured on the same sample and carbon contents calculated for each individual sample before averaging. The effect of this bias is likely to be smaller (but still greater than zero) when the primary objective is to determine stock changes over time.

Variance and mean carbon content are significantly and positively related to each other, although some sites showed much higher variability than predicted by this relationship, as a likely consequence of their particular site history, forest management, and micro-topography. Because of the proportionality between mean and variance, the number of samples required to detect a fixed change in soil carbon stocks varied directly with the site mean carbon content from less than 10 to several thousands across the range of carbon stocks normally encountered in temperate and Boreal forests. This raises important questions about how to derive an optimal sampling strategy across such a varied range of conditions so as to achieve the aims of the Kyoto Protocol.

Overall, on carbon-poor forest sites with little or no disturbance to the soil profile, it is possible to detect changes in total soil organic carbon over time of the order of 0.5 kg (C) m^{-2} with manageable sample sizes even using simple random sampling (i.e., about 50 samples per sampling point). More efficient strategies will reveal even smaller differences. On disturbed forest sites (ploughed, windthrow) this is no longer possible (required sample sizes are much larger than 100). Soils developed on coarse aeolian sediments (sand dunes), or where buried logs or harvest residues of the previous rotation are present, can also exhibit large spatial variability in soil carbon. Generally, carbon-rich soils will always require larger numbers of samples. On these sites, simple random sampling is unlikely to be the preferred method, because of its inherent inefficiency. More sophisticated approaches, such as paired re-sampling inside relatively small plots (see, for example, Ellert *et al.*, 2001) are likely to reduce sample size significantly and lead to detection of smaller differences in carbon stocks over time. However, it remains to be shown that at these sites the application of efficient sampling designs will result in the detection of differences relevant for the objectives of the Kyoto Protocol (cf., Conant *et al.*, 2003).

Finally, it should also be noted that, compared to the accuracy with which changes in atmospheric carbon content can be detected (less than 1 p.p.m. CO_2), changes in soil carbon stocks are very uncertain. A release of 0.5 kg (C) from 1 m^2 of soil surface is equivalent to an increase in CO_2 concentration of about 125 p.p.m. in the air column above the same area.

Acknowledgements

The EU-funded CarboAge project (contract ENV4-CT97-0577) and the NERC grant GR9/4806 provided the framework within which this study took place. Thanks are due to Mark Rayment for his encouragement to write this paper. Thanks are also due to Pete Smith (Aberdeen University, UK) and Pierre Bernier (Canadian Forest Service, Canada) for comments on an earlier version of the paper.

References

Anderson, V.L. and McLean, R.A. (1974) *Design of Experiments, A Realistic Approach*. Marcel Dekker, New York.

Conant, R.T. and Paustian, K. (2002) Spatial variability of soil organic carbon in grasslands: implications for detecting change at different scales. *Environmental Pollution* **116**: S127–S135.

Conant, R.T., Smith, G.R. and Paustian, K. (2003) Spatial variability of soil carbon in forested and cultivated sites: implications for change detection. *Journal of Environmental Quality* **32**: 278–286.

Ellert, B.H., Janzen, H.H. and McConkey, B.G. (2001) Measuring and comparing soil carbon storage. *In*: Lal, R., Kimble, J.M., Follet, R.F. and Stewart, B.A. (eds) *Assessment Methods for Soil Carbon*, pp. 131–146. CRC Press, Boca Raton.

Fernandez, I.J., Rustad, L.E. and Lawrence, G.B. (1993) Estimating total soil mass, nutrient content, and trace metals in soils under a low elevation spruce-fir forest. *Canadian Journal of Soil Science* **73**: 317–328.

Garten, C.T. Jr., Post, W.M.III., Hanson, P.J. and Cooper, L.W. (1999) Forest soil carbon inventories and dynamics along an elevation gradient in the southern Appalachian Mountains. *Biogeochemistry* **45**: 115–145.

Homann, P.S., Bormann, B.T. and Boyle, J.R. (2001) Detecting treatment differences in soil carbon and nitrogen resulting from forest manipulations. *Soil Science Society of America Journal* **65**: 463–469.

Huntington, T.G., Ryan, D.F. and Hamburg, S.P. (1988) Estimating soil nitrogen and carbon pools in a northern hardwood forest ecosystem. *Soil Science Society of America Journal* **52**: 1162–1167.

Johnson, C.E., Johnson, A.H. and Huntington, T.G. (1990) Sample size requirements for the determination of changes in soil nutrient pools. *Soil Science* **150**: 637–644.

Liski, J. (1995) Variation in soil organic carbon and thickness of soil horizons within a boreal forest stand – effect of trees and implications for sampling. *Silva Fennica* **29**: 255–266.

Palmer, C.J., Smith, W.D. and Conkling, B.L. (2002) Development of a protocol for monitoring status and trends in forest soil carbon at a national level. *Environmental Pollution* **116**: S209–S219.

Papritz, A. and Webster, R. (1995a) Estimating temporal change in soil monitoring. 1. Statistical theory. *European Journal of Soil Science* **46**: 1–12.

Papritz, A. and Webster, R. (1995b) Estimating temporal change in soil monitoring. 2. Sampling from simulated fields. *European Journal of Soil Science* **46**: 13–27.

Post, W.M., Izaurralde, R.C., Mann, L.K. and Bliss, N. (2001) Monitoring and verifying changes of organic carbon in soil. *Climatic Change* **51**: 73–99.

Schlesinger, W.H. (1997) *Biogeochemistry. An Analysis of Global Change*, 2nd edition. Academic Press, San Diego, CA.

Schwager, S.J. and Mikhailova, E.A. (2002) Estimating variability in soil organic carbon storage using the method of statistical differentials. *Soil Science* **167**: 194–200.

Watson, R.T., Noble, I.R., Bolin, B., Ravindranath, N.H., Verardo, D.J. and Dokken, D.J. (2000) *Land Use, Land-Use Change and Forestry*. Special Report, IPCC. Cambridge University Press, Cambridge, UK.

Weber, M. (1999) Carbon budget of a virgin *Nothofagus* forest in Tierra del Fuego. *Forstwissenschaftliches Centralblatt* **118**: 156–166. (In German.)

Webster, R. and Oliver, M.A. (1990) *Statistical Methods in Soil and Land Resource Survey*. Oxford University Press, Oxford.

Webster, R. and Oliver, M.A. (2001) *Geostatistics for Environmental Scientists*. John Wiley & Sons, Chichester, U.K.

Zar, J.H. (1999) *Biostatistical Analysis*, 4th edition. Prentice Hall International, NJ.

12

Fractional contributions by autotrophic and heterotrophic respiration to soil-surface CO$_2$ efflux in Boreal forests

Peter Högberg, Anders Nordgren, Mona N. Högberg,
Mikaell Ottosson-Löfvenius, Bhupinderpal-Singh,
Per Olsson and Sune Linder

1. Introduction

Soil-surface CO$_2$ flux from terrestrial ecosystems, representing the sum of decomposition of soil organic matter and root respiration, releases 20–40% of the total terrestrial CO$_2$ emissions to the atmosphere (Raich and Schlesinger, 1992). An estimate for European forests (Janssens *et al.*, 2001) found that although 80% of gross primary production was returned to the atmosphere through respiration, as much as 70% of the ecosystem respiration occurred below ground. In one case, a productive mixed conifer forest in central Sweden, the ecosystem was a net carbon source as an effect of high soil respiration caused by earlier drainage of wetter parts of the forest (Valentini *et al.*, 2000). The large impact of soil respiration on the ecosystem carbon balance calls for a closer examination of the associated compartmentalized fluxes and their controls.

The Boreal forest is the second largest terrestrial biome in the world, covering approximately 15.6×10^6 km^2 (cf., Gower *et al.*, 2001). The amount of soil carbon in this biome is disproportionately larger than that in other forest biomes, because of the deep, often frozen, organic deposits (Dixon *et al.*, 1994). The Boreal forest is expected to experience the greatest warming of all forest biomes, raising concern over the fate of the large quantities of soil carbon (see, for example, Cox *et al.*, 2000; Cramer *et al.*, 2001). The large extent of the boreal forest, the large amount of carbon contained in the soil, and the expected climatic warming, make the Boreal forest a key biome to be understood and correctly represented in global carbon models.

2. What is respiring: what is autotrophic and heterotrophic respiration in soils?

Soil-surface CO₂ efflux is commonly divided into autotrophic and heterotrophic respiration. The former should in a strict sense refer to the activity by plant roots (the activity by algal and bacterial autotrophs can usually be neglected in the context of boreal forest soils). In reality it is, however, quite complex and technically difficult quantitatively to make the separation between autotrophic and heterotrophic soil respiration. It is none-the-less of the utmost importance to separate these components quantitatively, because of their large impact on the ecosystem carbon balance.

Plant root respiration should be the sole and dominant autotrophic respiration component. In northern temperate forests, fine tree roots are, however, almost invariably covered by ectomycorrhizal fungi (Taylor *et al.*, 2000), which, like the plant roots, are fed by sugars derived from current or stored photosynthates (Smith and Read, 1997). The current understanding is that sucrose from the plant is hydrolysed by plant invertases, whereafter glucose is preferentially absorbed by the fungus (Smith and Read, 1997). The flux of carbon to the fungus is assumed to be maintained by high rates of fungal respiratory activity and by the conversion to sugars not commonly used by plants, e.g., glycogen, mannitol, and trehalose.

The ectomycorrhizal fungi form sheaths several tens of micrometres thick, from which an extramatrical mycelium extends into the surrounding soil (*Figures 1 and 2*). The latter may be finely divided and composed of single hyphae or may aggregate to form tube-like structures called rhizomorphs. There is evidence that ectomycorrhizal roots leak significant quantities of dissolved organic carbon (DOC) (Garbaye, 1994; Högberg and Högberg, 2002). Surfaces of ectomycorrhizal sheaths and extramatrical hyphal structures are, therefore, often covered by colonies of bacteria thought to use labile carbon compounds leaked by mycorrhizal fungi (see, for example, Garbaye, 1994; Timonen *et al.*, 1998). Mycorrhizal fungi are, like other fungi, classified as heterotrophs, but serve as an integral and functional extension of the root system of the autotrophic plant. We therefore suggest they be classified as autotrophs, in line with their function, rather than based on their taxonomic position.

To elaborate further, there are achlorophyllous plants, which are functional heterotrophs. Among these are mycoheterotrophs, which receive their carbon from

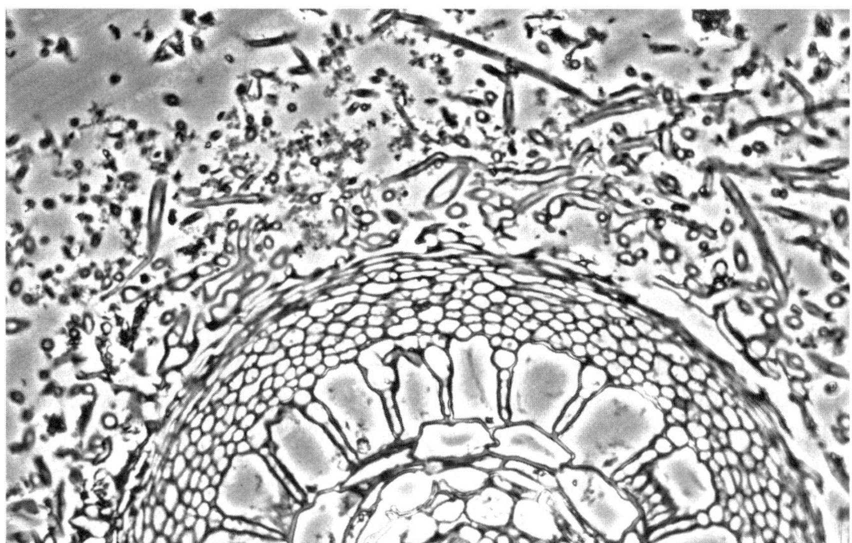

Figure 1. *Cross-section through an ectomycorrhizal root showing the fungal Hartig net penetrating between plant cortical cells, the fungal sheath (or mantle), and an extensive extraradical (or extramatrical) mycelium. Root diam. 0.2 mm. Photograph: P. Högberg.*

Figure 2. *Ectomycorrhizal Scots pine roots and their extraradical mycelium, which sometimes forms fungal rhizomorphs. Photograph: courtesy of K. Olofsson.*

autotrophic plants via a common mycorrhizal fungus (Björkman, 1960; Leake, 1994). We would include the below-ground respiration by such a complex system as part of autotrophic respiration. Moreover, complex organic molecules, containing carbon, are to a minor extent also taken up from the soil by autotrophic plants. For example, plant root uptake of amino acids (Näsholm *et al.*, 1998) represents heterotrophy, but is not of such significance that the general classification of chlorophyllous plants as autotrophs should be questioned.

As mycorrhizal roots leak labile carbon compounds to adjacent microbes, which by convention are defined as heterotrophs, the separation of autotrophic and heterotrophic soil respiration becomes even more enigmatic. In this context, there is a wide range of likely ambiguity from organisms that are totally dependent on sugars and other labile carbon compounds from the mycorrhizal plants to others that are less, or not at all, dependent on a direct flux of carbon from the plants. Because under natural conditions no unequivocal division can be made between autotrophs and heterotrophs in soils, it seems most appropriate to consider an autotroph–heterotroph continuum (*Figure 3*).

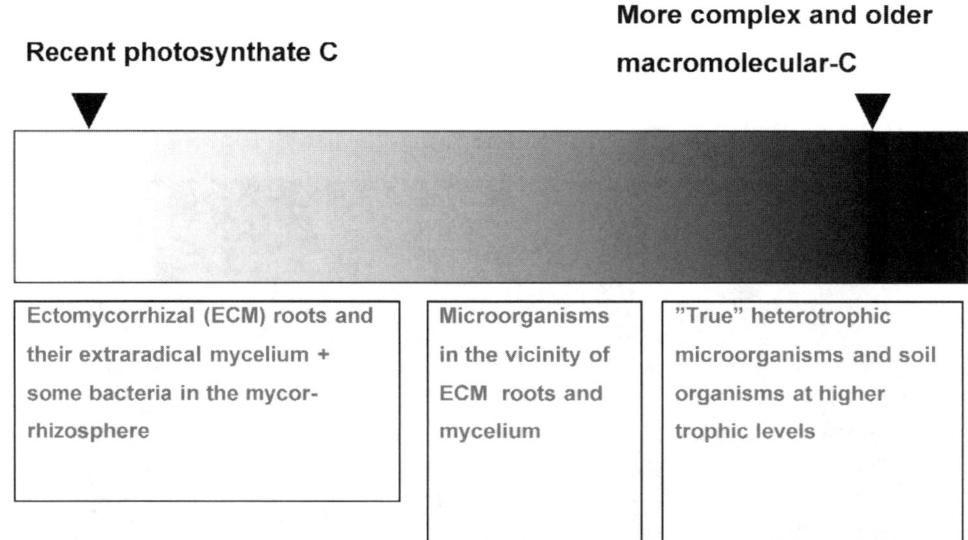

Figure 3. The 'autotroph–heterotroph continuum'. Note that the division between the functional groups may vary over time.

There is also a considerable seasonal variation in the contribution of the two major respiratory components to the total soil-surface CO$_2$ flux (see, for example, Hansen *et al.*, 1997). The reason for this is to be found in climatic constraints to the annual course of carbon uptake via photosynthesis (see, for example, Bergh *et al.*, 1998; Troeng and Linder, 1982) and a strong seasonality in above- and below-ground growth processes in trees (see, for example, Havranek and Tranquillini, 1995; Linder and Flower-Ellis, 1992; Majdi, 2001; McMurtrie *et al.*, 1994).

3. How do we separate autotrophic and heterotrophic soil activities?

Estimates of the contribution of autotrophic respiration to soil respiration vary widely (Hanson *et al.*, 2000). Part of this variation reflects true natural variability within and across ecosystems. It is, however, also most likely that much of the reported variability reflects methodological problems, which is to be expected given the complexity of the systems studied. For example, based on data available for *Liriodendron*-dominated stands in the southeast USA (Edwards and Harris, 1977; Edwards and Ross-Todd, 1979; Edwards and Sollins, 1973), the fractional contribution of autotrophic respiration can be estimated to vary between 0 and 80%, most likely as a consequence of the different methods used.

The methods used to separate autotrophic and heterotrophic soil respiration were classified into three broad groups by Hanson *et al.* (2000):

(i)   integration of components; the system is disintegrated physically, the respiration is measured for each component separately, and the sum of respiration by the various components is compared with that of the intact system;
(ii) use of physical barriers to exclude roots, or other types of comparison of respiration from soils with and without roots;
(iii) isotopic methods, which most often are based on labelling of the photosynthates by exposing the plant shoots to radioactive ^{14}C or stable ^{13}C.

For (i), it follows from the above (Section 2) that physical disruption of the delicate root–microbe–soil system is highly undesirable, primarily because roots and mycorrhizal hyphae will be disconnected from their principal carbon source (*Figures 1 and 2*). Moreover, one cannot assume that the heterotrophic component will respire at the same rate in isolation as in the presence of the autotrophic component. Although the interactions between the two components can be neutral in theory, there are reports of both negative (e.g., Gadgil and Gadgil, 1975; Koide and Wu, 2003) and positive (e.g., Dighton *et al.*, 1987; Entry *et al.*, 1991) effects of mycorrhizal roots on decomposition of organic matter. We should therefore stress the obvious risk in assuming that the summation procedure is a good validation tool for the method.

For (ii), the natural system of interest here is the soil of Boreal forests. It is a soil with roots (*Figure 2*), and one cannot assume that heterotrophic activity in naturally 'root-free' soil, or in soil from which roots are excluded by barriers (trenches, solid sheets, or fine meshes), is equivalent to the heterotrophic activity in root-free soil. Moreover, because active roots are excluded, the soils can become wetter since there is no plant uptake of water.

Isotopic methods (iii) do have the apparent advantage that they can be applied to complex natural systems, and do not involve physical disturbance of the system studied. The plant canopy can be exposed to labeled CO_2 in closed chambers (see, for example, Horwath *et al.*, 1994) or in free air carbon dioxide enrichment (FACE) experiments (see, for example, Andrews *et al.*, 1999). There are, however, problems with certain time lags in the system as the pulse of labelled carbon is distributed in a system with many pools of carbon, which have large differences in carbon turnover rates. For studies of the partial contribution of autotrophic activity to total soil respiration, one can use a single pulse-labelling, repeated pulse-labelling or continuous labelling. Problems of time lags are common to these approaches as one must take a decision about when the autotrophic respiration is completely labelled, whereas the heterotrophic activity should still not be labelled at all.

For trees, it may take one to several days for the carbon from current photosynthesis to appear in the soil-surface CO$_2$ efflux (e.g., Ekblad and Högberg, 2001; Högberg *et al.*, 2001; Horwath *et al.*, 1994). At the same time, there is rapid turnover of some components of root carbon. A recent report on the turnover of carbon in the extra-radical mycelium of arbuscular mycorrhizal fungi demonstrated a turnover time of five to six days, based on labelling of photosynthates with 'fossil' CO$_2$ depleted in ^{14}C (Staddon *et al.*, 2003). This rapid turnover of mycorrhizal carbon could be caused by a high respiratory rate and/or a rapid consumption of the mycelium by, for example, grazing soil animals. Another approach to estimate the fractional contribution of autotrophic soil activity is to use the enrichment in ^{14}C, which occurs as a result of atmospheric tests of nuclear weapons in the late 1950s and early 1960s. Since then, the enrichment has declined faster in the atmosphere than in recently formed soil organic matter. This leads to a lower 'ecosystem-age' of the soil CO$_2$ efflux than that of CO$_2$ respired as a result of decomposition of the organic matter. As a consequence, the fractional contribution of the different components of soil respiration can be modelled and calculated (V. Hahn, P. Högberg and N. Buchmann, unpublished data). However, although isotopic methods are attractive because they do not disturb the system studied, they require meticulous attention to the importance of time lags in the system. Another problem with many isotopic labelling methods is that they are comparatively expensive, especially at the scale of FACE experiments.

As an alternative one can sometimes use variations in natural δ^{13}C labelling of photosynthates (Ekblad and Högberg, 2001; Högberg *et al.*, 2004) that occur as a result of variations in ^{13}C fractionation during photosynthesis (Farquhar *et al.*, 1980). Ekblad and Högberg (2001) found that the δ^{13}C of the soil CO$_2$ efflux varied by 5‰, and correlated this variability with variations in the relative humidity (RH) of the air. If the ratio between the partial pressures of CO$_2$ inside the stomata (C_i), and in ambient air outside the leaf (C_a) is determined simultaneously, and the δ^{13}C of heterotrophic respiration (which is quite stable) is known, the fractional contribution of autotrophic respiration to total soil respiration can be estimated when the δ^{13}C of the soil CO$_2$ efflux differs significantly from that of heterotrophic respiration. This approach is attractive as it does not involve any disturbance of the system, but it does not allow calculation of the contribution of autotrophic respiration under moist conditions when the isotopic signatures of the two respiratory components overlap. This is critical, because we do not know if the values of the fractional contribution by autotrophic respiration to the total soil CO$_2$ efflux are representative for other than dry conditions. Moreover, the use of stored carbohydrates in roots may confound interpretations of the direct link between the isotopic composition of photosynthates and that of root respiration.

Another approach is to use soil carbon with a distinctly different carbon isotopic composition from that of the studied plant. An example is the use of the difference in δ^{13}C of about 14‰ between carbon derived from C3 and C4 plants (Smith and Epstein, 1971). In regions, where C3 plants occur naturally or are cropped, one can, in theory, plant C4 plants, where C3 plants have previously labelled the soil C. The fractional contribution of C4 autotrophic root respiration can then be distinguished from the background heterotrophic respiration resulting from decomposition of C3 plant organic matter (see, for example, Robinson and Scrimgeour, 1995). This approach cannot, however, be used in northern temperate forests, because soils there are naturally C3-carbon labelled and there are no C4 trees to be tested. As with other labelling approaches, there are also problems of interpreting the time lags in the system. We have tried injecting C4 sugar directly into the soil, anticipating that the heterotrophic organisms should shift from C3-carbon sources to C4-carbon sugar, leaving roots and their most closely associated microbiota as the sole source of C3-carbon-based respiration (Ekblad and Högberg, 2000; Högberg and Ekblad, 1996). In initial experiments with root-free soil in the laboratory, the C3-based respiration was eliminated a few days after additions of C4-carbon, whereas in the field the initial C3-carbon respiration was only reduced by 50%, suggesting that autotrophic respiration accounted for roughly half of the soil CO_2 efflux. With this approach there are problems of interpretation of time lags, and it is possible that some of the mycorrhizal fungi and other mycorrhizosphere organisms turn to the new C4-carbon source, while using recent plant photosynthates under natural conditions. Additionally, some heterotrophs obviously do not shift to the new C4-carbon source (K. Steinmann and P. Högberg, unpublished data). Hence, the approach cannot be recommended at this stage as a method to separate heterotrophic from autotrophic soil respiration.

4. Tree girdling: an alternative method to estimate autotrophic respiration

Several years of exploration led us to the idea of instantaneously and permanently inhibiting the carbon flow to below-ground autotrophic respiration by girdling trees over a large area. Once the pool of sugars in the roots is depleted, the fractional contribution of autotrophic respiration to total soil respiration can then be estimated (Högberg et al., 2001). Girdling in this context means stripping off the bark and phloem, down to the depth of the current xylem, around the circumference of the tree stem at a point below the green crown (*Figure 4*). We suggest that this physiological approach is added as a fourth group to the list of methods compiled by Hanson et al. (2000). So far, we have conducted three girdling experiments in Boreal forests. A short description of the experiments and some preliminary results are presented below.

Our first girdling experiment was done in a 50-year-old *Pinus sylvestris* L. forest on a poor sandy silt soil at Åheden, northern Sweden (Högberg et al., 2001; *Figure 4*). Three 900 m² plots were girdled in early June, here denoted early girdling (EG) plots, while three plots were girdled later in August, here denoted late girdling (LG) plots. Three additional plots were used as controls. EG plots lost their soil respiration activity relative to control plots slowly to begin with, but two months after girdling, the soil CO_2 efflux was more than 50% lower than from the control plots (*Figure 5*). In contrast, LG plots rapidly lost soil respiratory activity, and reached the level of EG plots in only two weeks. One apparent reason for the more rapid response to LG was that the amount of starch stored in the roots was low in late summer, whereas EG resulted in an initial depletion of a larger starch pool, thus delaying the response to the interrupted phloem transport of current photosynthates. Another potential reason for the more rapid response to LG is that more photosynthates are allocated below ground in late

Figure 4. *The forest girdling experiment in Scots pine forest at Åheden, northern Sweden. Photograph: P. Högberg.*

summer compared with early in the summer (Hansen *et al.*, 1997). We observed a decline in fine-root starch concentration (Högberg *et al.*, 2001) and a simultaneous transient increase in the $\delta^{13}C$ of the soil CO_2 efflux, which was interpreted as respiration of root sugars and decomposition of ectomycorrhizal mycelium (Bhupinderpal-Singh *et al.*, 2003). This would lead to an underestimate of the autotrophic component. The assumption that autotrophic respiration was underestimated was confirmed, as in the second summer after girdling the calculated autotrophic respiration accounted for up to 65% of total soil respiration (Bhupinderpal-Singh *et al.*, 2003).

Tree roots may reach far from the stems. *Picea abies* (L.) Karst. and *P. sylvestris* have been reported to have a radial spread of their roots of up to 20 m (Stone and Kalicz, 1991). Therefore, all trees on large plots were girdled to reduce interference by roots of non-girdled trees outside the plots. The precaution of using large plots was, however, effective as sporocarps of ectomycorrhizal fungi were virtually eliminated from the central 10 m × 10 m of the girdled plots. The concern that roots and mycorrhizal mycelium would quickly die on girdled plots, thereby enhancing the heterotrophic activity, appeared to be of minor importance because the effect of girdling resulted in a drastic reduction of the soil CO_2 efflux.

An advantage with girdling should be that it does not initially affect the soil properties or the micro-climate in the stand. Girdled trees will, however, reduce water uptake after some time and the soil will become wetter than in non-girdled control plots, especially during dry spells. In the longer term the girdled trees will reduce their leaf area, as an effect of increased litter fall and reduced foliage production, thus affecting the micro-climate in the stand. Again, the effect will be that the autotrophic component is underestimated. The first girdling experiment was performed in a *P. sylvestris* forest on a sandy silt soil, i.e., a soil with a comparatively limited

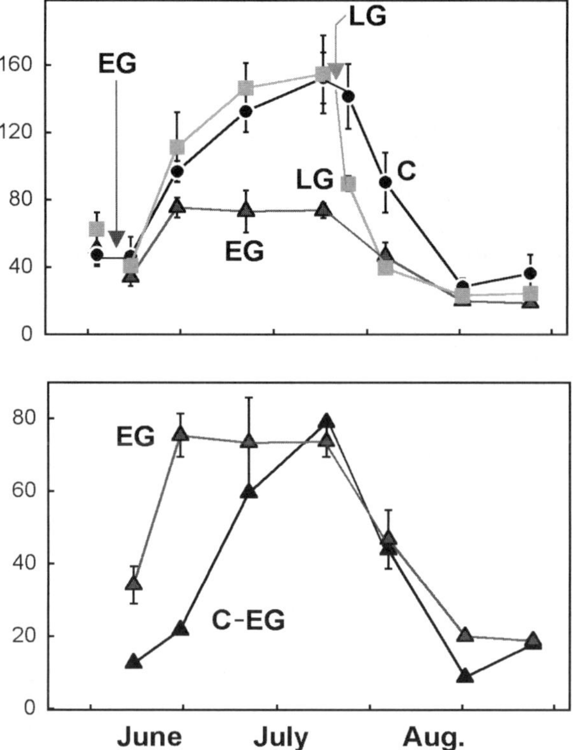

Figure 5. *Above: Soil respiration after early and late girdling (EG and LG, respectively) and in control plots (C) in the experiment at Åheden. Below: Autotrophic root respiration is calculated as the difference between CO_2 efflux on the C and EG plots. Reproduced from Högberg et al. (2001), with permission from Macmillan.*

water-holding capacity. As the first summers of the experiment were rainy, we did not observe any effect of girdling on the soil moisture content at this site.

The second girdling experiment was conducted at Stor Skogberget, also in the area of Umeå, in a mixed coniferous forest dominated by more than 100-year-old *P. abies* and *P. sylvestris*. This forest was on a till soil, and represented wetter and more fertile conditions. In particular, the organic mor layer was often more than 10 cm thick, compared with 2 cm at Åheden. The experiment consisted of three circular 700 m^2 (15 m radius) plots, which were girdled in early June 2002, and three control plots. As at Åheden, the response to girdling was considerable, and in August the calculated autotrophic respiration exceeded the heterotrophic respiration (A. Nordgren, P. Högberg, M. Ottosson-Löfvenius and Bhupinderpal-Singh, unpublished data).

A third girdling experiment was performed in a long-term nutrient optimization experiment at Flakaliden (64°07′ N; 19°27′ E; altitude 310 m above sea level) in a 40-year-old *P. abies* plantation on a till soil. Details about the nutrient optimization experiment can be found in Bergh *et al.* (1999) and Linder (1995). Girdling was performed in early June 2002 on 1000 m^2 plots in three unfertilized control (C) and three fertilized (F) stands, with three non-girdled control plots for each treatment. At the time of girdling, 15 years after the start of the nutrient treatments, there were major differences in stand characteristics between the control and fertilized stands. The C-plots had a standing volume of 52 m^3 ha^{-1} and a leaf area index (LAI) of 3.4, whereas the F-plots had a standing volume of 172 m^3 ha^{-1} and an LAI of 7.6. During the first summer after girdling, the response of the unfertilized stands (C) was similar to the early girdling at Åheden (Högberg *et al.*, 2001), with a reduction of soil CO$_2$ efflux of more than 60% (compared with the non-girdled control plots) The soil CO$_2$ efflux in the girdled F-plots was reduced by less. A reduced water uptake in girdled trees was found to affect the soil water content during dry periods with the most pronounced effect in the F-plots with high LAI, but the differences disappeared during periods with rain.

These three experiments strongly suggest that autotrophic respiration in Boreal forest accounts for up to 50–65% of the soil respiration during the summer. Data from Åheden, where we have records for several years, suggest that the cumulative contribution through the summer is about 50% (Bhupinderpal-Singh *et al.*, 2003). If the winter is included, the contribution of autotrophic respiration should be slightly smaller.

At present, we conclude that girdling provides a robust and inexpensive way of estimating, by subtraction, the contribution of autotrophic activity to total soil respiration. There are certainly problems associated with girdling, as with any other of the methods discussed, but the problems are not so great that they invalidate the method. Usually, the fractional contribution of autotrophic activity will be underestimated as a result of enhanced heterotrophic activity. Thus the figure for the former will be conservative. Currently there are several other girdling experiments elsewhere.

5.   Effects of temperature on components of soil respiration

Temperature is often used as a major driving variable in models of effects of global climate change on soil carbon dynamics (see, for example, McGuire *et al.*, 1992; Melillo *et al.*, 1993; Xiao *et al.*, 1998). It is therefore of significance that some authors have reported Q_{10} values up to twice as high for root respiration as for heterotrophic respiration (Boone *et al.*, 1998; Epron *et al.*, 2001). Effects of temperature are obviously difficult to disentangle from potential effects of seasonality of carbon allocation to roots (Hansen *et al.*, 1997). Moreover, one also has to consider time lags and acclimation in the response to temperature change (see, for example, Atkin *et al.*, 2000; Larigauderie and Körner, 1995), which are usually not included in models of global change.

Our girdling experiments provided an opportunity to follow the response of soil respiratory components to changes in soil temperature. Previous work on changes of the δ^{13}C of the soil CO$_2$ efflux (Ekblad and Högberg, 2001) and the girdling experiments (e.g., *Figure 5*) suggest a fast link between photosynthesis and root respiration, and that the seasonality of below-ground carbon allocation and phenology must be taken into account. Calculations based on our data support the assumption that Q_{10} is higher for autotrophic than for heterotrophic soil respiration. We think, however, that this is related to the fact that autotrophic respiration shows a stronger seasonality with large variations during a period of comparatively small changes in soil temperature (Bhupinderpal-Singh *et al.*, 2003). As indicated by the dramatic response to girdling, the flux from above-ground to below-ground appeared to be of utmost importance, and it seems unlikely that it should be much affected by minor changes in soil temperature. In particular, it is not obvious why autotrophic respiration, including that of ectomycorrhizal fungi, should be more sensitive to temperature than that of heterotrophs including saprotrophic fungi.

We thought that short-term temperature anomalies, especially periods of cold weather during the summer, could provide an opportunity to avoid confusing effects of temperature with effects of seasonality in below-ground carbon allocation. So far, we have been able to analyse three events of short-term (1–3 weeks long) temperature decreases at the beginning and middle of summer. In these cases (*Table 1*), the autotrophic respiration did not react to the drops in temperature, but sometimes even increased. Meanwhile, the heterotrophic activity clearly decreased in all three cases. Hence, this argues against the idea that autotrophic respiration should be more sensitive to temperature. In our view, autotrophic soil respiration should be modelled as driven by canopy photosynthesis and below-ground allocation of photosynthates, and the interactions of these processes with soil and air temperatures and with soil water content.

6. A new perspective on the debate about turnover rates of fine roots

The paradigm has been that the fine roots of trees turn over quickly, with lifespans of a couple of months at the most, although rates of turnover are highly variable and root turnover rates are difficult to measure in undisturbed natural systems (see, for example, Schoettle and Fahey, 1994). Recently, the paradigm of rapid root turnover has been challenged by observations in mini-rhizotrons (see, for example, Majdi and Kangas, 1997) and by observations of surprisingly high 'bomb ^{14}C' ages of roots (Gaudinski *et al.*, 2000, 2001). For Boreal *P. sylvestris*, root turnover was considered to be the major fate of the carbon allocated below ground by the trees (Ågren *et al.*, 1980; Persson, 1978). The fast and large decline in soil respiration after girdling is inconsistent with this view (Högberg *et al.*, 2002).

Table 1. Effects of temperature decreases during the summer on the autotrophic and heterotrophic components of soil respiration. Data for Åheden from Bhupinderpal-Singh et al. (2003) and for Flakaliden from P. Olsson, P. Högberg and S. Linder (unpublished).

Site	Tree species	Temperature change (°C, days)	Respiration change (%)	
			Autotrophic	Heterotrophic
Åheden	*Pinus sylvestris*	–6, 20*	+11	–50
Flakaliden				
Unfertilized	*Picea abies*	–0.9, 7**	+25	–14
Fertilized	*Picea abies*	1.1, 7**	+100	–20
Unfertilized	*Picea abies*	–2.4, 14**	±0	–33
Fertilized	*Picea abies*	–2.4, 14**	±0	–20

*Second summer after girdling; root starch reserves were depleted.
**The summer after girdling during a period of depletion of root sugar and starch reserves.

We thus used our estimate of a 50% contribution from root respiration to total soil respiration as a constraint in a model of partitioning between root growth and root respiration (Högberg *et al.*, 2002; *Figure 6*). The model suggested that about 75% of the carbon allocated below ground is used for respiration, and this results in a much more sensible partitioning between respiration and growth (50:50) for the whole tree than does the notion of very high rates of root turnover. As estimates of below-ground carbon allocation, autotrophic respiration, and root biomass become more precise, we are convinced that it will be possible to reconcile the contradicting views about fine-root turnover in trees.

7. Summary and concluding remarks

Soil-surface CO$_2$ efflux ('soil respiration') accounts for roughly two-thirds of forest ecosystem respiration, and can be divided into heterotrophic and autotrophic components. Conventionally, the latter is defined as respiration by plant roots. In Boreal forests, however, fine roots of trees are invariably covered by ectomycorrhizal fungi, which by definition are heterotrophs, but like the roots, receive sugars derived from photosynthesis. There is also a significant leaching of labile carbon compounds from the ectomycorrhizal roots. It is, therefore, more meaningful in the context of carbon balance studies to include mycorrhizal fungi and other mycorrhizosphere organisms, dependent on the direct flux of labile carbon from photosynthesis, in the autotrophic component. Hence, heterotrophic activity becomes reserved for the decomposition of more complex organic molecules in litter and other forms of soil organic matter. In reality, the complex situation is perhaps best described as a continuum from strict auto-

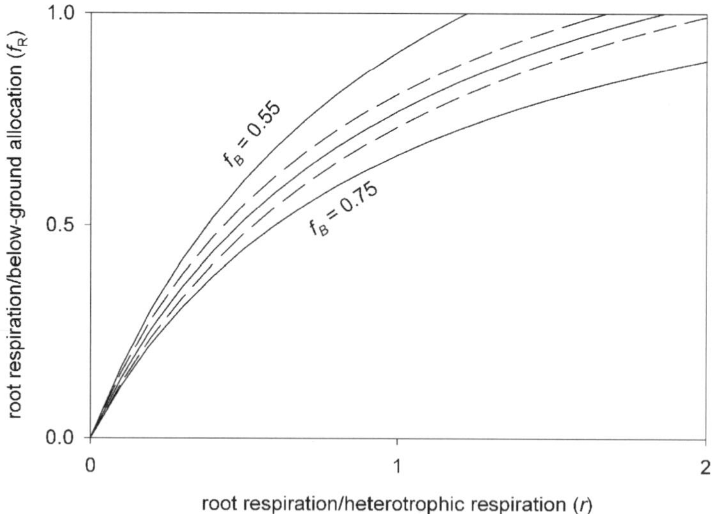

Figure 6. *Fraction of below-ground allocation which is used for root respiration (f_R) as a function of the ratio between root and heterotrophic respiration (r). The lines (f_B) correspond to different fractions (of photosynthates allocated below ground). Oecologia, "Carbon allocation ...," Högberg et al. Vol 132, pp. 579–582, Copyright (2002), with permission from Springer Verlag.*

trophy to strict heterotrophy. As a result of this, and associated methodological problems, estimates of the contribution of autotrophic respiration to total soil respiration have been highly variable. Based on recent stand-scale tree girdling experiments we have estimated that autotrophic respiration in boreal forest accounts for up to 50–65% of soil respiration during the snow-free part of the year. Girdling experiments and studies of the $\delta^{13}C$ of the soil CO_2 efflux show that there is a lag of a few days between the carbon uptake by photosynthesis and the release by autotrophic soil respiration of the assimilated carbon. In contrast, estimates of 'bomb ^{14}C' and other approaches have suggested that it takes years to decades between carbon uptake via photosynthesis and the bulk of soil heterotrophic activity. Temperature is normally used as a driver in models of soil processes and it is often assumed that autotrophic soil activity is more sensitive to temperature than is heterotrophic activity, but this is questionable.

It is inherently difficult to make a precise separation of autotrophic and heterotrophic respiration from soils. The partitioning between these two components is highly variable in space and time, and taxonomic autotrophs and heterotrophs may perform the function of the other group to some degree. Care should be taken to disturb as little as possible the delicate plant–microbe–soil system, and this speaks for non-intrusive isotopic methods. There are, however, problems in modelling the flux of isotopes through this complex system. Girdling of tree stands is a very robust alternative approach to make the distinction between autotrophic and heterotrophic activities, but ultimately kills the trees and cannot, therefore, always be used. A further development would be to block the phloem sugar transport reversibly. We propose that thus assumption needs further critical testing.

Acknowledgements

This study was made possible by financial support from the former Swedish Council of Forestry and Agricultural Research (SJFR), the former Swedish Natural Science Council (NFR), Swedish Research Council for Environment, Agricultural Sciences and Spatial Planning (FORMAS), the Swedish National Energy Administration (STEM), and the European Union through the Environment R and D Programme (ECOCRAFT, Contract No. ENV4-CT95-077; FORCAST, Contract No. EVK2-CT-1999-00035). This work contributes to the Global Change and Terrestrial Ecosystem (GCTE) core project of the International Geosphere–Biosphere Program (IGBP).

References

Ågren, G.I., Axelsson, B., Flower-Ellis, J.G., Linder, S., Persson, H., Staaf, H. and Troeng, E. (1980) Annual carbon budget for a young Scots pine. *Ecological Bulletins* **32**: 307–313.

Andrews, J.A., Harrison, K.G., Matamala, R. and Schlesinger, W.H. (1999) Separation of root respiration from total soil respiration using carbon-13 labeling during Free-Air Carbon dioxide Enrichment (FACE). *Soil Science Society of America Journal* **63**: 1429–1435.

Atkin, O.K., Edwards, E.J. and Loveys, B.R. (2000) Response of root respiration to changes in temperature and its relevance to global warming. *New Phytologist* **147**: 141–154.

Bergh, J., McMurtrie, R.E. and Linder, S. (1998) Climatic factors controlling the productivity of Norway spruce: A model-based analysis. *Forest Ecology and Management* **110:** 127–139.

Bergh, J., Linder, S., Lundmark, T. and Elfving, B. (1999) The effect of water and nutrient availability on the productivity of Norway spruce in northern and southern Sweden. *Forest Ecology and Management* **119:** 51–62.

Bhupinderpal-Singh, Nordgren, A., Ottosson-Löfvenius, M., Högberg, M.N., Mellander, P.-E. and Högberg, P. (2003) Tree root and soil heterotrophic respiration as revealed by girdling of boreal Scots pine forest: extending observations beyond the first year. *Plant, Cell and Environment* **26:** 1287–1296.

Björkman, E. (1960) *Monotropa hypopitys* L. an epiparasite on tree roots. *Physiologia Plantarum* **13:** 308–327.

Boone, R.D., Nadelhoffer, K.J., Canary, J.D. and Kaye, J.P. (1998) Roots exert a strong influence on the temperature sensitivity of soil respiration. *Nature* **396:** 570–572.

Cox, P.M., Betts, R.A., Jones, C.D., Spall, S.A. and Totterdell, I.J. (2000) Acceleration of global warming due to carbon-cycle feedbacks in a coupled climate model. *Nature* **408:** 184–187.

Cramer, W., Bondeau, A., Woodward, I., Prentice, I.C., Betts, R.A., Brovkun, V. *et al.* (2001) Global response of terrestrial ecosystem structure and function to CO_2 and climate change: results from six dynamic global vegetation models. *Global Change Biology* **7:** 357–373.

Dighton, J., Thomas, E.D. and Latter, P.M. (1987) Interactions between tree roots, mycorrhizas, a saprotroph and the decomposition of organic substrates in a microcosm. *Biology and Fertility of Soils* **4:** 145–150.

Dixon, R.K., Brown, S., Houghton, R.A., Solomon, A.M., Trexler, M.C. and Wisniewski, J. (1994) Carbon pools and flux of global forest ecosystems. *Science* **263:** 185–263.

Edwards, N.T. and Harris, W.F. (1977) Carbon cycling in a mixed deciduous forest floor. *Ecology* **58:** 431–437.

Edwards, N.T. and Ross-Todd, B.M. (1979) The effects of stem-girdling on biogeochemical cycles within a mixed deciduous forest in eastern Tennessee. I. Soil solution chemistry, soil respiration, litterfall and root biomass studies. *Oecologia* **40:** 247–257.

Edwards, N.T. and Sollins, P. (1973) Continuous measurement of carbon dioxide evolution from partitioned forest floor components. *Ecology* **54:** 406–412.

Ekblad, A. and Högberg, P. (2000) Analysis of $\delta^{13}C$ of CO_2 distinguishes between microbial respiration of added C4-sucrose and other soil respiration in a C3-ecosystem. *Plant and Soil* **219:** 197–209.

Ekblad, A. and Högberg, P. (2001) Natural abundance of ^{13}C in CO_2 respired from forest soils reveals speed of link between tree photosynthesis and root respiration. *Oecologia* **127:** 305–308.

Entry, J.A., Rose, C.L. and Cromack Jr, K. (1991) Litter decomposition and nutrient release in ectomycorrhizal mat soils of a Douglas-fir ecosystem. *Soil Biology and Biochemistry* **23:** 285–290.

Epron, D., Le Dantec, V., Dufrene, E. and Granier, A. (2001) Seasonal dynamics of soil carbon dioxide efflux and simulated rhizosphere respiration in a beech forest. *Tree Physiology* **21:** 145–152.

Farquhar, G.D., Von Caemmerer, S. and Berry, J.A. (1980) A biochemical model of photosynthetic carbon dioxide assimilation in leaves of 3-carbon pathway species. *Planta* **149**: 78–90.

Gadgil, R.L. and Gadgil, P.D. (1975) Suppression of litter decomposition by mycorrhizal roots of *Pinus radiata*. *New Zealand Journal of Forestry Science* **5**: 33–41.

Garbaye, J. (1994) Helper bacteria: a new dimension to the mycorrhizal symbiosis. *New Phytol.* **128**: 197–210.

Gaudinski, J., Trumbore, S.E., Davidson, E.A. and Zheng, S. (2000) Soil carbon cycling in a temperate forest: radiocarbon-based estimates of residence times, sequestration rates and partitioning of fluxes. *Biogeochemistry* **51**: 33–69.

Gaudinski, J., Trumbore, S.E., Davidson, E.A., Cook, A.C., Markewitz, D. and Richter, D.D. (2001) The age of fine-root carbon in three forests of the United States measured by radiocarbon. *Oecologia* **129**: 420–429.

Gower, S.T., Krankina, O., Olson, R.J., Apps, M., Linder, S. and Wang, C. (2001) Net primary production and carbon allocation patterns of boreal forest ecosystems. *Ecological Applications* **11**: 1395–1411.

Hansen, J., Turk, R., Vogg, G., Heim, R. and Beck, E. (1997) Conifer carbohydrate physiology: updating classical views. In: Rennenberg, W.E. and H. Ziegler, H. (eds) *Trees – Contributions to Modern Tree Physiology*, pp. 97–108. Backhuys, Leiden, The Netherlands.

Hanson, P.J., Edwards, N.T., Garten, C.T. and Andrews, J.A. (2000) Separating root and soil microbial contributions of soil respiration: a review of methods and observations. *Biogeochemistry* **48**: 115–146.

Havranek, W.H. and Tranquillini, W. (1995) Physiological processes during winter dormancy and their ecological significance. In: Smith, W.K. and Hinckley, T.M. (eds) *Ecophysiology of Coniferous Forests*, pp. 95–124. Academic Press, San Diego. CA.

Högberg, M.N. and Högberg, P. (2002) Extramatrical ectomycorrhizal mycelium contributes one-third of microbial biomass and produces, together with associated roots, half the dissolved organic carbon in a forest soil. *New Phytologist* **154**: 791–795.

Högberg, P. and Ekblad, A. (1996) Substrate-induced respiration measured *in situ* in a C3-plant ecosystem using C4-sucrose. *Soil Biol. Biochem.* **28**: 1131–1138.

Högberg, P., Nordgren, A. and Ågren, G.I. (2002) Carbon allocation between tree root growth and root respiration in boreal pine forest. *Oecologia* **132**: 579–582.

Högberg, P., Nordgren, A., Buchmann, N., Taylor, A.F.S., Ekblad, A., Högberg, M.N., Nyberg, G., Ottosson-Löfvenius, M. and Read, D.J. (2001) Large-scale forest girdling shows that current photosynthesis drives soil respiration. *Nature* **411**: 789–792.

Högberg, P., Ekblad, A., Nordgren, A., Plamboeck, A.H., Ohlsson, A., Bhupinderpal-Singh and Högberg, M.N. (2004) Factors determining the ^{13}C abundance of soil-respired CO_2 in boreal forests. In: Flanagan, L.B., Ehleringer, J.R. and Pataki, D.E. (eds) *Stable Isotopes and Biosphere–Atmosphere Interactions. Processes and Biological Controls*, pp. 47–68. Elsevier Academic Press, Oxford, U.K.

Horwath, W.R., Pregitzer, K.S. and Paul, E.A. (1994) ^{14}C allocation in tree–soil systems. *Tree Physiology* **14**: 1163–1176.

Janssens, I.A., Lankreijer, H., Matteucci, G., Kowalski, A.S., Buchmann, N., Epron, D. *et al.* (2001) Productivity overshadows temperature in determining soil and ecosystem respiration across European forests. *Global Change Biology* **7**: 269–278.

Koide, R.T. and Wu, T. (2003) Ectomycorrhizas and retarded decomposition in a *Pinus resinosa* plantation. *New Phytologist* **158**: 401–407.

Larigauderie, A. and Körner, C. (1995) Acclimation of leaf dark respiration to temperature in alpine and lowland plant species. *Annals of Botany* **76**: 245–252.

Leake, J.R. (1994) The biology of myco-heterotrophic (saprophytic) plants. *New Phytologist* **127**: 171–216.

Linder, S. (1995) Foliar analysis for detecting and correcting nutrient imbalances in Norway spruce. *Ecological Bulletins* **44**: 178–190.

Linder, S. and Flower-Ellis, J.G.K. (1992) Environmental and physiological constraints to forest yield. In: Teller, A., Mathy, P. and Jeffers, J.N.R. (eds) *Responses of Forest Ecosystems to Environmental Changes*, pp. 149–164. Elsevier. The Hague.

Majdi, H. (2001) Fine root production and longevity in response to water and nutrient availability in a Norway spruce stand in northern Sweden. *Tree Physiology* **2**: 1057–1061.

Majdi, H. and Kangas, P. (1997) Demography of fine roots in response to nutrient applications in a Norway spruce stand in SW Sweden. *Ecoscience* **4**: 199–205.

McGuire, A.D., Melillo, J.M., Joyce, L.A., Kicklighter, D.W., Grace, A.L., Moore, B.III. and Vorosmarty, C.J. (1992) Interactions between carbon and nitrogen dynamics in estimating net primary production for potential vegetation in North America. *Global Biogeochemical Cycles* **6**: 101–124.

McMurtrie, R.E., Gholz, H.L., Linder, S. and Gower, S.T. (1994) Climatic factors controlling the productivity of pine stands: A model-based analysis. *Ecological Bulletins* **43**: 173–188.

Melillo, J.M., McGuire, A.D., Kicklighter, D.W., Moore III, B., Vorosmarty, C.J. and Schloss, A.L. (1993) Global climate change and terrestrial net primary production. *Nature* **363**: 234–240.

Näsholm, T., Ekblad, A., Nordin, A., Giesler, R., Högberg, M. and Högberg, P. (1998) Boreal forest plants take up organic nitrogen. *Nature* **392**: 914–916.

Persson, H. (1978) Root dynamics in a young Scots pine stand in Central Sweden. *Oikos* **30**: 508–519.

Raich, J.W. and Schlesinger, W.H. (1992) The global carbon dioxide flux in soil respiration and its relationship to vegetation and climate. *Tellus Series B – Chemical and Physical Meteorology* **44**: 81–99.

Robinson, D. and Scrimgeour, C.M. (1995) The contribution of plant C to soil CO$_2$ measured using δ^{13}C. *Soil Biology and Biochemistry* **27**: 1653–1656.

Schoettle, A.W. and Fahey, T.J. (1994) Foliage and fine root longevity of pines. *Ecological Bulletins* **43**: 136–153.

Smith, B.N. and Epstein, S. (1971) Two categories of ^{13}C/^{12}C ratios for higher plants. *Plant Physiology* **47**: 380–384.

Smith, S.E. and Read, D.J. (1997) *Mycorrhizal Symbiosis*, 2nd edition. Academic Press, London.

Staddon, P.L., Bronk Ramsey, C., Ostle, N., Ineson, P. and Fitter, A.H. (2003) Rapid turnover of hyphae of mycorrhizal fungi determined by AMS microanalysis of ^{14}C. *Science* **300**: 1138–1140.

Stone, E.L. and Kalicz, P.J. (1991) On the maximum extent of tree roots. *Forest Ecology and Management* **46**: 59–102.

Taylor, A.F.S., Martin, F. and Read, D.J. (2000) Fungal diversity in ectomycorrhizal communities of Norway spruce (*Picea abies* [L.] Karst.) and beech (*Fagus sylvatica* L.) along north–south transects in Europe. *Ecological Studies* **142**: 343–365.

Timonen, S., Jorgensen, K.S., Haahtela, K. and Sen, R. (1998) Bacterial community structure at defined locations of *Pinus sylvestris–Suillus bovinus* and *Pinus sylvestris–Paxillus involutus* mycorrhizospheres in dry pine forest humus and nursery peat. *Canadian Journal of Microbiology* **44**: 499–513.

Troeng, E., and Linder, S. (1982) Gas exchange in a 20-year-old stand of Scots pine. I. Net photosynthesis of current and one-year-old shoots within and between seasons. *Physiologia Plantarum* **54**: 7–14.

Valentini, R., Matteucci, G., Dolman, A.J., Schulze, E.-D., Rebmann, C., Moors, E.J. *et al.* (2000) Respiration as the main determinant of carbon balance in European forests. *Nature* **404**: 861–865.

Xiao, X., Melillo, J.M., Kicklighter, D.W., McGuire, A.D., Prinn, R.G., Wang, C., Stone, P.H., Sokolov, A. (1998) Transient climate change and net ecosystem production of the terrestrial biosphere. *Global Biogeochemical Cycles* **12**: 345–360.

13

Trace gas and CO$_2$ contributions of northern peatlands to global warming potential

Tuomas Laurila, Mika Aurela, Annalea Lohila and Juha-Pekka Tuovinen

1. Introduction

Natural mires are unique habitats where plants have adapted to the wet and, commonly, nutrient-poor environment (Crum, 1988). They form usually in lowlands in climatic conditions in which precipitation exceeds evaporation. Mires have slowly accumulated carbon in peat deposits because decomposition in anoxic conditions below the water table is slow. This accumulation takes place despite the low photosynthetic CO$_2$ uptake rates of mire vegetation compared with most other upland ecosystems. Among terrestrial ecosystems, northern peatlands are one of the largest reservoirs of carbon that has been taken from the atmosphere, mostly during the Holocene. Gorham (1991) estimated that they comprise as much as one-third of the carbon stored globally in soils. Extensive natural mires still exist in uninhabited areas of Eurasia and North America. However, in populated regions, most wetlands have been drained for agriculture, forestry, and energy production.

The three most important greenhouse gases (GHGs) that are well mixed in the atmosphere and have long lifetimes are CO$_2$, methane (CH$_4$), and nitrous oxide (N$_2$O). There are sources and sinks of all three of these gases in peatland ecosystems. Natural mires affect concentrations of radiatively active trace gas concentrations in the atmosphere predominantly in two ways. Firstly, they have been long-term continuous sinks of CO$_2$, thus reducing atmospheric CO$_2$ content and hence its warming effect. Secondly, methane-producing microbes are usually found in the anoxic organic soil layer below the water table, resulting in significant CH$_4$ emissions to the

---

atmosphere (Whiting and Chanton, 1993). On the other hand, microbial N_2O production rates in natural mires tend to be low because these environments are usually too anoxic, acid, and nutrient-poor for denitrification and nitrification bacteria (Regina *et al.*, 1996). Liming and addition of nutrients are common practices for making the soil more fertile. Together with the lowered water table, these additions change the GHG budgets profoundly.

It is generally assumed that drained peatlands are sources of CO_2 (Gorham, 1991) and small sinks of CH_4 (Maljanen *et al.*, 2002), whereas those in agricultural use are sources of N_2O (Regina *et al.*, 1996). However, emissions and uptake rates of GHGs vary greatly, depending on land use and local environmental conditions, and are highly dependent on vegetation, hydrology, nutrient status, and temperature. Climate change will modify these variables and most probably also the composition of mire flora (Oechel *et al.*, 2000). Thus, it is not only local human intervention that may affect GHG fluxes, but we can expect large-scale changes in GHG fluxes to result from the indirect human actions that lead to global warming.

Terrestrial vegetation in temperate and Boreal regions has been recognized as a major sink of atmospheric CO_2 (IPCC, 2001a). The net CO_2 uptake by terrestrial vegetation is considered so important that it has been included as a component in the international protocols for reducing atmospheric concentrations of greenhouse gases and climate change (Schulze *et al.*, 2002). A country participating in the Kyoto protocol can benefit from these natural carbon sinks and reduce its abatement commitment to anthropogenic emissions, if it can show, in a verifiable way, changes in the carbon balance of relevant land-use classes, compared with the reference year, 1990. However, peatlands pose particular scientific and practical challenges for calculating GHG budgets and sink capacities. In this presentation, we do not aim at a full description of GHG budgets of different peatlands, but, using enlightening examples, we will demonstrate which gases are important in natural northern mires and how their exchanges are modified by human intervention and land-use change. We will present examples from sites at which CO_2 exchanges have been measured using micrometeorological techniques, thus avoiding many problems related to studies that rely on chamber measurements. Unlike observations of long-term carbon accumulation rates, these studies yield present-day gas fluxes. Our focus is on the northern Boreal and Arctic regions, where most peatlands are located.

2. Global emissions and radiative forcing

The three gases considered here have both anthropogenic and natural sources. Natural ecosystems are generally sinks of CO_2 as part of the carbon fixed in photosynthesis is stored in the soil. On geological time-scales, these carbon deposits may have turned into the oil and gas reserves that we are now using as a massive energy source. It is estimated that at present 1.4 Pg (C) per year is stored in the terrestrial ecosystems (IPCC, 2001b). Unfortunately, quantitative data are missing about which land-use classes are sinks and which are sources, and how these are geographically distributed. The estimation of current carbon budgets of peatlands is hampered by the lack of a general ecosystem model for simulating GHG budgets on a global scale. Even on a plot scale, there remain interesting open questions, such as whether, on average, natural mires maintain, under present climate conditions, the same carbon net uptake rates as during the Holocene.

The latest IPCC Scientific Assessment (IPCC, 2001b) summarizes the global CH_4 and N_2O emissions, and their considerable uncertainties. The total emissions of methane are 600 Tg (CH_4) per year, of which wetlands are the largest source category (137 Tg (CH_4) per year). As the total natural emissions are 172 Tg (CH_4) per year, wetland emissions would have been predominant before human influence. Despite the different vegetation, rice fields have similar microbial methanogenic processes to those in mires and also have high methane emission rates (88 Tg (CH_4) per year).

Nitrogen fertilizers and livestock grazing turn cultivated lands and pastures into large emitters of N_2O compared with natural soils. Agricultural soils emit 4.2 Tg (N_2O) per year of the total emissions of 17.7 Tg (N_2O) per year. Natural mires generally have very low emission rates, but if the soil is drained and fertilized for better growth, microbial conditions in the soil change dramatically. Mineralization of the nitrogen bound in organic compounds in peat may be very high. Among the peatlands, it is those in agricultural use, covering globally 10 000 km^2 (Immirzi *et al.*, 1992), that are the most important sources. On the global scale, the emissions from cultivated histosols are difficult to estimate but most probably their proportion is small.

Infra-red absorption properties of the three greenhouse gases considered here vary greatly. In the present atmosphere, one added CH_4 molecule absorbs infrared radiation about 23 times as efficiently as one added CO_2 molecule, and one N_2O molecule about 296 times a molecule of CO_2 on a 100 year time horizon (IPCC 2001b). Once emitted, CH_4 is oxidized by hydrogen radicals in the atmosphere and consumed by methanotrophic bacteria in soils, leading to an average global lifetime of 12 years. Almost the sole sink of N_2O is photo-dissociation in the stratosphere, and the atmospheric lifetime of N_2O is 114 years. To compare the effectiveness of emissions of different GHGs, we have to take into account both their lifetime in the atmosphere and an appropriate time horizon over which to make the assessment. Best estimates of the time-integrated warming effects of GHG emissions, relative to CO_2, the so-called global warming potential (GWP), are given in the IPCC reports for different time horizons (IPCC, 2001b). We have used the time horizons given there of 20, 100, and 500 years for our GWP calculations. The 100 year time horizon is mandatory for Kyoto Protocol reporting. We suggest that the slow processes occurring in natural mires might support the use of longer time-scales.

3. Northern peatlands

Peatlands may be classified in various ways. One classification is based on the process of their formation. In principle, there are three processes for the initiation of peatland formation: primary formation, paludification, and terrestrialization. In primary mire formation, mire vegetation begins to grow on virgin mineral soil that has emerged from water or ice. This process is common in the land uplift areas of Scandinavia and the Hudson Bay area in Canada. Paludification is a process in which peatland vegetation starts to grow on mineral forest soil. In terrestrialization, floating peat spreads out on ponds and small lakes. Expanding mire vegetation, particularly *Sphagnum* mosses, gradually changes hydrology and soil properties, favouring wetland plants and further succession towards a mire.

The most extensive areas of northern peatlands are located in Canada and Russia. In some countries, the proportion of mires is very large. In Finland, the peatland area is about 89 000 km², which is 30% of the total land area (Finnish Forest Research Institute, 2001). Northern peatlands encompass an area of 3.5 million km² and contain a carbon pool of 273 Pg (Turunen *et al.*, 2002). These mires have developed during the Holocene. Turunen *et al.* (2002) assumed an average age of 4200 years, and this leads to an annual carbon sink of 65 Tg.

Vegetation in mires varies a great deal, depending mainly on hydrological conditions and nutrient availability (Eurola *et al.*, 1984). The plant species composition provides a criterion for mire type classification. A broad division of mire types is usually made between bogs and fens. Ombrotrophic bogs receive water and nutrients by precipitation, and nutrients to some extent also by dry deposition, creating an acidic habitat, with low concentrations of nutrients. *Sphagnum* mosses are among the few plant species adapted to this nutrient-poor environment and are by far the most common species in raised bogs. Usually ground or surface water from the surrounding mineral soil areas, where the vegetation is mesotrophic, influences the border areas of a raised bog. Fens also receive nutrients and neutralizing alkaline materials in surface and ground water from the surrounding uplands. They may be oligotrophic or mesotrophic, varying from acidic to alkaline. Vegetation is usually abundant and species rich, and sedges are common. Young mires are often fens. The bog phase begins when the peat layer grows so that the groundwater from the surroundings no longer reaches the mire center.

Climate is one of the most important factors for development of the type of mire (Crum, 1988). *Sphagnum* mosses do not survive over extended periods in the inundated conditions that are common during extended spring floods in the northernmost areas. In Finland, the *aapa* mires, which are fens in which wet flarks are separated by hummocky strings, are numerous in the northern part of the country. In the more southern raised-bog areas, the effective temperature sum above a mean daily temperature of 5 °C exceeds 1000 degree days (Turunen *et al.*, 2002). In the *aapa* mire zone, the annual maximum of the water-equivalent of snow cover in spring exceeds 140 mm. Towards the north, the surface structure of *aapa* mires becomes more patterned. Soil frost is a major factor forming hummocks. In the sub-Arctic areas, where the annual average temperature is below 0 °C, we may find palsas. These are peat mounds in which an ice core is present all year round. In the permafrost Arctic areas, fens and polygon mires with wet hollows 10–30 m in diameter are common. A cool maritime climate, for example in Ireland and Scotland, supports formation of blanket bogs.

The distinction between bogs and fens is important for the estimation of greenhouse gas budgets (Whiting and Chanton, 2001). *Sphagnum* peat in bogs decomposes very slowly, leading to generally higher carbon accumulation rates than in fens where more decomposable plant litter is provided by sedges and other vascular plants. For CH_4 emission to the atmosphere, it is not only the CH_4 production rate that is important: gas transfer from the CH_4-producing layers in the peat to the atmosphere also needs to be efficient. In bogs, there is often a superficial oxic moss layer in which most of the CH_4 is oxidized to CO_2. Sedges, which are common in fens, have in their stems a tissue comprising air-filled spaces (aerenchyma) through which oxygen diffuses to the roots and CH_4 produced in the peat diffuses directly to the atmosphere (Whiting and Chanton, 1993).

An average long-term rate of carbon accumulation (LORCA) can be calculated from the total carbon mass and the basal age of the mire. Turunen *et al.* (2002) calculated the average LORCA in the bog and *aapa* mire regions of Finland. The average in the bog areas (26.1 g (C) m^{-2} per year) was higher than that in the *aapa* mire region (17.3 g (C) m^{-2} per year). It is important to emphasize that LORCA is different from the actual rate of carbon accumulation (ARCA) and that present-day carbon budgets have to be estimated using other methods, such as gas flux measurements.

4. Carbon dynamics in pristine mires

The carbon budget of a mire is a balance between gaseous fluxes to the atmosphere, net transport of dissolved carbon by the soil water, and peat accumulation (*Figure 1*). Part of the CO_2 carbon fixed by photosynthesis is used for growth of plants and the rest is emitted back to the atmosphere as autotrophic respiration from above-ground and below-ground parts of the plants, or is deposited as 'leaf' litter onto the surface or as 'root' litter below ground. Decomposition of this detritus by heterotrophic organisms, which takes place mostly in the oxic upper layers (acrotelm), produces a further respiratory CO_2 emission. Old litter layers that gradually become inundated if the water table is close to the surface form peat.

The anoxic layer (catotelm) is of vital importance for carbon accumulation and CH_4 production; decomposition of dead plant structures forming peat is very slow in this layer. In anoxic conditions methanogenic bacteria decay peat producing CH_4 in significant amounts (see, for example, Le Mer and Roger, 2001). The capacity for net primary productivity (NPP), temperature, water table level, nutrient availability, and vegetation type are important variables affecting CH_4 production (Nykänen *et al.*, 1998; Schimel, 1995). The depth and structure of the acrotelm strongly affect CH_4 emissions because a substantial part of the CH_4 emitted from the catotelm is oxidized by methanotrophic bacteria before emission to the atmosphere. In bogs, the *Sphagnum* layer above the water table curbs the CH_4 flux to the atmosphere. The highest CH_4 emission rates are usually observed in fens in which sedge-type plants are common, because the aerenchyma provides a direct gas transfer route between the root zone and the atmosphere (Schimel, 1995), bypassing oxidation in the acrotelm.

Lateral and vertical fluxes of carbon dissolved in soil water are significant because carbon densities in peatlands are high. The rate of net lateral loss is usually relatively low (Sallantaus, 1992), but Turunen *et al.* (1999) reported downward leaching rates to the underlying mineral soil that are comparable in magnitude to the carbon exchange between the mire and the atmosphere. However, neither of these leaching rates is commonly measured, and in most studies they remain unknown.

5. Measurements of greenhouse gas fluxes

For the measurement of carbon budgets there are two different approaches. Traditional biomass inventories are most suitable for observing accumulation rates

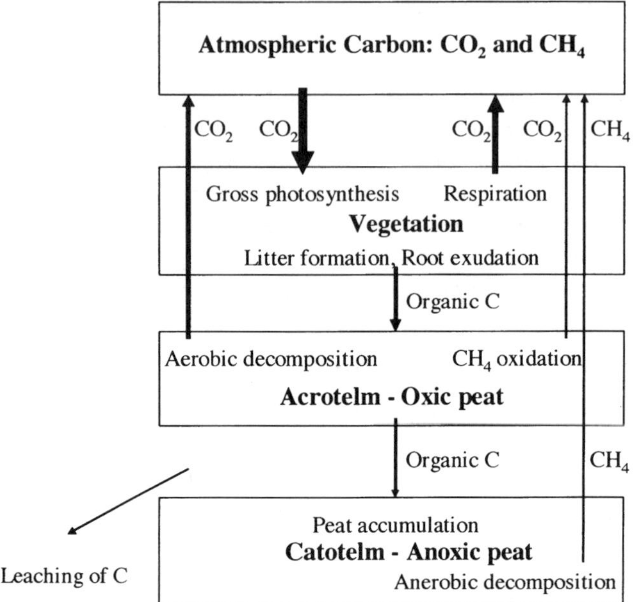

Figure 1. The carbon pools of a mire consist of vegetation and oxic and anoxic peat layers. Photosynthesis of CO₂ by vegetation is the primary source of carbon to the ecosystem. The photosynthetic products are used for growth and maintenance respiration. Above- and below-ground litter production provide carbon to peat formation in the oxic and anoxic layers. Decomposition of organic material in the acrotelm is a source of CO₂ to the atmosphere, but decomposition in the anoxic catotelm by methanogenic bacteria produces CH₄ which has to pass through the oxic layers before it is emitted to the atmosphere. Part of the carbon dissolved in the water is leached to the soil or transported away from the wetland.

over long periods such as decades or centuries, and the widely used LORCA is a measure of carbon accumulation over millennia. An advantage of biomass harvesting is that we obtain detailed information on the accumulation rates of different plants. A disadvantage is that it is very laborious to obtain statistically reliable values for short-term stock changes, especially for the below-ground components. Another way to investigate CO_2 flows is to measure the short-term gas flux; for CH_4 and N_2O this is the only possibility we have.

Gas fluxes provide information at the high temporal resolution that is necessary for investigating processes in plants and ecosystems and for determining GHG budgets on different temporal scales. In ecosystem research there are two widely used approaches to flux measurements: chamber methods and micrometeorological methods (Fowler *et al.*, 2001). In chamber methods, a part of the system is isolated in a chamber and the rate of change of gas concentration is measured. An advantage of chamber methods is that the subject of investigation is known exactly and environmental variables controlling the fluxes can be varied experimentally for the determination of response functions. A disadvantage is that the conditions inside the chamber always differ from those in the ambient air and that, for ecosystem scale studies in particular, major spatial and temporal extrapolation based on small samples is often necessary. In the case of tall vegetation, such as forest stands, for example, it is very difficult to obtain reliable ecosystem budgets by using chamber methods (see Chapters 7, 8, and 15, this volume, for example).

Micrometeorological methods are based on a completely different approach. A net mass balance over the target area is measured by observing turbulent flow velocities and gas concentrations. In the method, which is commonly used today for CO_2 fluxes, vertical wind speed and fluctuations in gas concentration are recorded at a rate of several times a second (Baldocchi, 2003; Chapters 7, 8, and 9, this volume). Typically, an average flux is calculated as their covariance over a 30 minute period, and consequently the method is generally known as the eddy covariance method. Processing of the data is complicated by additional corrections attributable to imperfect responses of the instruments, atmospheric density fluctuations, and flow alignment (Aubinet et al., 2000). The method relies on observations of turbulent transfer, so during periods of low wind speed, reliable measurement may not be obtained. For these periods, so called gap-filling algorithms are applied, and this results in additional uncertainty (Falge et al., 2001). As an automated system, however, micrometeorological instrumentation provides a continuous time-series of fluxes, and this is a clear advantage over most chamber-based systems. For ecosystem-scale gas budget studies, there are some further advantages, as the measurement system does not influence the ecosystem in any way, and the result represents an area-integrated average net flux that is an appropriate quantity for calculating average budgets over longer periods of days to years. None-the-less, there are some disadvantages related to the method: the area of vegetation from which the measured signal comes is not known exactly, and photosynthesis and respiration rates have to be extracted from the observations by using a simple model, as it is their net effect that is measured. However, for annual budgets on an ecosystem scale, eddy covariance fluxes are superior to chamber measurements. The CO_2 budgets, which we present in the following section, have mostly been measured by the eddy covariance method, with appropriate post-collection data-processing procedures.

As an example of the possibilities of eddy covariance flux studies, we first show flux measurements from a sub-Arctic fen site located at Kaamanen in northern Finland that has been operating since 1997 (Aurela et al., 2001). The site is briefly described in the next section. The daily average CO_2 budgets (*Figure 2*) show that all seasons (winter, thaw, pre-leaf, summer, and fall) contributed significantly to the annual net CO_2 exchange. Both the timing of these periods and the CO_2 balance over the growing season showed significant inter-annual variation. The flux measurements were able to resolve small annual CO_2 budgets of only 0.5 mol m^{-2} per year, on average, and their significant variation from year to year (Aurela et al., 2004). Such results indicate that measurements over several years are necessary for estimating the annual budget of this kind of mire. This example demonstrates the power of the eddy covariance method for obtaining reliable, ecosystem-scale carbon budgets in environmental conditions in which it is difficult to employ other methods to measure gas fluxes continuously, through the Arctic winter, the spring flooding period, and the growing season.

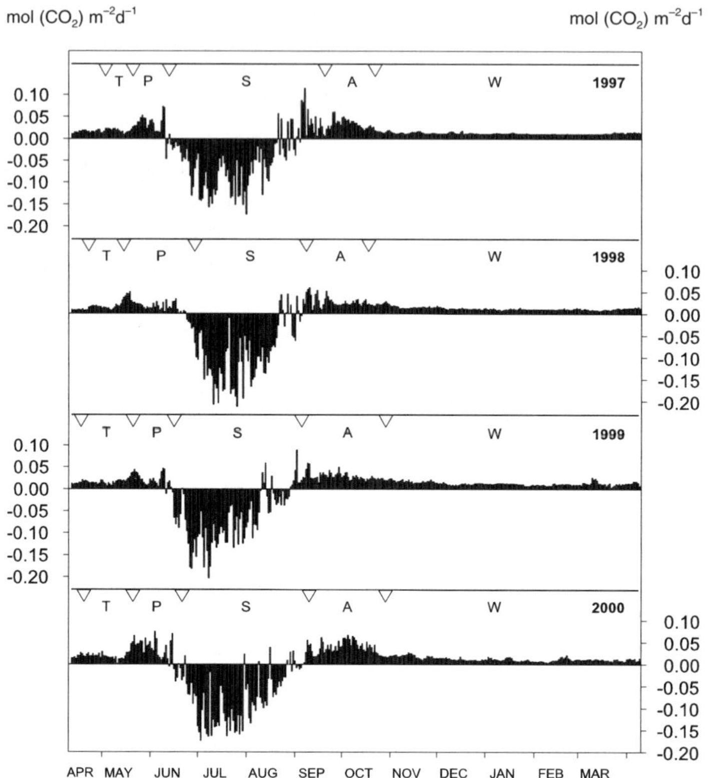

Figure 2. *Daily average fluxes of CO₂ measured using the micrometeorological eddy covariance method at a sub-Arctic aapa mire in northern Finland (Aurela et al., 2004). The thaw (T), pre-leaf (P), summer (S), fall (A), and winter (W) seasons are also indicated. Net CO₂ uptake (negative numbers) takes place during the short summer, beginning in late June and lasting until late August. Most of the inter-annual variation of the carbon balances results from the timing of the periods in spring and early summer. The effluxes during winter are fairly constant.*

6. Natural bogs and fens

The gas exchanges and greenhouse warming potentials of natural mires are conveniently presented using three common northern wetland types as examples. These are a Boreal bog in the southern Boreal region, a northern *aapa* mire in the border area between the north Boreal and sub-Arctic regions, and a fen in the permafrost region. At all these sites, extensive eddy covariance flux measurements have been made, providing reliable estimates of the CO₂ budgets.

The most southern of the example sites is the Mer Bleue bog in Ontario, Canada (45°24′ N, 75°30′ W), which is described by Lafleur *et al.* (2003). The peatland is 8500 years old with the bog phase starting 6000 years ago. Peat depth in the center of this large oval-shaped bog is 5–6 m, decreasing to about 2 m at the edges. The vegetation largely consists of *Sphagnum* mosses and low-growing ericaceous and deciduous shrubs. The mean annual temperature is 5.8 °C and precipitation 910 mm per year.

The open flark fen at Kaamanen (69°08′ N, 27°17′ E) in northern Finland is a typical example of the northern *aapa* mire ecosystems in the region (Aurela *et al.*, 2002;
Heikkinen *et al.*, 2002). There are series of dry hummocky strings (0.3–0.8 m high)
and wet hollows. The peat layer is about 1 m deep. The fen is about 7000 years old and
has a LORCA of 11 g (C) m^{-2} per year (J. Turunen, personal communication). The
hollows are covered by different sedges (*Carex* spp.) and some moss species, whereas
the higher strings are dominated by various shrubs, such as *Ledum palustre*,
Empetrum nigrum, *Rubus chamaemorus*, and *Betula nana*. The maximum single-
sided leaf area index (LAI) is estimated to be 0.7 (Aurela *et al.*, 2001). The mean annual
temperature is −1.3 °C and precipitation 395 mm per year. Only the topsoil is frozen
during the winter because the snow cover effectively insulates it.

The third example is an Arctic fen in the permafrost region on the northern part
of the eastern coast of Greenland (74°28′ N, 20°34′ W) where extensive eddy covariance CO_2 and CH_4 flux studies have been made (Friborg *et al.*, 2000; Nordstroem *et
al.* 2001; Soegaard and Nordstroem, 1999). The vegetation comprises *Eriophorum*
spp., *Dupontia psilosontha*, and *Arctagrostis latifolia*, with the hummocks and the
more elevated areas dominated by *Salix arctica* and mosses, overlying a peat layer of
0.2–0.3 m. The maximum LAI in August is 1.1. The mean annual temperature within
the Zackenberg Ecological Research Operations (ZERO) study area is −9.6°C. The
mean July temperature is 4.7 °C and the active peat layer then extends to a depth of
0.2–0.8 m.

At the Mer Bleue bog in Canada, the annual average CO_2 uptake rate first measured by Lafleur *et al.* (2001) using the eddy covariance method was −5.6 mol m^{-2} per
year, representing substantial carbon uptake from the atmosphere (*Figure 3*). (Note:
the conventional micrometeorological sign convention is for upward fluxes from vegetation to the atmosphere to be treated as positive and downward removals from the
atmosphere to vegetation as negative.) The growing season is very long, extending
over 200 days and resulting in an average annual carbon uptake rate for the four measurement years of −4.5 mol m^{-2} per year. In normal weather conditions, the net
exchange is relatively stable (Lafleur *et al.*, 2003). However, during one year with an
exceptionally dry summer, net uptake was only −0.77 mol m^{-2} per year. Losses of carbon as CH_4 were small because of the thick, oxic *Sphagnum* layer. Methane emissions
and dissolved organic and inorganic carbon exports amounted to 0.22 mol m^{-2} per
year in total (Fraser *et al.*, 2001). Thus the LORCA is only about one-third of the
annual net uptake indicated by the recent gas flux measurements (Moore *et al.*, 2002).

At the *aapa* mire in northern Finland, the growing season is about three months.
During the 71 day summer period in 1998 (see *Figure 2*), the average net CO_2 uptake
was −6 mol m^{-2} (Aurela *et al.*, 2002). Respiration during the other seasons, including
the winter, resulted in an average annual net uptake of −1.6 mol m^{-2}. Measurements by
Aurela *et al.* (2004) in later years showed significant inter-annual variation of the
budget, ranging from zero up to three times that observed in 1998, and related to
weather conditions, particularly the timing of snow melt, in a highly systematic way.
Based on eddy covariance flux measurements at the site, Hargreaves *et al.* (2001) estimated the annual CH_4 emission as 0.35 mol m^{-2}. A crude estimate of the leaching of
total organic carbon is 0.6 mol m^{-2} per year (Aurela *et al.*, 2002). Summation of these
values, suggests that, within the measurement errors, the present carbon budget averaged over six years, is similar to the LORCA (Aurela *et al.*, 2004).

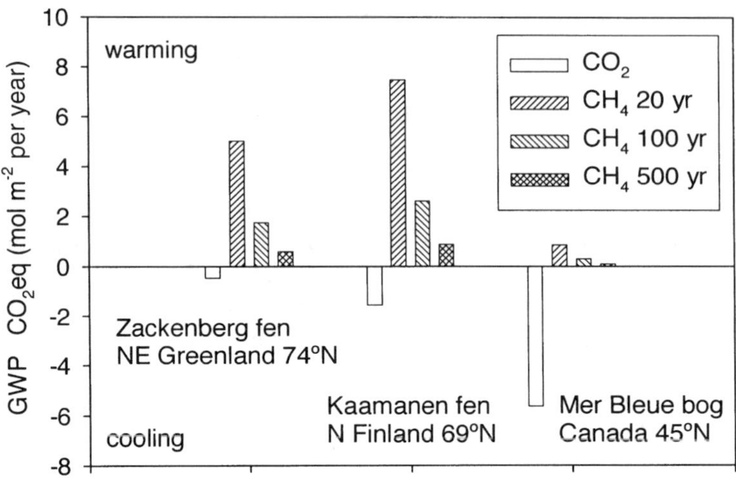

Figure 3. *GWPs calculated from annual CO_2 balances and CH_4 emissions for an Arctic fen in the permafrost region in Greenland (Friborg et al., 2000; Nordstroem et al., 2001), a sub-Arctic fen in northern Finland (Hargreaves et al., 2001; Aurela et al., 2004), and a bog in Canada (Fraser et al., 2001; Lafleur et al., 2001). The results show higher CO_2 uptake by the bog ecosystem and higher CH_4 emissions by the fens.*

At the Arctic Zackenberg fen site in Greenland, the high-summer CO_2 uptake rates were comparable to those observed in warmer climates (Laurila *et al.*, 2001; Nordstroem *et al.*, 2001), but because of the very short growing season, lasting only six weeks, the average annual net rate of uptake was very low (Nordstroem *et al.*, 2001). This low net uptake rate and the very low emission rates during the long winter season when the soil is frozen resulted in a small positive annual carbon budget of about 0.5 mol m⁻². The CH_4 emission rates in midsummer were relatively high, considering the low soil temperatures (Friborg *et al.*, 2000). The annual emissions, however, were again small because the period when the topsoil temperatures are above 0 °C was short.

Summarizing the annual carbon budgets of the different mires, the CO_2 uptake rate tends to be higher in climates with a longer growing season, and at bog sites compared with fen sites. On the other hand, the CH_4 emission rates are higher at the typically wetter fen ecosystems. There is a clear dependence of the GWPs on latitude (*Figure 3*), even though this data set is very limited. In the southern bog, the CO_2 uptake dominated the GWP budget and the site has an unambiguous net cooling effect. As we move northwards, the relative importance of CH_4 emissions increases, and in the sub-Arctic fen the sign of the total GWP depends on the time horizon selected. The high-Arctic fen has a net warming effect even on the 500 year time horizon. The N_2O emissions are assumed to be insignificant at these pristine and nutrient poor mires. These results are in a good agreement with those of Whiting and Chanton (2001), who emphasized the influence of the time horizon in their study on a more southern transect (30°–54° N).

7. Peatlands drained for forestry

In countries with extensive mire areas, there is pressure to develop the timber production capacity. In Finland, one-third of the forestry land area is classified as peatland. Widespread ditching took place mostly during the 1960s, and currently over half of the mires are drained (Finnish Forest Research Institute, 2001). Water-table drawdown is necessary for sufficient aeration of the roots of trees and most other vascular plants. Decomposition of peat in the thicker oxic layer also accelerates nitrogen mineralization. The soil is deficient in potassium and phosphorus, which should be added for healthy tree growth. Relatively fertile forested mires are most suitable for timber production (see also Chapter 8, this volume). In poor ombrotrophic bogs, economic timber production is difficult.

Long-term forest growth experiments initiated in the 20th Century provide valuable information on subsequent changes in the vegetation and carbon stocks of drained peatlands. After ditching, the peatland ecosystem enters a dynamic state of change that may last for decades (Laine *et al.*, 1995); tree growth accelerates and succession of the ground flora towards plant species common on upland forest soils begins. More carbon is accumulated into above-ground and below-ground tree biomass (Laiho *et al.*, 2003). Usually, after a short period, the build-up of carbon into plant tissues exceeds the losses resulting from the increased heterotrophic respiration. On very long time-scales, it is difficult to estimate whether litter production of the new forest ecosystem is large enough to sustain soil carbon accumulation. Despite the much higher rate of carbon fixation of drained, forested, peatlands, peat accumulation during the past millennia has been an attribute of water-logged ecosystems.

Hargreaves *et al.* (2003) have investigated the effects of afforestation on carbon budgets of peatlands in Scotland, where an area of about 5000 km^2 has been ploughed, drained, and planted with Sitka spruce (*Picea sitchensis* (Bong.) Carr.). The annual carbon budgets of a 3–5 m thick ombrotrophic blanket peat and a chronosequence of young Sitka spruce plantations up to 26 years of age were estimated using eddy-covariance CO_2 flux measurements and modelling. Their results show that the afforested ecosystem was a net carbon source of 16–33 mol m^{-2} per year for two to four years after draining and planting, after which it became a carbon sink of 25 mol m^{-2} per year for four to eight years, increasing up to 42 mol m^{-2} per year thereafter. The estimated peat decomposition rate below the forest canopy was 8 mol m^{-2} per year or less. The natural, undisturbed mire used for comparison accumulated carbon at a rate of 2 mol m^{-2} per year. Using these results, Hargreaves *et al.* (2003) estimated that afforested peatlands in Scotland accumulate more carbon in plants, soil, and products than the amount that is lost from the peat for 90–190 years, i.e., during the first or second timber production rotations, the afforested peatland is a net sink of carbon. However, on longer time-scales it will become beneficial to restore wetland vegetation, because carbon stocks in most peatlands exceed the stocks that can be added by the growing of trees. However, considering the total GHG budget, the cessation of CH_4 emissions after aeration of the topsoil favours the draining of peatlands.

Minkkinen *et al.* (2002) estimated the effects of drainage on the carbon budget and radiative forcing of forest peatlands over a 200-year-period, including all greenhouse gases, for a comprehensive national study in Finland (*Figure 4*). Land-use changes in Finnish mires were taken from the national forest inventory statistics, and the changes in peatland carbon budgets of undrained and drained mires of different mire types in different climate regions were estimated from measurements of peat carbon stocks. The estimation of annual CH_4 and N_2O emissions, the computation of tree biomass growth using a stand simulator, and the calculation of radiative forcings are described by Minkkinen *et al.* (2002). Their national-scale results show several interesting dynamics. The most remarkable changes took place after the 1950s, when extensive draining started, and the trees growing on the mires began to absorb significant amounts of carbon. However, during the 21st Century, when tree removals began, the net CO_2 uptake by trees decreases. Carbon accumulation in the peat continues through the simulation period because the amount of litter produced is large and its high lignin content makes it resistant to decomposition. The total effect during this 200 year simulation period is a net cooling, over half of which results from the strongly reduced CH_4 emissions. In contrast, N_2O emissions are expected to increase from the most fertile mires.

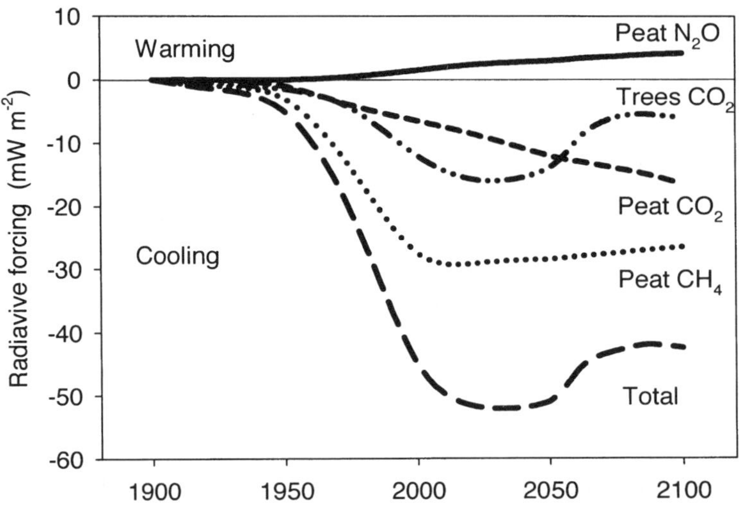

Figure 4. Radiative forcing induced by the draining of Finnish peatlands for forestry in 1900–2100. The 'Trees CO₂' line also includes wood products. Redrawn from Minkkinen et al. (2002). The largest effect on the total radiative forcing has been the decrease of methane emissions resulting from the widespread draining in the 1960s and 1970s. Draining results in accelerated peat decomposition and increased plant growth and litter production. As an average, the last of these counteracting processes is greater over all drained peatlands. The carbon stored in the tree biomass has a net cooling effect which is lower after the first rotation. Draining may increase the N₂O emissions in some fertile peatlands, and this has a warming effect.

8. Peatlands in agricultural use

In populated regions, mires have been drained for agriculture for many centuries. The oldest records even in Finland are from medieval times (Myllys, 1996). Peatlands were easy to clear for agriculture, but the poor nutrient status of the resulting peatfields was a problem. The oldest fields in Europe have already changed from histosols to mineral soils as a result of the long-term decomposition of organic matter. Globally, the area of cultivated peatlands is a few percent of the total farmed area but it contains about half of the carbon content of agricultural soils in the world (Paustian *et al.*, 1998). In Russia 71 000 km^2 (about 12% of the total peatland area) (Kosov and Krestapova, 1996) and in Finland 7000–10 000 km^2 have been drained for agriculture (Myllys, 1996). Presently the area of cultivated peatlands in Finland is much less, about 2000 km^2, or 10% of the total area of agricultural fields (Myllys, 1996). Less productive soils have been forested or abandoned, and some original peatfields are no longer classified as histosols because their organic matter content is less than 40%.

Mechanical disturbance of a peatfield by ploughing activates peat decomposition, and the microbial processes change considerably from those in a pristine mire when a peatfield is limed and fertilized, leading to high heterotrophic respiration rates (Nykänen *et al.*, 1995). This kind of histosol is also ideal for intense microbial nitrification and denitrification processes, producing high rates of N_2O and NO emissions. An original wetland surface emitting CH_4 turns into an oxic soil which instead becomes a CH_4 sink. CH_4 emissions occur only during possible seasonal flooding events.

The rate of carbon lost from peatfields may be estimated by comparing soil carbon stocks over long time periods, e.g., 10–20 years. By analysing the decrease in depth of soil, Armentano and Menges (1986) estimated that Boreal agricultural peat soils loose carbon at a rate of 18 mol m^{-2} per year under cereal cropping and 2.5 mol m^{-2} per year under pasture. The IPCC default emission factors for cultivated organic soils are 8.3 and 83 mol m^{-2} per year in cold and warm temperate climates, respectively. In Finland, for GHG budget reporting for the Kyoto Protocol, we use emission factors of 33 mol m^{-2} per year and 16 mol m^{-2} per year for crops and pasture, respectively (Pipatti, 2001; see also Chapter 1, this volume). Direct gas-flux measurements are needed to determine these emission factors, and to obtain estimates of other GHG emissions. There exist only a few studies on peatfields and these are mostly based on the chamber technique, which makes the estimation of annual budgets problematic.

To our knowledge, the first study using the eddy covariance method for CO_2 fluxes from peatfields was made at Jokioinen (60°54′ N, 23°31′ W) in southern Finland. The flux study site is located in the southern Boreal zone and has an annual mean air temperature of 3.9 °C and precipitation of 580 mm per year. Lohila *et al.* (2003, 2004a) have described the site and measurement methods. The soil comprises well-decomposed peat and is classified as a Terric histosol. The depth of the peat layer close to the micrometeorological mast was 0.5–0.6 m. The measurement period extended over three winters and two growing seasons. During the first summer, barley with under-sown grass was grown and during the second summer forage grass, a common agricultural practice. The fluxes of N_2O and CH_4 were measured in the same field by using chambers, and from those measurements annual flux averages were calculated as described by Martikainen *et al.* (2002) and Regina *et al.* (2004).

From the CO_2 flux measurements, Lohila *et al.* (2004a) obtained a carbon budget of 17 mol m^{-2} per year and 6.6 mol m^{-2} per year for the barley and grass fields, respectively (*Figure 5*). The peatfield was a smaller annual source of CO_2 to the atmosphere when growing grass because of a much longer growing season; the field was a net CO_2 sink for only 40 days when growing barley. These CO_2 budgets represent the overall CO_2 flux between the peatfield and the atmosphere, but if we consider the annual carbon budget of the system, then we have to take into account the amounts of harvested biomass, which are regarded as short-lived products, rapidly returned as CO_2 to the atmosphere. The amounts of carbon that were transported away from the field were larger for the grass harvest than for the barley grain harvest, resulting in total annual carbon losses of 38 and 28 mol m^{-2}, respectively. These amounts are larger than those based on long-term soil loss by Armentano and Menges (1986). The emission factors used for Kyoto Protocol reporting in Finland are larger than the measured CO_2-exchange budgets but lower than the estimate of total carbon losses. An interesting point is that the summed estimates of the decline in soil carbon stock, loss of carbon to the atmosphere and loss of carbon as harvested products are larger when the field is growing grass than barley (Lohila *et al.*, 2004a). This contrasts both with the results of Armentano and Menges (1986) and with the IPCC methodology.

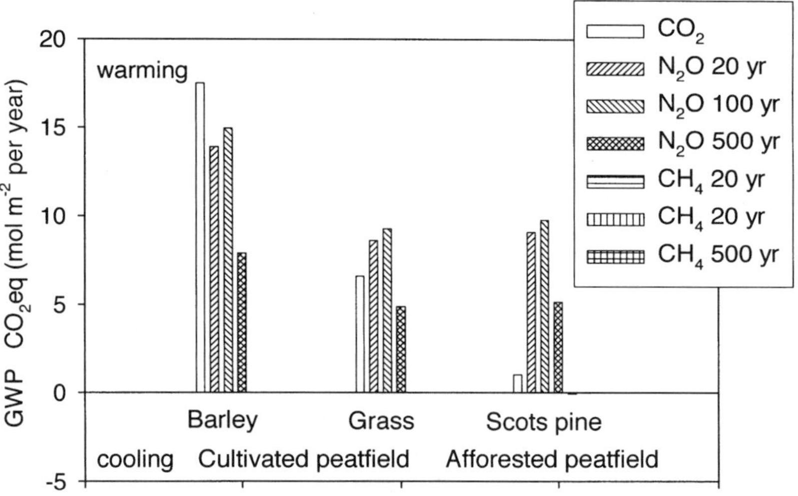

Figure 5. *GWPs estimated for a peatfield growing barley and grass (Regina et al., 2004; Lohila et al., 2004a) and an afforested peatfield (Lohila et al., 2004b; Pihlatie et al., 2004) in southern Finland. In drained peatfields, the CO_2 emissions are high but the CH_4 emissions are low owing to the lowered water table. In the afforested peatfield, carbon stored in the tree biomass nearly balances the carbon lost as a result of the peat decomposition. High N_2O emissions are typical for peatlands that have been in agricultural use.*

At Jokioinen, the chamber measurements showed that CH_4 flux rates were very low, and that the field was an extremely small methane sink, whereas the fluxes of both CO_2 and N_2O emissions contributed more or less equally to a large GWP. Soil temperature, moisture, and nutrient status are important determinants for N_2O emissions (Regina *et al.*, 1996). However, it is difficult to model emissions based on these environmental variables because emissions are often sporadic and high emission rates have been observed, for example, when soil temperatures are close to zero (Regina *et al.*, 2004). In general, the total GWP of agricultural peatlands is very large, and this suggests that land-use change to forestry, for example, is a mitigation option (see *Figure 5*).

9. Afforested peatfields

The area of peatlands in agricultural use is diminishing in many countries as a result of actions to cut down agricultural over-production. Less productive peatfields are often forested or abandoned. Recently, afforestation of peatfields has also been considered as an option to mitigate greenhouse gas emissions from cultivated peaty soils that are known to emit large amounts of CO_2 and N_2O into the atmosphere (Kasimir-Klemedtsson *et al.*, 1997; Lohila *et al.*, 2004b; Nykänen *et al.*, 1995). However, comprehensive studies of greenhouse gas budgets of afforested peatfields have not previously been made. Here we present the first eddy covariance, year-round, CO_2 flux measurements above an afforested peatfield, together with estimates of the annual N_2O and CH_4 emissions (*Figure 5*). We also compare the results to those obtained earlier from a cultivated peatfield.

The annual carbon budget was measured using the eddy covariance method above a Scots pine forest in Karvia, western Finland (62°12′ N, 22°42′ E) (Lohila *et al.*, 2004b). There has been agricultural activity on the site from 1940 until 1969. After clearing the site in 1970, mineral soil was added for conditioning the peat soil. In 1971, the field was planted with Scots pine (*Pinus sylvestris* L.) using 3.5–8 m row spacing. At present (2004), the mean height of the trees is approximately 13 m. The depth of the peat is more than 1 m. During the growing season, the water table level fluctuated between 0.40 and 0.55 m. The CO_2 flux data cover the period from October 2002 to September 2003. The emissions of N_2O and CH_4 were measured with the static chamber technique in May to November 2003. Sampling was weekly during the growing season and fortnightly during the fall.

The eddy covariance measurements show that the forest acted as a source of CO_2 from October to the beginning of May (Lohila *et al.*, 2004b). Net CO_2 uptake was observed between May and September, but there were considerable day-to-day variations depending on the weather, in particular reflecting susceptibility to conditions conducive to low rates of photosynthesis and high rates of soil respiration. Averaged over the year, the forest was a small source of CO_2 (about 1 mol m^{-2} per year). The total respiration rates were comparable to those observed in the peatfield at Jokioinen (*Figure 5*) (Lohila *et al.*, 2004a). Thus the peat decomposition rate was high and the annual net carbon balance brought close to zero by a high rate of carbon fixation by the growing trees and understorey vegetation. At this site, mineral soil was added as a soil conditioner when the field was in agricultural use, and this may explain the high decomposition and mineralization rates. For comparison, a Scots pine forest of about the same age growing on mineral soil at Hyytiälä in the same region has an annual net carbon uptake of −16 mol m^{-2} per year (Suni *et al.*, 2003).

The N_2O fluxes measured with the chamber method were relatively high during the measurement period (Pihlatie *et al.*, 2004). We extrapolated these measurements to obtain an estimate of annual N_2O emissions. Methane fluxes were small, the annual average flux for the year being close to zero. The GWP of the N_2O emissions is similar to the values observed in a peatfield with active cultivation (*Figure 5*).

Mineralization of the peat has not declined after afforestation. The forest stand seems to absorb CO_2 at much the same rate as it is released from the decomposing peat so that there is virtually no net impact on the atmospheric CO_2 concentration. Thus, afforestation seems beneficial from the atmospheric perspective but does not help in conserving the soil carbon.

10. Cut-away and restored peatlands

The energy use of peat is common in countries that do not have fossil fuel resources but do have abundant peat resources. Of the total global area of 1650 km^2 on which peat harvesting occurred in 1997, 750 km^2 are to be found in Ireland and 530 km^2 in Finland (Crill *et al.*, 2000). Russia, Belarus, and Sweden also have some commercial energy production by combustion of peat. The total area of peat harvesting for horticultural purposes in 1997 was 1034 km^2, the most intensive harvesting taking place in Germany, Canada, and Russia. It is estimated that in Finland the area of peatlands expected to be released annually during the next few years after peat harvesting has concluded will be about 27 km^2 (Crill *et al.*, 2000).

After harvesting, the bare, usually well-aerated peat surface decomposes at a high rate. Natural plant colonization in the harsh environment is very slow (Tuittila, 2000) and active measures are needed to facilitate rewarding after-use of these land areas. Alternatives for the after-use include afforestation, production of biofuel by short-rotation coppice, agriculture, and restoration of native vegetation. Afforestation has been the most widely used option in Finland, but for successful timber production it is important to take proper care of drainage, selection of tree species, and nutritional status of the soil (Kaunisto and Aro, 1996). By contrast, restoration aims at bringing back a naturally functioning ecosystem that reinitiates peat formation. A prerequisite for successful restoration is to manage the level of the water table so that it is close to the surface but the vegetation is not inundated for long periods. In many cases reintroduction of *Sphagnum* mosses is unsuccessful for this reason, and sedges are the preferred choice as pioneer plants (Tuittila, 2000).

Tuittila *et al.* (1999a, b) compared the CO_2 and CH_4 dynamics of the bare surfaces of cut-away peat lands to those after successful restoration in an investigation in the central part of Finland. The trace gas fluxes of *Eriophorum* sedges and *Sphagnum* plantations of various ages were measured by using static chambers.

The annual average fluxes (Tuittila, 2000) indicate a high decomposition rate of peat in the oxic layer of the bare peat surface. Methane fluxes are so low that they do not contribute to the annual GWPs calculated from the fluxes (*Figure 6*). After re-wetting and successful reintroduction and colonization by wetland plants, the wetland accumulates carbon. Methane production also recommences and the aerenchyma tissue of sedges provides an effective transport route for CH_4 direct to the atmosphere. The total GWP of the restored wetland is slightly negative (i.e., a slight climate cooling effect), but this depends on the time horizon used for the estimation of GWP.

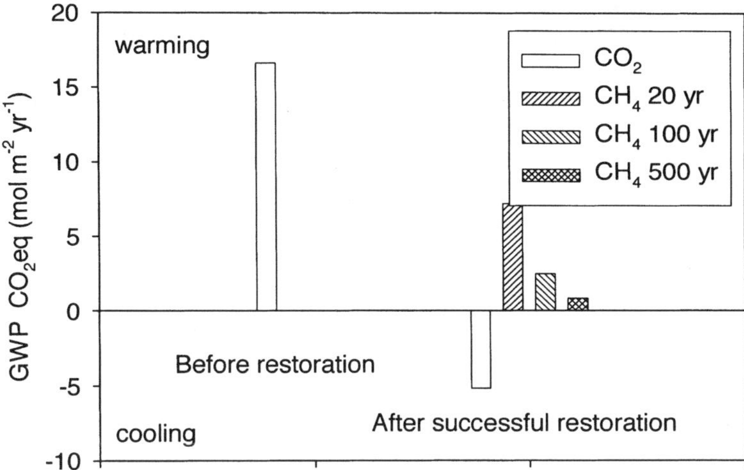

Figure 6. GWPs estimated for cut-away and restored cut-away peatlands (based on Tuittila, 2000) in central Finland. Before the restoration, peat decomposes at a high rate. After raising the water table and reintroducing sedge vegetation, the CO_2 emission turns to a net uptake and the CH_4 emissions increase to the levels typical for natural fens (Figure 3).

11. Climate change

Wetland ecosystems are sensitive to hydrological and temperature conditions, and this is manifest in the variations of mire type dominance in different climatic zones and in the temporally changing mire types and peat accumulation rates during the Holocene. In the Arctic areas, the rate of peat accumulation has been higher during warmer climate intervals, but in the southernmost bog regions carbon accumulation seems to be particularly susceptible to summer droughts (Gorham, 1991). However, drier conditions improve growth of forests and this may have a substantial impact on carbon uptake. Gorham (1991) also supposed that CH_4 emissions would increase with a warming climate, as the emission rate is dependent on soil temperature. More recent data mostly support these conclusions. Based on inter-annual variations in the measured CO_2 budget, Aurela *et al.* (2004) show that in the sub-Arctic region, where the short summer period limits vegetation growth, the annual net CO_2 uptake rate is highest during years when the spring is warm and the growing season begins early (see also Chapter 8, this volume). The inter-annual variability observed on a sub-Arctic fen (Griffis *et al.*, 2000) and in the bog zone (Arneth *et al.*, 2002) indicates that net carbon uptake is small during unusually dry summers, and a mire may turn from a sink to a source of CO_2. This is also supported by chamber measurements made on an ombrotrophic bog (Alm *et al.*, 1999). A modelling study by Griffis and Rouse (2001) also indicates that carbon accumulation of a sub-Arctic fen would benefit from a wet and warm climate.

Oechel *et al.* (1993) reported that extensive areas in Alaska have turned from sinks to sources of CO_2 as a result of the warming and drying of the mire soil, although this conclusion was partly revised when considering the role of autogenic adaptation of the vegetation in more recent years (Oechel *et al.*, 2000). In contrast to Oechel *et al.* (1993), Camill *et al.* (2001) found that peat accumulation rates after the thaw of permafrost were mostly increasing. However, local-scale responses to changes in environmental variables and plant species are complex, and responses at a landscape scale may differ from those found at flux site or chamber scales.

On decennial and longer time-scales, the autogenic development of wetland vegetation in response to environmental variables will be critical. The studies made at sites drained for forestry purposes many decades ago show that after ditching and drawdown of the water table, both ground flora and tree growth start to resemble that of upland forests (Laiho *et al.*, 2003). Rates of carbon fixation on forested peatlands are much higher than rates on natural mires. For many forested mires, there is additional carbon accumulation because above-ground and below-ground biomass stocks increase more than heterotrophic respiration losses from the peaty soil (Minkkinen *et al.*, 2002). Once the forest is established, high transpiration rates in summer tend to sustain a lower water table level, and although that may stimulate heterotrophic respiration and emissions of CO_2, it also strongly diminishes emissions of CH_4, with its much higher GWP.

There are currently many uncertainties about the impacts of climate change on trace gas emissions. For example, the NPP of vegetation providing root exudates to the soil and the occurrence of those plant species that are efficient in transporting CH_4 from anoxic soil layers to the atmosphere would seem to be of vital importance for CH_4 emissions (Schimel, 1995). As CH_4 emissions are lower from bogs than from fens, and taking into account that soil drying rather than soil wetting may follow from global warming, we may speculate that these effects together will surpass the effects of rising temperatures and length of the warm season on CH_4 emission rates. However, one should be exceptionally careful in crude generalizations in regions where permafrost may thaw. Christensen *et al.*, (2004) have reported on widespread wetting and changes in vegetation following recent permafrost thaw of mires in sub-Arctic Sweden where the annual mean air temperature is –0.7 °C. They estimated that the landscape-scale CH_4 emissions from mires increased by 22–66% over the period 1970–2000.

Finally, the current decline in the rate of growth of atmospheric CH_4 concentrations at northern latitudes (Dlugokencky *et al.*, 2003) does not support the likelihood of CH_4 emissions increasing in the future, but this remains a key question!

12. Conclusions

Peat formation and methane production in inundated conditions are intimately connected (Whiting and Chanton, 2001). Bogs, which typically have a negative GWP (i.e., a cooling effect), are mostly located in the southern and central parts of the Boreal region, whereas fens, which have high CH_4 emission rates and lower carbon accumulation rates, are usually found in the more northern latitudes. If there is a northward translocation of *Sphagnum* moss vegetation replacing sedge fens as a response to global warming, and if we assume that there is enough time for ecosystems to adapt, the total GWP effect in that region would be one of cooling. However, GWP depends very much on the time horizon considered. The two usually counteracting GHGs, CO_2 and CH_4, are important on different time-scales. Most wetlands have a warming effect on short time-scales, but for longer time horizons CO_2 budgets dominate and all carbon accumulating mires have a cooling effect on climate. One critical question is how peatlands will be sustained in the southern margin of the bog region. It might be expected that in this zone evapotranspiration will increase more than precipitation, and this may lead to extended summer drought periods, which would have deleterious effects on annual carbon accumulation. In these populated regions, there will also be a large demand for water resources for agriculture and other human activities, which may diminish the areas reserved as natural wetlands.

Peatlands are usually drained if they are used for economic purposes such as forestry, agriculture, and energy production, thus seriously reducing CH_4 emissions from the peat. From the GWP point of view, this helps to balance the large effect that warming has on CO_2 emissions from the decomposing peat. Successful afforestation may well have a cooling effect on climate, at least for some decades. However, if soil is losing carbon over many forestry rotations, this cooling effect is likely eventually to turn to warming, despite the tree growth.

Conditioning of soil for agriculture results in high rates of N_2O emissions, which continue even after afforestration. Afforestation of cultivated peatfields has a cooling effect, but it may be impossible to cut down their net warming effect totally because of the continuing N_2O emissions. Peat production for energy and horticultural purposes devastates the wetland mire ecosystem. After peat has been extracted, large areas of peat surface are left exposed and decomposing at a high rate. Successful restoration of vegetation can reduce significantly the net warming effect provided that active measures are taken. Restoration of cut-away peatlands is a good example of the adaptive capacity of wetland vegetation, and this gives rise to optimism for the autogenic adaptation of natural mires in the changing climate.

Nations with commitments resulting from their participation in the Kyoto Protocol (see Chapter 1, this volume) should collect appropriate information and develop tools for national GHG inventories *that should include peatlands* in the various related land-use categories (see Chapter 4, this volume, for example). For this undertaking, quantitative information on the responses of greenhouse gas fluxes to land-use changes is still insufficient. More ecosystem-scale flux studies are essential because the effects of land-use changes on the GWP may be positive or negative, depending on the original site type, management history, climate change, and the measures taken. Because modelling is an inherent part of national inventories, lack of a comprehensive modelling approach to GHG fluxes on peatland ecosystems, and to related land-use changes, severely hampers construction of emission inventories.

Acknowledgements

We thank the Academy of Finland, the European Commission, and the Ministry of Trade and Industry (Finland) for financial support.

References

Alm, J., Schulman, L., Walden, J., Nykänen, H., Martikainen, P.J. and Silvola, J. (1999) Carbon balance of a boreal bog during a year with an exceptionally dry summer. *Ecology* **80**: 161–174.

Armentano, T.V. and Menges, E.S. (1986) Patterns of change in the carbon balance of organic soil – wetlands of the temperate zone. *Journal of Ecology* **74**: 755–774.

Arneth, A., Kurbatova, J., Kolle, O., Shibistova, O.B., Lloyd, J., Vygodskaya, N.N., and Schulze, E.-D. (2002) Comparative ecosystem-atmosphere exchange of energy and mass in a European Russian and a central Siberian bog II. Interseasonal and interannual variability of CO_2 fluxes. *Tellus Series B – Chemical and Physical Meteorology* **54**: 514–530.

Aubinet, M., Grelle, A., Ibrom, A., Rannik, Ü., Moncrieff, J., Foken, T. *et al.* (2000) Estimates of the annual net carbon and water exchange of forests: The EUROFLUX methodology. *Advances in Ecological Research* **30**: 113–175.

Aurela, M., Laurila, T. and Tuovinen, J.-P. (2001) Seasonal CO_2 budgets on a sub-arctic mire. *Journal of Geophysical Research* **106**: 1623–1637.

Aurela, M., Laurila, T. and Tuovinen, J.-P. (2002) Annual CO_2 balance of a subarctic fen in northern Europe: Importance of the wintertime efflux. *Journal of Geophysical Research* **107**: 4607, doi:10.1029/2002JD002055.

Aurela, M., Laurila, T. and Tuovinen, J.-P. (2004) The timing of snow melt controls the annual CO_2 balance in a subarctic fen. *Geophysical Research Letters* **31**: L16119, doi:10.1029/2004GLO20315.

Baldocchi, D.D. (2003) Assessing the eddy covariance technique for evaluating carbon dioxide exchange rates of ecosystems: past, present and future. *Global Change Biology* **9**: 479–492.

Camill, P., Lynch, J.A., Clark, J. S., Adams, J.B. and Jordan B. (2001) Changes in biomass, aboveground net primary production, and peat accumulation following permafrost thaw in the boreal peatlands of Manitoba, Canada. *Ecosystems* **4**: 461–478.

Christensen, T.R., Johansson, T., Åkerman, H.J., Mastepanov, M., Malmer, N., Friborg, T., Crill, P. and Svensson, H. (2004) Thawing sub-arctic permafrost: Effects on vegetation and methane emissions. *Geophysical Research Letters* **31**: L04501, doi:10.1029/2003GL018680.

Crill, P., Hargreaves, K. and Korhola, A. (2000) The role of peat in Finnish greenhouse gas budgets. *Studies and Reports 10/2000*. Ministry of Trade and Industry. 71 pp.

Crum, H.A. (1988) *A Focus on Peatlands and Peat Mosses*. 306 pp. University of Michigan Press. Ann Arbor, Michigan.

Dlugokencky, E.J., Houweling, S., Bruhwiler, L., Masarie, K.A., Lang, P.M., Miller, J.B. and Tans, P.P. (2003) Atmospheric methane levels off: Temporary pause or a new steady-state? *Geophysical Research Letters* **30**: 1992. doi:10.1029/2003GL018126.

Eurola, S., Hicks, S. and Kaakinen, E. (1984) Key to Finnish mire types. In: Moore, P. D. (ed.) *European Mires*, pp. 11–117. Academic Press, London.

Falge, E., Baldocchi, D.D., Olson, R., Anthoni, P., Aubinet, M., Bernhofer, Ch. *et al.* (2001) Gap filling strategies for defensible annual sums of net ecosystem exchange. *Agricultural and Forest Meteorology* **107**: 43–69.

Finnish Forest Research Institute (2001) *Finnish Statistical Yearbook of Forestry*. Vammalan kirjapaino. 374 pp.

Fowler, D., Coyle, M., Flechard, C., Hargreaves, K., Nemitz, E., Storeton-West, R., Sutton, M. and Erisman, J.-W. (2001) Advances in micrometeorological methods for the measurement and interpretation of gas and particle nitrogen fluxes. *Plant and Soil* **228**: 117–129.

Fraser, C.J.D., Roulet, N.T. and Moore, T.R. (2001) Dissolved organic carbon biogeochemistry and hydrology at a temperate bog. *Hydrological Processes* **15**: 3151–3166.

Friborg, T., Christensen, T.R., Hansen, B.U., Nordstrøm, C. and Søgaard, H. (2000) Trace gas exchange in a high-arctic valley. 2. Landscape CH_4 fluxes measured and modeled using eddy correlation data. *Global Biogeochemical Cycles* **14**: 715–723.

Gorham, E. (1991) Northern peatlands: Role in the carbon balance and probable responses to climatic warming. *Ecological Applications* **1**: 182–195.

Griffis, T.J., Rouse, W.R. and Waddington, J.M. (2000) Interannual variability in net ecosystem CO_2 exchange at a subarctic fen. *Global Biogeochemical Cycles* **14**: 1109–1121.

Griffis, T.J. and Rouse, W.R. (2001) Modelling the interannual variability of net ecosystem CO_2 exchange at a subarctic sedge fen. *Global Change Biology* **7**: 511–530.

Hargreaves, K.J., Fowler, D., Pitcairn, C.E.R. and Aurela, M. (2001) Annual methane emission from Finnish mire estimated from eddy covariance campaign measurements. *Theoretical and Applied Climatology* **70**: 203–213.

Hargreaves, K.J., Milne, R. and Cannell, M.G.R. (2003) Carbon balance of afforested peatland in Scotland. *Forestry* **76**: 299–317.

Heikkinen, J.E., Maljanen, P.M., Aurela, M., Hargreaves, K.J. and Martikainen, P.J. (2002) Carbon dioxide and methane dynamics in a sub-arctic peatland in northern Finland. *Polar Research* **21**: 49–62.

Immirzi, C.P., Maltby, E. and Clymo, R.S. (1992) The global status of peatlands and their role in carbon cycling. A report for the Friends of the Earth by the Wetland Ecosystem Research Group, University of Exeter, London.

IPCC (2001a) *Climate Change 2001: Impacts, Adaptation, and Vulnerability.* Contribution of Working Group II to the Third Assessment Report of the Intergovernmental Panel on Climate Change, McCarthy, J.J., Canziani, O.F., Leary, N.A., Dokken, D.J. and White, K.S. (eds), pp. 237–342. Cambridge University Press, Cambridge.

IPCC (2001b) *Climate Change 2001: The Scientific Basis.* Contribution of Working Group I to the Third Assessment Report of the Intergovernmental Panel on Climate Change, Houghton, J.T., Ding, Y., Griggs, D.J., Noguer, M., van der Linden, P.J., Dai, X., Maskell, K. and Johnson, C. A. (eds). 881 pp. Cambridge University Press, Cambridge.

Kasimir-Klemedtsson, Å., Klemedtsson, L., Berglund, K., Martikainen, P.J., Silvola, J., and Oenema, O. (1997) Greenhouse gas emissions from farmed organic soils: a review. *Soil Use and Management* **13**: 245–250.

Kosov, V.I. and Krestapova, V.N. (1996) The peat resources of Russia and their utilization. In: Lappalainen, E. (ed) *Global Peat Resources*, pp. 127–131. International Peat Society, Jyväskylä.

Kaunisto, S. and Aro, L. (1996) Forestry use of cut-away peatlands. In: Vasander, H. (ed) *Peatlands in Finland*, pp. 130–134. Finnish Peatland Society, Helsinki.

Lafleur, P. M., Roulet, N. T. and Admiral, S. (2001) The annual cycle of CO_2 exchange at a boreal bog peatland. *Journal of Geophysical Research* **106**: 3071–3081.

Lafleur, P.M., Roulet, N.T., Bubier, J.L., Frolking, S. and Moore, T.R. (2003) Interannual variability in the peatland-atmosphere carbon dioxide exchange at an ombrotrophic bog. *Global Biogeochemical Cycles* **17**: 1036. doi: 10.1029/2002GB001983.

Laiho, R., Vasander, H., Penttila, T. and Laine, J. (2003) Dynamics of plant-mediated organic matter and nutrient cycling following water-level drawdown in boreal peatlands. *Global Biogeochemical Cycles* **17**: 1053, doi:.10.1029/2002GB002015.

Laine, J., Vasander, H. and Laiho, R. (1995) Long-term effects of water level drawdown on the vegetation of drained pine mires in southern Finland. *Journal of Applied Ecology* **32**: 785–802.

Laurila, T., Soegaard, H., Lloyd, C.R., Aurela, M., Tuovinen, J.-P. and Nordstroem, C. (2001). Seasonal variations of net CO_2 exchange in European Arctic ecosystems. *Theoretical and Applied Climatology* **70**: 183–201.

Le Mer, J. and Roger, P. (2001) Production, oxidation, emission and consumption of methane by soils: A review. *European Journal of Soil Biology* **37**: 25–50.

Lohila, A., Aurela, M., Regina, K. and Laurila, T. (2003) Soil and total ecosystem respiration in agricultural fields: effect of soil and crop type. *Plant and Soil* **251**: 303–317.

Lohila, A., Aurela, M., Tuovinen, J.-P. and Laurila, T. (2004a) Annual CO_2 exchange on peat field growing spring barley or perennial forage grass. *Journal of Geophysical Research* **109**: D18116. doi:10.1029/2004JD004715.

Lohila, A., Aurela, M., Aro, L. and Laurila, T. (2004b) Effects of afforestation on the CO_2 balance of an agricultural peat soil. In: Päivänen, J. (ed) *Wise Use of Peatlands, Proceedings on the 12th International Peat Congress, Volume 1*, pp. 145–149. International Peat Society, Jyväskylä.

Maljanen, M., Martikainen, P.J., Aaltonen, H. and Silvola, J. (2002) Short-term variation in fluxes of carbon dioxide, nitrous oxide and methane in cultivated and forested organic soils. *Soil Biology and Biochemistry* **34**: 577–584.

Martikainen, P.J., Regina, K., Syväsalo, E., Laurila, T., Lohila, A., Aurela, M. *et al.* (2002) Agricultural soils as sinks and sources of greenhouse gas: A research consortium (AGROGAS). In: Käyhkö, J. and Talve, L. (eds). *Understanding the Global System – the Finnish Perspective*, pp. 55–68. Painosalama, Turku.

Minkkinen, K., Korhonen, R., Savolainen, I. and Laine, J. (2002) Carbon balance and radiative forcing of Finnish peatlands 1900–2100 – the impact of forestry drainage. *Global Change Biology* **8**: 785–799.

Moore, T.R., Bubier, J.L., Frolking, S.E., Lafleur, P.M. and Roulet, N.T. (2002) Plant biomass and production and CO_2 exchange in an ombrotrophic bog. *Journal of Ecology* **90**: 25–36.

Myllys, M. (1996) Agriculture on peatlands. In: Vasander, H. (ed) *Peatlands in Finland*, pp. 66–74. Finnish Peatland Society, Helsinki.

Nordstroem, C., Soegaard, H., Cristensen, T.R., Friborg, T. and Hansen, B.U. (2001) Seasonal carbon dioxide balance and respiration for a high-arctic fen ecosystem in NE-Greenland. *Theoretical and Applied Climatology* **70**: 149–166.

Nykänen, H., Alm, J., Lång, K., Silvola, J., and Martikainen, P.J. (1995) Emissions of CH_4, N_2O and CO_2 from a virgin fen and a fen drained for grassland in Finland. *Journal of Biogeography* **22**: 351–357.

Nykänen, H., Alm, J., Silvola, J., Tolonen, K. and Martikainen, P.J. (1998) Methane fluxes on boreal peatlands of different fertility and the effect of long-term experimental lowering of the water table on flux rates. *Global Biogeochemical Cycles* **12**: 53–69.

Oechel, W.C., Hastings, S.J., Vourlitis, G.L., Jenkins, M.A., Riechers, G. and Grulke, N. (1993) Recent change of Arctic tundra ecosystems from a net carbon dioxide sink to a source. *Nature* **361**: 520–526.

Oechel, W.C., Vourlitis, G.L., Hastings, S.J., Zulueta, R.C., Hinzman, L. and Kane, D. (2000) Acclimation of ecosystem CO_2 exchange in the Alaskan Arctic in response to decadal climate warming. *Nature* **406**: 978–981.

Paustian, K., Cole, C.V., Sauerbeck, D. and Sampson, N. (1998) CO_2 mitigation by agriculture: an overview. *Climate Change* **40**: 135–162.

Pihlatie, M., Rinne, J., Lohila, A., Laurila, T., Aro, L. and Vesala, T. (2004) Nitrous oxide emissions from an afforested peat field using eddy covariance and enclosure techniques. In: Päivänen, J. (ed) *Wise Use of Peatlands, Proceedings of the 12th International Peat Congress, Volume 2*, pp. 1010–1014. International Peat Society. Jyväskylä.

Pipatti, R. (2001) Greenhouse gas emissions and removals in Finland. *VTT Research Notes 2094*. Technical Research Centre of Finland, Espoo.

Regina, K., Nykänen, H., Silvola, J. and Martikainen, P.J. (1996) Fluxes of nitrous oxide from boreal peatlands as affected by peatland type water level and nitrification capacity. *Biogeochemistry* **35**: 401–418.

Regina, K., Syväsalo. E., Hannukkala, A. and Esala M. (2004) Fluxes of N_2O from farmed peat soils in Finland. *European Journal of Soil Science*. In Press.

Sallantaus, T. (1992) Leaching in the material balance of peatlands – preliminary results. *Suo* **43**: 253–358.

Schimel, J.P. (1995) Plant transport and methane production as controls on methane flux from arctic wet meadow tundra. *Biogeochemistry* **28**: 183–200.

Schulze, E.-D., Valentini, R. and Sanz, M.-J. (2002) The long way from Kyoto to Marrakesh: Implications of Kyoto Protocol negotiations for global ecology. *Global Change Biology* **8**: 505–518.

Soegaard, H. and Nordstroem, C. (1999) Carbon dioxide exchange in a high-arctic fen estimated by eddy covariance measurements and modelling, *Global Biogeochemical Cycles* **5**: 547–562.

Suni, T., Rinne, J., Reissel, A., Altimir, N., Keronen, P., Rannik, Ü., Dal Maso, M., Kulmala, M. and Vesala, T. (2003) Long-term measurements of surface fluxes above a Scots pine forest in Hyytiälä, southern Finland, 1996–2001. *Boreal Environment Research* **8**: 287–301.

Tuittila, E.-S., Komulainen, V.-M., Vasander, H. and Laine, J. (1999a) Restored cut-away peatland as a sink for atmospheric CO_2. *Oecologia* **120**: 563–574.

Tuittila, E.-S., Komulainen, V.-M., Vasander, H., Nykänen, H., Martikainen, P.J. and Laine, J. (1999b) Methane dynamics of a restored cut-away peatland. *Global Change Biology* **6**: 569–581.

Tuittila, E.-S. (2000) Restoring vegetation and carbon dynamics in a cut-away peatland. *Publications in Botany from the University of Helsinki* **30**: 1–38.

Turunen, J., Tolonen, K., Tolvanen, S., Remes, M., Ronkainen, J. and Jungner, H. (1999) Carbon accumulation in the mineral subsoil of boreal mires. *Global Biogeochemical Cycles* **13**: 71–79.

Turunen, J., Tomppo, E., Tolonen, K. and Reinikainen, A. (2002) Estimating carbon accumulation rates of undrained mires in Finland – application to boreal and subarctic regions. *Holocene* **12**: 79–90.

Whiting, G.J. and Chanton, J.P. (1993) Primary production control of methane emission from wetlands. *Nature* **364**: 794–795.

Whiting, G.J. and Chanton, J.P. (2001) Greenhouse carbon balance of wetlands: methane emission versus carbon sequestrion. *Tellus Series B – Chemical and Physical Meteorology* **53**: 521–528.

14

Contribution of trace gases nitrous oxide (N_2O) and methane (CH_4) to the atmospheric warming balance of forest biomes

Rainer Brumme, Louis V. Verchot, Pertti J. Martikainen and Christopher S. Potter

1. Introduction

Forest ecosystems play an important role in the global carbon cycle and have fixed a large amount of carbon from the atmosphere, which is stored in vegetation and soils. Until a steady state between uptake and release of CO_2 is reached, these ecosystems accumulate carbon. Nowadays, forest ecosystems are hardly in steady state because of their intensive use for wood and other products, a changing chemical and physical environment, and intensive forest management (IPCC, 2001; Kauppi *et al.*, 1992). These human activities have to be considered when evaluating whether forests are acting as a sink or source of carbon. Actually, it is becoming increasingly evident that forests, especially in the temperate region, act as sinks for atmospheric CO_2 (Jarvis *et al.*, 2001; Kauppi *et al.*, 1992; Malhi *et al.*, 1999). Non-CO_2 greenhouse gases (GHGs) such as nitrous oxide (N_2O) or methane (CH_4) also contribute to the atmospheric warming balance of forest ecosystems. N_2O is especially troublesome because, in addition to being an efficient greenhouse gas, it is involved in the destruction of stratospheric ozone (Andreae and Schimel, 1989; IPCC, 2001). The relative radiative effects of N_2O and CH_4 are about 296 and 23 times that of CO_2 (global warming potential: GWP) when compared on a mass basis and using a 100-year horizon (IPCC, 2001). Considering that soils contribute to 57% of the total global N_2O

The Carbon Balance of Forest Biomes, edited by H. Griffiths and P.G. Jarvis.
© 2005 Taylor & Francis Group

sources (IPCC, 2001), it is clear how important soils are for atmospheric warming. On the other hand, forest soils are generally net sinks for atmospheric CH_4. CH_4 oxidation in soils amounts only to about 5% of the total global sources of this gas, but is about double the annual increase in the atmospheric burden, indicating the importance of CH_4 oxidation in soils.

N_2O is produced during nitrification and denitrification processes in soils and is controlled by several factors that determine the rates of these processes (Firestone and Davidson, 1989). Factors such as oxygen, nitrogen, and carbon availability are most important at the process level, whereas humus type, species composition, soil conditions, and climate regulate N_2O production at the ecosystem level (Brumme *et al.*, 1999). This complex regulation system leads to diurnal, day-to-day, seasonal, and year-to-year variation by modulating nitrification and denitrification at the process scale, and to stand-to-stand variation by regulation at the ecosystem scale.

Soils both produce and consume CH_4. The net soil–atmosphere CH_4 flux is the result of the balance between the two offsetting processes of methanogenesis (microbial production) and methanotrophy (microbial consumption), which is largely controlled by the redox status of the soil. Methanotrophy is a process that involves the oxidation of CH_4 by bacteria; CH_4 is the carbon source, or the electron donor, in the respiration reaction. Methanotrophs in soil are obligate aerobes, whereas methanogens are anaerobes. In well-drained upland soils, CH_4 oxidation generally exceeds CH_4 production and there is a net uptake by the soil of CH_4 from the atmosphere. Methane oxidation is also an important process in wetland soils at the aerobic soil–water interface.

The regulation of CH_4 oxidation in soils is less complex than that of N_2O. Although CH_4 oxidation undergoes seasonal variation, control at the process scale is generally mediated by soil water content, whereas soil temperature is of minor importance (Smith *et al.*, 2000). For our purposes, the most important concern is regulation at the ecosystem scale. Soil texture and pH, humus type, species composition, and land-use history have been found to be responsible for large variations of annual CH_4 uptake in the landscape.

The objective of this paper is to update the estimate of N_2O source and CH_4 sink strength in the global forest biomes. We will apply some innovations, using our understanding of the mechanisms controlling the rate of N_2O emission and CH_4 consumption by soils, to develop a meaningful stratification scheme for forests that should give a better estimate than that currently available.

2. Approaches to estimate the biome fluxes and data sources

Extrapolations of GHG fluxes are often done by surveying the literature, averaging means from several studies, and multiplying by the area covered by the biome. The 'stratify-and-multiply' approach implies several uncertainties, particularly if the sites measured are not representative of the variation within the biome. Some effort has been made to improve the biome estimates by modelling the fluxes using information about the landscape from geographical information systems (GISs) (Bouwman *et al.*, 1993; Nevison *et al.*, 1996; Potter *et al.*, 1996a, b). Other approaches use key factors regulating the fluxes at the ecosystem scale, together with GIS, to calculate the biome fluxes (Matson and Vitousek, 1990).

Matson and Vitousek (1990) used soil fertility in tropical rain forests, assuming that the production of N_2O increases with increasing nitrogen mineralization. The same approach was used by Breuer *et al.* (2000) to update the estimate. Schulte-Bisping *et al.* (2003) used plant species, humus type, and climate criteria to estimate the total N_2O flux of German forest soils. This stratification was based on the observation that only broadleaved forest soils with moder or mor humus had high, seasonal fluctuating emissions (seasonal emission pattern: SEP) while needle-leaved forests and broadleaved forests with mull humus had low and constant emissions (background emission pattern: BEP) (Brumme *et al.*, 1999). Some new approaches to estimate the N_2O and CH_4 fluxes on the biome scale have been developed for this paper and will be described in detail.

2.1 *Nitrous oxide emission*

The N_2O emissions in the tropics undergo strong seasonal patterns expressed as high fluxes in the wet season and low fluxes in the dry season (*Figure 1*). The driving force behind this pattern, precipitation, was used to estimate fluxes of the tropical forest biome. The whole-year measurements of Keller and Reiners (1994) and Verchot *et al.* (1999) indicate that the threshold value of precipitation between high and low N_2O emissions is about 250 mm per month (*Figure 2*). The areas with precipitation totals larger or smaller than 250 mm per month were calculated from a 1° gridded dataset (Legates and Willmott, 1990). We multiplied the computed areas with the mean emissions for the dry and wet seasons to estimate the tropical N_2O flux ('dry/wet-season' approach). We used the following sources of data to estimate the mean N_2O emissions in the dry and wet seasons: Breuer *et al.* (2000); García-Méndez *et al.* (1991); Goreau and deMello (1987); Ishizuka *et al.* (2002); Keller *et al.* (1983, 1986, 1988); Keller and Reiners (1994); Kiese and Butterbach-Bahl (2002); Livingston *et al.* (1988); Luízao *et al.* (1989); Matson and Vitousek (1987); Matson *et al.* (1989); Matson and Vitousek (1990); Melillo *et al.* (2001); Verchot *et al.* (1999); Weitz *et al.* (1998).

Another approach for estimating the emission of the tropical biome was based on a non-linear regression between monthly precipitation and N_2O emissions (precipitation approach), using data from the studies of Keller and Reiners (1994) and Verchot *et al.* (1999) (*Figure 2*):

$$y = 1.0197e^{(0.0034x)},$$

where y is the N_2O emission in nanograms of $N_2O - N$ per square centimetre per hour, and x is the monthly precipitation in millimeters.

The 'stratify-and-multiply' approach for the tropical biome has a larger dataset because all data published in the literature review by Breuer *et al.* (2000) were included ($n = 49$).

The N_2O emission estimate for the temperate biome was based on the approach used by Schulte-Bisping *et al.* (2003). Similar to the tropics, N_2O emissions in some temperate forests undergo strong seasonal variation (*Figure 1*). Differences exist in the variables and threshold values driving these patterns. Generally, soil temperature of 8–10 °C (Brumme, 1995) and precipitation of more

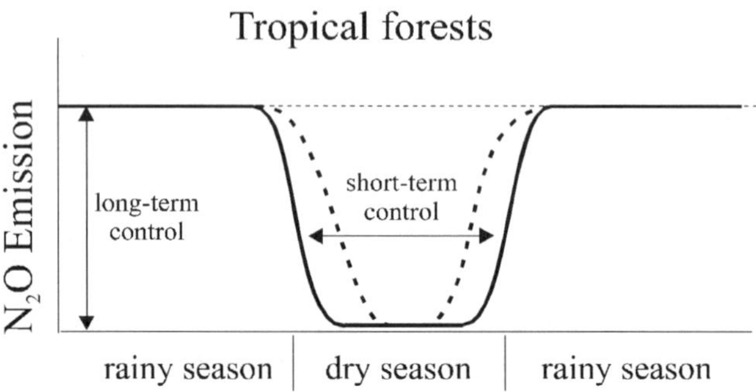

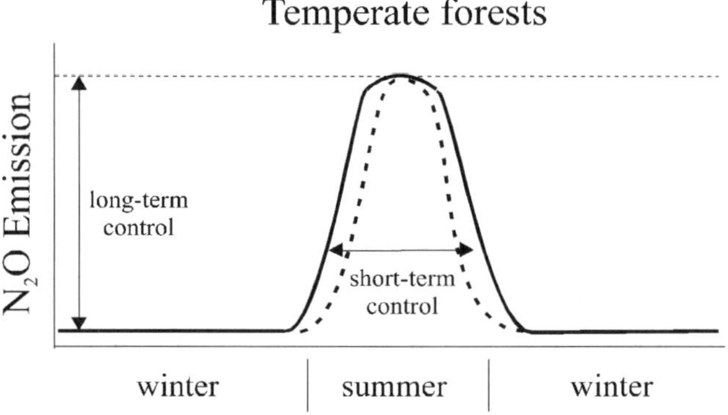

Figure 1. Short-term control by climate (duration of dry and rainy seasons in tropical forests and warm/wet and cold periods in temperate forests) and long-term control by state variables (annual climate, soil and vegetation properties, management) on N_2O emissions in the tropical and temperate biomes.

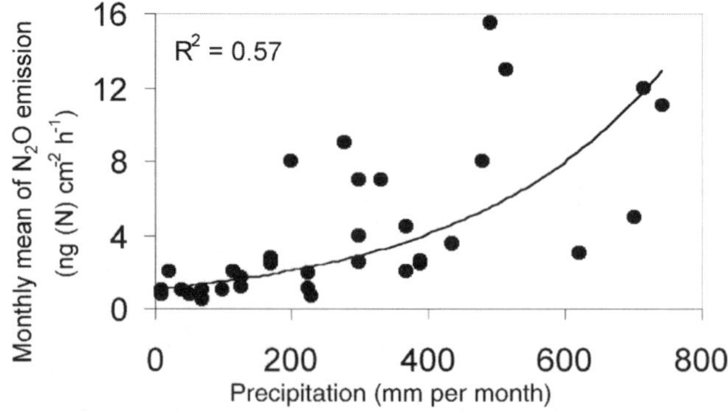

Figure 2. Relationship between monthly precipitation and monthly N_2O emissions in the tropics based on data from Keller and Reiners (1994) and Verchot et al. (1999).

than 100 mm per month has been found to discriminate between high summer and low winter fluxes. These threshold values were derived from a 13-year N_2O emissions record at Solling, Germany (Teepe, R. and R. Brumme, unpublished data). As long as the temperature remains below the threshold or, at higher temperatures, the precipitation remains below the threshold, N_2O emissions are low. We calculated the global forest area that falls within these climate criteria on a monthly basis by using the Legates and Willmott dataset (1990). According to Schulte-Bisping et al. (2003), only 4% of the total forest area in Germany falls within the range for forests with SEP. To calculate the temperate biome flux of N_2O, we assumed that this proportion is valid for the entire global temperate forest biome, and multiplied the appropriate area by the mean monthly fluxes of the period with high SEP. The remaining area was multiplied by the mean monthly fluxes of the BEP scenario. Mixed forests (broadleaved or needle-leaved forests mixed with other species) were assumed to have fluxes in between SEP and BEP. We used the following sources of data to estimate the mean N_2O emissions for the temperate biome: Ambus and Christensen (1995); Ambus et al. (2001); Bowden et al. (1990); Brumme et al. (1999); R. Brumme (unpublished, two annual datasets); Goodroad and Keeney (1984); Klemedtsson et al. (1997); MacDonald et al. (1997); Mogge et al. (1998); Papen and Butterbach-Bahl (1999); Schmidt et al. (1988); Teepe et al. (2000); Teepe et al. (2004b); R. Teepe and R. Brumme (unpublished, three annual datasets), Zechmeister-Boltenstern and Meger (1997); Zechmeister-Boltenstern et al. (2002).

Most flux measurements in the Boreal forest biome have been made during summer, and winter exchanges have been ignored. A study by Sommerfeld et al. (1993) showed that emission of N_2O and uptake of CH_4 continued throughout the period of snow cover indicating that the snow was porous and microbial activity did not cease when the soil surface was frozen. More recent studies have shown that even cold winter periods contribute significantly to the net annual fluxes (Flessa et al., 1995; Teepe et al., 2000). The few winter flux measurements made in Boreal forests indicate that about 40% of the annual soil emissions occur during the winter (Alm et al., 1999; Maljanen et al., 2003; Regina et al., 1998). We used the 'stratify-and-multiply' approach to estimate the N_2O emissions from the Boreal forest biome (*Table 1*) by using the measured summer fluxes and the assumed winter proportion. Spruce forest fens and bogs present a special case, with much higher fluxes compared with the overall mean (Huttunen et al., 2002, 2003b). The area of spruce and alder fens and bogs is 5.9×10^4 km^2 in the Boreal and sub-Arctic parts of the former Soviet Union (Botch et al., 1995). The area of Boreal forest fens and bogs in the rest of Europe and North America was estimated from the total peatland area of 1.9×10^6 km^2 (Kosov and Kreshtapova, 1966; Lappalainen, 1996; Lappalainen and Zurek, 1996; Rubec, 1996). Assuming that the total area of Boreal and sub-Arctic spruce/alder swamps is 1% of this area (J. Turunen, personal communication), the actual area is then 7.8×10^4 km^2. We used the following sources of data to estimate the mean N_2O emissions in the Boreal biome: Blew and Parkinson (1993); Corre et al. (1999); Huttunen et al. (2002); Klemedtsson et al. (1997); Maljanen et al. (personal communication); Martikainen et al. (1994); Saari A., Heiskanen J. and Martikainen, P.J. (personal communication); Schiller and Hastie (1996); Simpson et al. (1997).

Table 1. *Global estimates of N_2O–N emission from tropical (20×10^6 km^2), temperate (6.75×10^6 km^2), and Boreal (23.5×10^6 km^2) forest biomes (all approaches except IPCC (2001) were converted to the areas given in this table) (ranges or standard deviation in parenthesis).*

Approach	Tropical forest biome		Temperate forest biome		Boreal forest biome	
	kg ha^{-1} per year	Tg per year	kg ha^{-1} per year	Tg per year	kg ha^{-1} per year	Tg per year
Matson and Vitousek, 1990	1.6	3.2	—	—	—	—
Bouwman *et al.*, 1993	1.3	2.4	0.20	0.15	—	—
Potter *et al.*, 1996a	1.2	2.4	0.49	0.33	0.18	0.58
IPCC, 2001	—	3.0 (2.2–3.7)	—	1.0 (0.1–2.0)	—	—
Breuer *et al.*, 2000	2.4	4.8	—	—	—	—
'Stratify-and-multiply' approach	1.9 (2.0)	3.8	1.0 (1.5)	0.70	0.27 (0.52)	0.62
'Dry/wet-season' approach	2.0	4.0	—	—	—	—
Precipitation approach	1.8	3.8	—	—	—	—
SEP/BEP approach	—	—	0.43	0.30	—	—

SEP: seasonal emission pattern; BEP: background emission pattern.

2.2 *Methane uptake*

Although biome characteristics, such as temperature and length of growing season, are important variables in controlling the magnitude of microbial processes over the time-frame of a year, our mechanistic understanding of CH_4 uptake at the ecosystem to landscape scale suggests that other variables that affect the diffusivity of gases within soils, such as soil texture and soil water content, may be more important controlling factors (Smith *et al.*, 2000; Striegl, 1993; Verchot *et al.*, 2000). Here, we examine the potential for explaining variability across datasets by including additional variables that are reported in most studies and that have mechanistic significance, namely: soil texture, mean temperature, latitude, and annual rainfall. The relationships developed from this analysis will be used to estimate the CH_4 fluxes in forests throughout the world.

We compiled a database from 46 studies that reported CH_4 fluxes from forests (*Table 3*). We included data from studies that reported soil texture, annual rainfall, mean temperature and latitude coordinates. We analysed the data to explore the relationships between CH_4 flux and environmental variables through regression analysis, using SAS statistical software (SAS Institute, 1992). Soil texture was divided into classes according to the USDA classification system (Brady, 1974). Generally, these studies presented data that were complete enough to calculate the total flux from the site for the frost-free season. However, Mosier and others (Maljanen *et al.*, 2003; Mosier *et al.*, 1993, 1996, 2002; Sommerfeld *et al.* (1993)), working in alpine tundra and Boreal ecosystems, showed that even in snow-covered or frozen soils, CH_4 fluxes were maintained, albeit at lower rates than during the frost-free seasons. Thus, when studies reported fluxes only for the frost-free part of the year, we assumed that these fluxes represented 70% of the annual flux.

3. Results and discussion

3.1 *Nitrous oxide emission of forest biomes*

Nitrous oxide emission increased from the Boreal, to the temperate, to the tropical forest in all the approaches used for estimating the biome N_2O fluxes (*Table 1*). This suggests that temperature regime is the dominating factor on the biome scale. *Figure 3* indicates that sites with low annual N_2O emissions exist in all biomes and that higher fluxes are restricted to the temperate and tropical biomes. The homogeneous distribution of the fluxes in the tropical biome compared with the temperate biome suggests that the processes responsible for the formation of N_2O may be less restricted in the tropical biome. A low mean annual temperature of 2.3 °C at those Boreal sites with flux measurements is most likely responsible for the low mean annual flux of 1.4 µg (N) m^{-2} h^{-1} (standard deviation (SD) 2.3 µg (N) m^{-2} h^{-1}, $n = 15$). Even in the summer season, when long, warm summer days increase soil temperature in the upper soil layers to values known in temperate forests, N_2O emissions are still low (3.0 µg (N) m^{-2} h^{-1}, SD 5.9 µg (N) m^{-2} h^{-1}, $n = 18$). Thus variables other than temperature probably limit the summer fluxes in the Boreal zone. Lundgren and Söderström (1983) reported that bacterial activity correlated best with the rainfall, and Paavolainen *et al.* (2000) found that nitrification and denitrification were highly stimulated by artificial addition of water. Boreal forests receive relatively little rainfall and the morainic soils have a large capacity for water filtration, i.e., they are well aerated. Therefore soil water may well be a very important factor for oxygen status and availability of mineral nitrogen in Boreal forest soils. Most of the Boreal and sub-Arctic biomes have low atmospheric nitrogen deposition, and they are highly nitrogen limited; this may be an additional factor limiting N_2O emissions. Only spruce fens and bogs on organic soils with the water table below the soil surface had high N_2O emissions of 11 µg (N) m^{-2} h^{-1} (SD 11 µg (N) m^{-2} h^{-1}, $n = 3$). Nitrification and denitrification activities in the Boreal organic forest soils are higher than those in upland forest soils as a result of higher nitrogen content in the organic soils. As discussed above, Boreal soils are also capable of N_2O production during winter and thus can have relatively high N_2O emission rates during that season.

In contrast to the Boreal biome, a high mean annual air temperature of 24 °C at tropical sites resulted in high N_2O emissions of 43 µg (N) m^{-2} h^{-1} (SD 55 µg (N) m^{-2} h^{-1}, $n = 24$) during the wet season. During the dry season N_2O emissions were lower at 17 µg (N) m^{-2} h^{-1} (SD 15 µg (N) m^{-2} h^{-1}, $n = 20$), primarily caused by less water-filled pore space (WFPS) and more aerobic soil conditions. In total, the mean emission from the tropical biome is 22 µg (N) m^{-2} h^{-1} (SD 23 µg (N) m^{-2} h^{-1}, $n = 49$).

The temperate biome represents an intermediate between the tropical and Boreal biomes with a mean annual temperature of 7.6 °C at sites with flux measurements. The temperate biome has been stratified into soils having very low N_2O emissions during the whole year of 4.8 µg (N) m^{-2} h^{-1} (SD 3.6 µg (N) m^{-2} h^{-1}, $n = 33$) (BEP) and those showing a seasonal course of emissions (SEP) (Brumme *et al.*, 1999). Forests with SEP had winter fluxes similar to BEP whereas the summer fluxes at SEP were similar to the wet season in the tropics, and estimated to be 90 µg (N) m^{-2} h^{-1} (SD 29 µg (N) m^{-2} h^{-1}, $n = 5$). In total, the mean emission from the temperate biome is 8.9 µg (N) m^{-2} h^{-1} (SD 11 µg (N) m^{-2} h^{-1}, $n = 38$). Not included in the total mean were

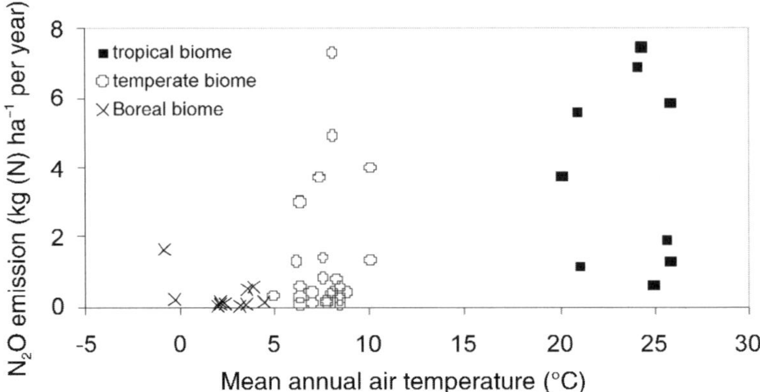

Figure 3. *Site-specific annual N_2O emissions in the Boreal, temperate, and tropical biome dependent on the mean annual air temperature.*

winter emissions and forests with organic soils. Forests with organic soils cover only a small area, but could be very strong sources of N_2O of up to 7.3 kg (N) per hectare per year (1 hectare (ha) = 10^4 m^2) (Brumme *et al.*, 1999), when the water table is below the surface.

The tropical biome emits by far the largest amount of N_2O into the atmosphere, amounting to between 2.4 and 3.2 Tg (N) per year (Bouwman *et al.*, 1993; Matson and Vitousek, 1990; Potter *et al.*, 1996a) (*Table 1*). The different approaches were based on the same few field studies and were recently updated by Breuer *et al.* (2000) using new field measurements in the tropics. They used all published flux data, independent of the duration of measurements and stratified the sites into soils of different fertility according to the approach of Matson and Vitousek (1990). The basis of their approach is the observation that N_2O emissions increase with increasing nitrogen mineralization. Breuer *et al.* (2000) suggested a 50% higher biome flux of 4.8 Tg (N) per year, compared with the upper limit of Matson and Vitousek (1990) of 3.2 Tg (N) per year. The 'stratify-and-multiply' approach applied here gave a value about 20% lower than the estimate of Breuer *et al.* (2000), and this may indicate that fertile soils are under-represented in the field studies.

We used two new approaches to estimate the tropical biome flux based on the seasonal pattern of N_2O emission and the climate (*Figure 1*). Almost all studies in the tropics indicated a seasonal pattern of N_2O flux with low emissions in the dry and high emissions in the wet season. *Figure 1* generalizes the seasonal N_2O pattern in the tropics and illustrates that the annual emissions depend on the amount and duration of high emissions in the wet season (Brumme *et al.*, 1999). The high emissions during the wet season are determined by factors at the ecosystem scale, such as annual climate, soil properties or forest management (i.e., long-term control). Thus, in natural conditions, we would not expect any large changes in the amount of N_2O emission during the wet season at a specific site. The inter-annual variation of N_2O emission is determined by the relative duration of the wet season relative to the dry season (short-term control), as indicated by the studies of Keller and Reiners (1994) and Verchot *et al.* (1999), for example. Our estimate for this biome is 4 Tg (N) per year (*Table 1*), a value that is similar to the result of 3.8 Tg (N) per year that was obtained by computing the relationship between monthly precipitation and N_2O emissions (*Figure 2*) for the tropical biome (precipitation approach, *Table 1*). These estimates are similar to the 'stratify-and-multiply' approach, but somewhat lower than the 4.8 Tg (N) per year of Breuer *et al.* (2000). The comparable results of these partly independent approaches, the 'dry/wet season', 'precipitation', and 'stratify-and-multiply' approaches, suggest that the IPCC value is too low. The updated estimate for the tropical biome lies in the range between 3.8 and 4.8 Tg (N) per year.

The tropical biome has been further stratified into tropical seasonal forest and rain forest, according to Holdridge (1967). The tropical seasonal forests account for about 57% and the rain forests for about 43% of the tropical N_2O flux (*Table 2*). The tropical rain forests receive more than 250 mm per month during 40% of a year, in contrast to the tropical seasonal forests, which receive that much rainfall during only 13% of the year. The higher total flux in the tropical seasonal forest is because they cover a larger area.

The annual contribution of the temperate forest biome to the atmospheric N_2O load has been calculated to be between 0.15 and 0.33 Tg (N) per year (Bouwman *et al.*, 1993; Potter *et al.*, 1996a). Considering the latest data sources, the updated 'stratify-and-multiply' approach provides an estimate of 0.7 Tg (N) per year, which is double the previous estimates. This higher biome flux estimate is attributable to sites with high fluxes, as high as are observed in the tropics, which were first discovered on an acid, moder-humus soil at Solling, Germany (Brumme and Beese, 1992). Broad-leaved forests with moder or mor humus and forests with organic soils may have distinct seasonal emissions with high fluxes in warm and wet periods, as discussed by Brumme *et al.* (1999). *Figure 1* illustrates that the annual N_2O flux of temperate forest with SEP is primarily determined by a short period with relatively high temperature and sufficient precipitation. The contribution of this period of high emission to the total biome flux is about 5%. This rather small proportion is likely to increase if spring temperatures and summer precipitation increase in the future. Not included in our estimate are emissions in winter and emissions exacerbated by forest harvesting. Forest harvesting could increase the compacted area and convert forest with low BEPs into forest with high SEPs (Teepe *et al.*, 2004). Winter emissions significantly increase the temperate biome flux. It has been shown that winter emissions occurred even at sites with low BEP and could increase the annual flux by more than 50% (Teepe *et al.*, 2000).

Table 2. *N_2O–N emission estimates from tropical seasonal (12.8 × 10⁶ km²) and tropical rain forests (7.2 × 10⁶ km²) for the dry (less than 250 mm per month) and the wet season (more than 250 mm per month) (standard deviations in parentheses).*

Tropical seasonal forest				Tropical rain forest			
Dry season		Wet season		Dry season		Wet season	
kg ha⁻¹ per year	Tg per year	kg ha⁻¹ per year	Tg per year	kg ha⁻¹ per year	Tg per year	kg ha⁻¹ per year	Tg per year
1.5 (1.3)	1.7	3.8 (4.8)	0.62	1.5 (1.3)	0.65	3.8 (4.8)	1.1

The total temperate forest biome flux (SEP/BEP approach) amounts to about 0.30 Tg (N) per year, which is similar to the estimate of Potter *et al.* (1996a), but only half the flux estimated by the 'stratify-and-multiply'- approach (*Table 1*). We found no indication in the data that the higher estimate by the 'stratify-and-multiply' approach results from harvesting or winter emissions and consequently we assume that the data are not fully representative of the temperate biome because of a large proportion of sites with SEP emissions patterns. Nevertheless, considering the uncertainties that may arise from possible harvesting and winter effects, we conclude that the temperate forest biome flux is likely to lie between 0.30 and 0.70 Tg (N) per year, which is within the range of the IPCC estimate.

The only estimate for the Boreal biome by Potter *et al.* (1996a) was confirmed by the 'stratify-and-multiply' approach and amounted to 0.62 Tg (N) per year (*Table 1*). Fens and bogs contribute only about 0.008 Tg (N) per year, because of the relatively small area that they occupy, although the flux per unit area is about four times the flux from northern upland forests. A large part of the Boreal forest is managed for forestry and many of the management interventions increase the N_2O emissions, for example, clear-felling and fertilization. However, the N_2O emissions from the managed Boreal coniferous forests are generally less than 1 kg (N) ha^{-1} per year (Martikainen, 1996).

There are several uncertainties when predicting possible changes in the N_2O emissions in northern latitudes resulting from the projected global warming. Firstly, the increase in temperature associated with global climate change is projected to be highest in the northern latitudes (IPCC, 2001) where winter snow cover is of great importance for the annual N_2O emissions (e.g., Brooks *et al.*, 1997). Reduction in the snow cover is likely to decrease N_2O emissions during winter, and hence also the annual N_2O emissions. Secondly, the N_2O emissions from organic soils with high water tables are closely linked to the hydrological conditions (Martikainen *et al.*, 1993). The projected changes in hydrology are more important than the changes in temperature for predicting N_2O emissions from natural organic soils.

3.2 *Methane uptake of forest biomes*

Of the 46 studies that contained data appropriate for our analysis, 20 were in temperate forests, 17 were in Boreal forest, and only 9 were in the tropics (*Table 3*). The temperate forest biome has a much higher sampling intensity, and is the smallest forest biome. Tropical forests are under-represented in the global dataset, and the studies are mostly in the neotropics. The data were not normally distributed (Shapiro–Wilk test, $W = 0.798$, $p < 0.0001$, *Figure 4*), and logarithmic transformation did not improve the nature of the distribution. There was a break in the linearity of the probability plot between 6.50 and 8.50 kg (CH_4) ha^{-1} per year, and the values for the 12 annual fluxes that exceeded 8.50 kg (CH_4) ha^{-1} per year fall on a line with steeper slope. Nine of these high fluxes were measured in temperate forests. At the lower end of the flux range, there were four observations of net emission (i.e., negative uptake values), all from temperate forests. Variances were also not homogenous across biomes or texture classes (Levene's test for homogeneity of variance: $p = 0.0079$ for biomes and $p = 0.0194$, for texture class). Therefore, we used the Kruskal–Wallis test, the non-parametric analogue of an analysis of variance, and the Mann–Whitney U-test, the non-parametric analogue of the t-test, to make comparisons.

Table 3. *Original data for methane uptake in Boreal, temperate, and tropical forests with environmental variables used in this analysis.*

Reference	Biome	Latitude (decimal degrees)	CH$_4$ (kg ha^{-1} per year)	Rainfall (mm per year)	Mean annual temperature (°C)	Soil texture class[1]
Billings *et al.*, 2000	Boreal	64.00	1.26	269	−3.5	M
Burke *et al.*, 1997	Boreal	55.91	3.23	585	−3.9	C
Burke *et al.*, 1997	Boreal	55.91	1.97	585	−3.9	F
Castro *et al.*, 1994	Boreal	44.30	1.61	867	5.9	O
Castro *et al.*, 1994	Boreal	43.30	2.66	990	4.9	O
Castro *et al.*, 1994	Boreal	44.30	2.82	1668	5.7	O
Castro *et al.*, 1994	Boreal	44.33	2.82	846	6.9	O
Castro *et al.*, 1994	Boreal	44.50	4.27	1141	4.4	O
Gulledge and Schimel, 2000	Boreal	64.75	1.78	269	−3.5	M
Gulledge *et al.*, 1997	Boreal	64.75	2.37	269	−3.5	M
Gulledge *et al.*, 1997	Boreal	64.75	2.06	269	−3.5	M
Huttunen *et al.*, 2003a	Boreal	67.00		507	−0.8	O
Huttunen *et al.*, 2003a	Boreal	60.21	0.22	701	4.5	O
Huttunen *et al.*, 2003a	Boreal	61.23	2.12	607	4.3	O
Huttunen *et al.*, 2003a	Boreal	60.21	1.66	701	4.5	O
Huttunen *et al.*, 2003a	Boreal	61.23	1.6	607	4.3	O
Kasimir-Klemedtsson and Klemedtsson, 1997	Boreal	58.04	1.4			
Maljanen *et al.*, 2003	Boreal	62.31	5.22	643	2.6	O
Maljanen *et al.*[2]	Boreal	64.86	2.46	517	2.1	
Maljanen *et al.*[2]	Boreal	65.00	4.3	517	2.1	
Maljanen *et al.*[3]	Boreal	61.20	11.33	631	3.9	
Maljanen *et al.*[2]	Boreal	65.00	3.99	517	2.1	
Maljanen *et al.*[2]	Boreal	64.64	1.35	517	2.1	
Maljanen *et al.*[2]	Boreal	65.00	1.94	517	2.1	
Martikainen *et al.*, 1994	Boreal	63.55	3.47	561	2.4	C
Martikainen *et al.*, 1994	Boreal	63.55	4.64	561	2.4	O
Martikainen *et al.*, 1996	Boreal	62.46	5.84	667	2.1	O
Saari *et al.*, 1998	Boreal	61.51	6.54	639	3.2	C
Saari *et al.*[4]	Boreal	61.51	5.68	581	3.5	C
Savage *et al.*, 1997	Boreal	55.40	3.99			
Savage *et al.*, 1997	Boreal	55.40	3.07			
Schiller and Hastie, 1996	Boreal	49.05	0.3	656	0.1	O
Whalen and Reeburgh, 1991	Boreal	65.44	0.89	208	−5.3	M
Whalen and Reeburgh, 1991	Boreal	65.44	0	208	−5.3	C
Whalen and Reeburgh, 1991	Boreal	65.44	0	208	−5.3	C
Whalen and Reeburgh, 1991	Boreal	65.44	0	208	−5.3	M
Whalen and Reeburgh, 1991	Boreal	65.44	0	208	−5.3	C
Whalen and Reeburgh, 1991	Boreal	65.44	0	208	−5.3	M
Whalen and Reeburgh, 1991	Boreal	65.44	0.7	208	−5.3	C
Whalen and Reeburgh, 1991	Boreal	65.44	1.15	208	−5.3	M

Table 3. continued

Reference	Biome	Latitude (decimal degrees)	CH_4 (kg ha^{-1} per year)	Rainfall (mm per year)	Mean annual temperature (°C)	Soil texture class[1]
Whalen and Reeburgh, 1991	Boreal	65.44	0.82	208	−5.3	M
Whalen and Reeburgh, 1991	Boreal	65.44	0	208	−5.3	C
Whalen and Reeburgh, 1991	Boreal	65.44	0.54	208	−5.3	M
Ambus and Christensen, 1995	Temperate	55.55	0.49	773	8.1	C
Ambus and Christensen, 1995	Temperate	55.55	1.21	773	8.1	C
Born et al., 1990	Temperate	49.35	3.1	577	8.6	M
Born et al., 1990	Temperate	49.35	12.6	577	8.6	C
Born et al., 1990	Temperate	49.35	12.7	577	8.6	C
Born et al., 1990	Temperate	49.30	0.9	577	8.6	F
Born et al., 1990	Temperate	49.30	3.7	577	8.6	F
Brumme and Borken, 1999	Temperate		0.75	750	8.0	M
Brumme and Borken, 1999	Temperate		1.34	650	8.5	M
Brumme and Borken, 1999	Temperate		2.48	680	7.8	M
Brumme and Borken, 1999	Temperate		0.3	650	8.5	M
Brumme and Borken, 1999	Temperate		0.19	650	8.5	M
Brumme and Borken, 1999	Temperate		0.73	1090	6.4	M
Brumme and Borken, 1999	Temperate		0.1	1090	6.4	M
Brumme[5]	Temperate		0.69	613	8.5	C
Brumme[6]	Temperate		0.63	1230	6.2	M
Brumme[6]	Temperate		2.86	1230	6.2	M
Dobbie and Smith, 1996	Temperate	56.08	9.42	723	6.9	C
Dong et al., 1998	Temperate	50.43	6.4	540	8.6	C
Dörr et al., 1993	Temperate	48.10	1.01	914	7.9	M
Dörr et al., 1993	Temperate	49.17	2.32	577	8.6	F
Dörr et al., 1993	Temperate	28.28	5.55	47	19.4	C
Goldman et al., 1995	Temperate	41.33	18.25	1041	9.6	C
Goldman et al., 1995	Temperate	41.00	19.71	1041	9.6	C
Goldman et al., 1995	Temperate	40.50	13.87	1080	11.8	C
Lessard et al., 1994	Temperate	45.37	1.86	865	4.7	M
Phillips et al., 2001	Temperate	35.97	2.12	1140	15.5	C
Priemé and Christensen, 1997	Temperate		3.4			C
Steudler et al., 1989	Temperate	42.00	14.72	1080	7.3	M
Steudler et al., 1989	Temperate	42.00	12.43	1080	7.3	M
Teepe et al., 2000	Temperate		0.25	670	7.6	M
Teepe and Brumme[7]	Temperate	52.75	2.28	1000	7.0	M
Teepe and Brumme[7]	Temperate	52.75	4.18	1000	7.0	M
Teepe et al., 2004a	Temperate		10.4	680	7.8	C
Teepe et al., 2004a	Temperate		1.95	680	7.8	M
Yavitt et al., 1995	Temperate	42.87	−2.63	1230	5.2	C
Yavitt et al., 1993	Temperate	42.87	−2.23	1230	5.2	C
Hudgens and Yavitt, 1997	Temperate	42.50	2.73	899	7.7	F
Hudgens and Yavitt, 1997	Temperate	42.50	0.91	899	7.7	F
Hudgens and Yavitt, 1997	Temperate	42.50	1.56	899	7.7	F

Yavitt *et al.*, 1990	Temperate	38.62	0	999	10.0	C
Yavitt *et al.*, 1990	Temperate	38.62	−0.44	999	10.0	M
Yavitt *et al.*, 1990	Temperate	38.62	0.02	999	10.0	M
Yavitt *et al.*, 1990	Temperate	38.62	0.07	999	10.0	M
Ishizuka *et al.*, 2002	Tropical	5.00	1.86	2060	24.0	C
Keller *et al.*, 1990	Tropical	9.00	1.28	2600	27.0	F
Keller *et al.*, 1990	Tropical	9.00	2.85	2600	27.0	F
Keller *et al.*, 1993	Tropical	10.20	5.62	4000	26.0	C
Keller *et al.*, 1993	Tropical	10.43	5.83	3962	25.8	C
Keller and Reiners, 1994	Tropical	10.43	4.6	3962	25.8	C
Keller and Reiners, 1994	Tropical	10.43	4.38	3962	25.8	C
Keller *et al.*, 1986	Tropical	2.93	1.37	1551	26.7	F
Steudler *et al.*, 1996b	Tropical	10.50	5.93	2200	25.5	C
Steudler *et al.*, 1996b	Tropical	10.50	3.42	2200	25.5	C
Verchot *et al.*[8]	Tropical	5.03	5.6	2518	24.0	M
Verchot *et al.*[8]	Tropical	5.03	5.06	2518	24.0	M
Verchot *et al.*, 2000	Tropical	2.98	2.4	1860	24.6	F
Verchot *et al.*, 2000	Tropical	2.98	2.1	1860	24.6	F
dos Santos, 1997	Tropical	22.47	8.75	2100	21.0	C
dos Santos, 1997	Tropical	22.47	9.31	2100	21.0	C

[1]Soil texture classes: C, coarse; M, medium; F, fine; O, organic.
[2]Maljanen *et al.* manuscript. See References.
[3]Maljanen *et al.* manuscript. See References.
[4]Saari A., Heiskanen, J. and Martikainen, P.J., unpublished data.
[5]Brumme, R. and Beese, F., unpublished data.
[6]Brumme, R. and Teepe, R., unpublished data.
[7]Teepe, R. and Brumme, R., unpublished data.
[8]Verchot, L.V. Hutabarat, L. and Martikainen, P.J., unpublished data.

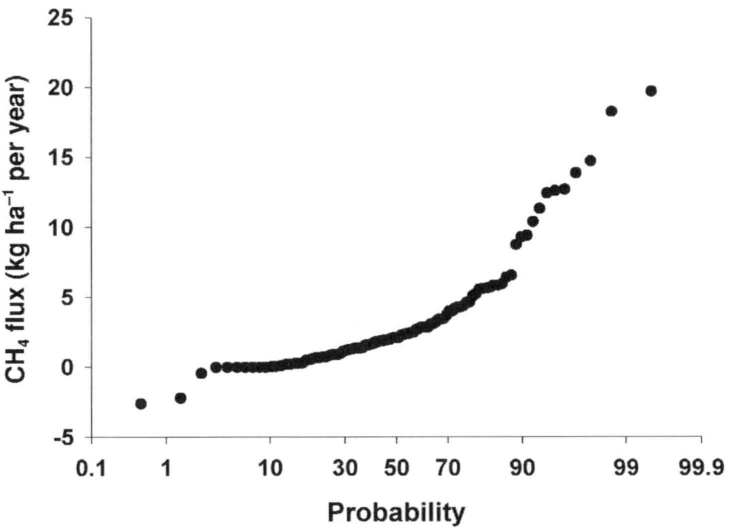

Figure 4. *Probability plot of CH₄ fluxes from the global dataset showing non-normality by deviations from a linear relationship at very low values and at values above 8.5 kg ha⁻¹ per year.*

We used a stepwise regression technique to explore the potential of five parameters that are frequently reported in studies of CH_4 fluxes and that have a mechanistic relationship with the flux: biome (coarse-scale integrator of productivity, growing season length and temperature), annual rainfall, mean annual temperature, soil texture, and latitude (proxy for the length of the growing season). The analysis showed that two variables, biome and soil texture class, accounted for 18.1% of the variation in the global dataset (p = 0.0032). Individually, texture class explained 11.9% of the variation, whereas biome explained 4.3%. As would be expected, mean CH_4 uptake rates were highest in tropical forests, followed by temperate forests, and Boreal forests (*Table 4*). These differences were significant at p = 0.026 (Kruskal–Wallis test). Globally, coarse-textured soils had much higher average uptake rates compared with organic soils, medium-textured soils, and fine-textured soils (*Table 4*). This difference was significant at p = 0.032 (Kruskal–Wallis test). Stratifying the world's forests by soil texture produced a much lower global estimate than that produced by stratifying by biome. This can be attributed to two factors: (i) this stratification scheme excluded the area covered by lithosols, very thin soils, and rock outcrops not suitable for plant growth; and (ii) it gave more weight to the importance of medium- and fine-textured soils, which are more common globally than coarse-textured soils, and have lower CH_4 flux rates. We found no data for CH_4 consumption in lithosols. Lithosols make up a significant portion of the global forest biome, accounting for 8.04×10^6 km^2 globally, and for as much as 40% of the tundra forests and 22% of the Boreal forest biome. Even in the tropics these soils account for 3–5% of the forest biome. We assumed that because these soils are generally unproductive and have a small organic matter content, biological activity must be low. We did not attempt to estimate the flux from the area with these soils. Although it is probably not reasonable to expect that the flux from these soils is zero, in the absence of data the implicit assumption in our global estimate is that these soils do not contribute to the net forest flux.

The data were poorly balanced and this precluded analysing a full model based on biome and soil texture (*Table 4*). For example, organic soils only appeared in the database in the Boreal biome, there were only two observations on medium-textured soils in the tropics, etc. Because the two most important variables identified in the analysis above were biome and texture, we stratified the data first by biome and analysed a reduced model within each biome according to natural groupings of texture classes represented in the dataset.

For the Boreal forest biome, there were 34 observations with adequate data for this analysis. We grouped coarse and organic soils together and fine and medium-textured soils together. The coarse-textured and organic soils have a mean uptake rate of 2.5 kg (CH_4) ha^{-1} per year, whereas the medium- and fine-textured soils have a mean uptake rate of 1.2 kg (CH_4) ha^{-1} per year. The difference between these two groupings was not significant (Mann–Whitney test, p = 0.120). Grouping the soils in this way explained 11.9% of the variation in uptake within the biome.

For the temperate forest biome, there were 44 observations with adequate data for this analysis. A similar grouping that combined medium- and fine-textured soils accounted for the highest portion of the variability (16.3%). Coarse soils were treated as a separate category, since there were no temperate organic soils in the dataset. The mean uptake rate for coarse-textured soils was 6.6 kg (CH_4) ha^{-1} per year, and that for

Table 4. *Estimates of CH₄ uptake by soils with different stratification schemes. The first estimate is made without stratification. The global dataset is then stratified by biome or by soil texture individually. Finally, the whole dataset is stratified by both biome and soil texture.*

Stratification level	n	Mean (kg (CH_4) ha^{-1} yr^{-1})	Variance	Area (10^6 km^2)	Total flux (Tg per year)
None	102	3.4	163 668	50.23	17.0
Forest biome					
Boreal	43	2.41	50 099	23.48	5.7
Temperate	44	4.0	301 861	6.75	2.7
Tropical	15	4.4	59 220	20.00	8.8
Total				**50.23**	**17.2**
Soil texture					
Coarse	35	5.2	297 438	6.82	3.5
Medium	34	2.2	104 885	24.10	5.3
Fine	12	2.0	7 199	9.27	1.9
Lithosols	0	n.d.	n.d.	8.03	—
Organic	13	2.8	31 818	1.96	0.5
Total				**50.19**	**11.2**
Biome and soil texture					
Boreal					
Coarse + organic	22	2.5	45 317	5.15	1.3
Medium + Fine	12	1.2	6 289	12.05	1.5
Lithosols	0	n.d.	n.d.	6.24	—
Temperate					
Organic	0	n.d.	n.d.	0.08	—
Coarse	17	6.6	497 140	1.00	0.7
Medium + Fine	27	2.3	120 674	4.74	1.1
Lithosols	0	n.d.	n.d.	0.94	—
Tropical					
Organic	0	n.d.	n.d.	0.12	—
Coarse + Medium	11	5.8	45 199	11.76	6.9
Fine	5	2.0	4 524	7.26	1.5
Lithosols	0	n.d.	n.d.	0.86	—
				50.19	**12.9**

n.d., no data.

medium and fine-textured soils was 2.3 kg (CH_4) ha^{-1} per year. The difference between these two groups was significant at $p = 0.085$ (Mann–Whitney test).

In the tropical biome, there were 16 observations with adequate data for the analysis. The aggregation that accounted for the largest percentage of the variation (47.1%) grouped coarse- and medium-textured soils together. In the tropical biome, there were only two observations on medium-textured soils, and these were both sandy–clay–loam soils, on the coarser end of the spectrum within the medium-texture classification. Coarse- and medium-textured soils had a mean uptake rate of 5.89 kg

(CH_4) ha^{-1} per year, whereas fine-textured soils had a mean uptake of 2.0 kg (CH_4) ha^{-1} per year. The difference between these two groups was significant at $p = 0.006$ (Mann–Whitney test).

That soil texture proved to be the stratification variable that accounted for the largest part of the variation in the global dataset is consistent with current thinking on the biogeochemical controls of CH_4 oxidation in soils (Dörr et al., 1993; Striegl, 1993; Verchot et al., 2000). Verchot et al. (2000) summarized the data for the neotropics and noted that fine-textured soils generally consumed between 1.5 and 2.0 kg (CH_4) ha^{-1} per year, while medium- and coarse-textured soils consumed more than 4.0 kg (CH_4) ha^{-1} per year. As Striegl et al. (1993) noted earlier, because gas transport of O_2 and CH_4 is the most important factor affecting CH_4 fluxes, and soil texture strongly affects diffusivity of gases within soils, Verchot et al. (2000) concluded that texture class is more important than other factors at the biome scale. Dörr et al. (1993) drew a similar conclusion from a comparison of several sites in central Europe. The conclusions of Verchot et al. (2000) and Dörr et al. (1993) drawn at a regional scale, and the conclusion that we draw here at the global scale, are consistent in their affirmation of soil texture class as one of the key factors controlling the magnitude of CH_4 fluxes in soils.

The question that remains is whether the stratification that we used was successful in providing a better estimate of the sink strength. Stratification reduced the variance in most strata each time. In the stratification by climate zones, the variances in the Boreal and tropical strata are lower than the global variance, but variance in the temperate stratum was larger than the global variance. As we noted above, the sample in the tropical forest biome was relatively small compared with the other biomes, and all data were from humid environments. We conclude that a large part of the variation in that biome simply has not yet been sampled. Using the global mean to extrapolate the global forest CH_4 sink gives a value of 17.0 Tg (CH_4) per year. Stratifying the dataset into biomes produces a virtually identical value of 17.2 Tg (CH_4) per year (*Table 4*). Thus, we gain little by this stratification with the current paucity of data.

Stratification by soil texture resulted in reduced variance in three out of the four texture classes and produced a much reduced estimate of the total global sink strength of 11.2 Tg (CH_4) per year, because of the elimination of large areas of the forest biomes for which there were no data. The variance remained high in coarse-textured soils. The double stratification scheme, using texture class within biome, reduced the variance below that of the variance in the biome-only stratification scheme in five out of the six strata (*Table 4*). The variance remained high in the temperate forests, and particularly high in those with coarse-textured soils in that biome. The stratification by climate zone and texture class gave a global estimate of 12.9 Tg (CH_4) per year, which is probably a reasonable lower bound.

Whereas several researchers have made estimates of global budgets using 'biome' to stratify their data, our approach suggests that more robust estimations can be made if additional factors with mechanistic significance are taken into consideration. Indeed, stratification by biome reduced the variance in the whole dataset and further stratification by soil texture class within biomes reduced the variance even more (*Table 4*). A large part of the lower global flux estimate using soil texture as a stratification variable resulted from the decrease in area over which mean fluxes were multiplied. There is no basis for expectation that the consumption rates in forests with lithosols would be similar to that observed on more developed soils. Yet, stratification schemes based only on climatic zone assume a similar rate on all soils.

Our estimates for the different forest biomes of the world are higher than the modelled estimates of Potter *et al.* (1996b) who found the global CH_4 forest sink to be 7.4 Tg (CH_4) per year. Generally, 'stratify-and-multiply' estimates have given higher results than the modelled estimates of Potter *et al.* (1996b). In the same paper, the authors presented the results of a 'stratify-and-multiply' estimate of 14.7 Tg (CH_4) per year for the global forests. Their modelled estimate for the global sink of all biomes, including grasslands and deserts, was 17.0 Tg (CH_4) per year. Our 'stratify-and-multiply' estimate for all biomes was somewhat higher, 20.6 Tg (CH_4) per year. Dörr *et al.* (1993) and Smith *et al.* (2000) used the 'stratify-and-multiply' approach and both estimated a global total soil CH_4 sink of 29 Tg (CH_4) per year.

4. Contribution of CH_4 and N_2O to the atmospheric warming balance of forests biomes

The atmospheric warming balance of forest biomes depends on the net ecosystem exchanges (NEEs) of CO_2 and the GHGs, N_2O and CH_4. The emission of N_2O from soils to the atmosphere increases the warming potential of forest biomes, whereas the flux of CH_4 from atmosphere to soils reduces the warming potential. As a result of the much higher GWP of N_2O compared with CH_4 (IPCC, 2001), the emission of N_2O is of much more concern than the uptake of CH_4 (Andreae and Schimel, 1989; Bouwman, 1990; Granli and Bøckman, 1994; van Ham *et al.*, 2002). Assuming a mean annual N_2O efflux from the biomes of 2.0, 0.72, and 0.27 kg (N) ha^{-1} per year and a 100 year integration period, the annual atmospheric impacts are 930, 335, and 126 kg (CO_2 equivalents) ha^{-1} in the tropical, temperate, and Boreal forest biomes, respectively (*Table 5*). Taking the average values weighted by soil texture class, CH_4 uptake amounted to 4.4, 3.1, and 1.6 kg (CH_4) ha^{-1} per year in the tropical, temperate, and Boreal forest biomes, respectively, and produced a CO_2-equivalent reduction in the atmospheric load of 101, 71, and 37 kg (CO_2) ha^{-1} per year. These figures only take into account the direct warming effects of CH_4. Indirect effects of CH_4, such as tropospheric ozone production, stratospheric water vapour production, CO_2 production, and the temporal changes in the CH_4 adjustment time resulting from its coupling with OH$^-$, are not included in this estimate. Estimates for these indirect effects are less certain than those for the direct effects, but may amount to as much as 30% of the direct effects (IPCC, 2001).

Thus, the net atmospheric impact of exchanges of CH_4 and N_2O by forests is 829, 264, and 89 kg (CO_2 equivalents) ha^{-1} per year, for the tropical, temperate, and Boreal forest biomes, respectively. The numbers for the Boreal and temperate forest biomes are small (*Table 5*) compared with the magnitude of the NEE of CO_2 (Jarvis *et al.*, 2001) and do not significantly influence the atmospheric warming balances in these biomes. In the tropics, the NEE seems to be lower but this estimate is based on measurements at only a few sites (Malhi *et al.*, 1999). Thus, the net atmospheric impact of N_2O and CH_4 may significantly reduce the cooling effect of the NEE of CO_2 in the case of the tropical forest biome.

Table 5. Atmospheric impact of N_2O emission and CH_4 uptake of soils (kilograms (CO_2 equivalents) per hectare per year, using a 100 year time horizon) compared with the net ecosystem exchange of CO_2 (NEE, kilograms (CO_2) per hectare per year) of forest biomes (see text for data source of NEE).

Biome	N_2O	CH_4	Sum of non-CO_2 GHGs	NEE	NEE of non-CO_2 GHGs (%)
	kg (CO_2) ha^{-1} per year				
Tropical forest	930	−101	829	2 200	38%
Temperate forest	335	−71	264	7 300 to 26 000	3.6% to 1.0%
Boreal forest	126	−37	89	0 to 4 800	0 to 1.9%

Acknowledgements

We thank Anne Saari, Marja Maljanen, and Jari Huttunen for providing unpublished data for methane and nitrous oxide emissions, and Jukka Turunen for providing information on the distribution of Boreal organic forest soils. We thank Arvin Mosier for review.

References

Alm, J., Saarnio, S., Nykänen, H., Silvola, J. and Martikainen, P.J. (1999) Winter CO_2, CH_4 and N_2O fluxes on some natural and drained boreal peatlands. *Biogeochemistry* **44**: 163–-186.

Ambus, P. and Christensen, S. (1995) Spatial and seasonal nitrous oxide and methane fluxes in Danish forest-, grassland-, and agroecosystems. *Journal of Environmental Quality* **24**: 993–1001.

Ambus, P., Jensen, J.M., Prieme, A., Pilegaard, K. and Kjoller, A. (2001) Assessment of CH_4 and N_2O fluxes in a Danish beech (*Fagus sylvatica*) forest and an adjacent N-fertilised barley (*Hordeum vulgare*) field: effects of sewage sludge amendments. *Nutrient Cycling in Agroecosystems* **60**: 15–21.

Andreae, M.O. and Schimel, D.S. (1989) *Exchange of Trace Gases between Terrestrial Ecosystems and the Atmosphere.* John Wiley, New York.

Billings, S.A., Richter, D.D. and Yarie, J. (2000) Sensitivity of soil methane fluxes to reduced precipitation in boreal forest soils. *Soil Biology and Biochemistry* **32**: 1431–1441.

Blew, R.D. and Parkinson, D. (1993) Nitrification and denitrification in a white spruce forest in southwest Alberta, Canada. *Canadian Journal of Forest Research* **23**: 1715–1719.

Born, M., Dörr, H. and Levin, I. (1990) Methane consumption in aerated soils of the temperate zone. *Tellus Series B – Chemical and Physical Meteorology* **42**: 28.

Botch, M.S., Kobak, K.I., Vinson, T.S. and Kolchugina, T.P. (1995) Carbon pools and accumulation in peatlands in the Former Soviet Union. *Global Biogeochemical Cycles* **9**: 37–46.

Bouwman, A.F. (1990) *Soils and the Greenhouse Effect.* John Wiley, New York.

Bouwman, A.F., Fung, I., Matthews, E. and John, J. (1993) Global analysis of the potential for N_2O production in natural soils. *Global Biogeochemical Cycles* **7**: 557–597.

Bowden, R.D., Steudler, P.A. and Melillo, J.M. (1990) Annual nitrous oxide fluxes from temperate forest soils in the northeastern United States. *Journal of Geophysical Research* **95**: 13997–14005.

Brady, N.C. (1974) *The Nature and Property of Soils*. Macmillan, New York.

Breuer, L., Papen, H. and Butterbach-Bahl, K. (2000) N_2O emission from tropical forest soils of Australia. *J. Geophysical Research* **105**: 26353–26367.

Brooks, P.D., Schmidt, S.K. and Williams, M.W. (1997) Winter production of CO_2 and N_2O from alpine tundra: Environmental controls and relationship to inter-system C and N fluxes. *Oecologia* **110(3)**: 403–413 APR 1997.

Brumme, R. (1995) Mechanisms for carbon and nutrient release and retention in beech forest gaps, 3. Environmental regulation of soil respiration and nitrous oxide emissions along a microclimatic gradient. *Plant and Soil* **168–169**: 593–600.

Brumme, R. and Beese, F. (1992) Effects of liming and nitrogen fertilization on emissions of CO_2 and N_2O from a temperate forest. *Journal of Geophysical Research* **97**: 12851–12858.

Brumme, R. and Borken, W. (1999) Site variation in methane oxidation as effected by atmospheric deposition and type of temperate forest ecosystem. *Global Biogeochemical Cycles* **13**: 493–501.

Brumme, R., Borken, W. and Finke, S. (1999) Hierarchical control on nitrous oxide emission in forest ecosystems. *Global Biogeochemical Cycles* **13**: 1137–1148.

Burke, R.A., Zepp, R.G., Tarr, M.A., Miller, W.L. and Stocks, B.J. (1997) Effect of fire on soil-atmosphere exchange of methane and carbon dioxide in Canadian boreal forest sites. *Journal of Geophysical Research* **102 (D24)**: 29289–29300.

Castro, M.S., Peterjohn, W.T., Melillo, J.M. and Steudler, P.A. (1994) Effects of nitrogen fertilization on the fluxes of N_2O, CH_4 and CO_2 from soils in a Florida slash pine plantation. *Canadian Journal of Forestry Research* **24**: 9–13.

Corre, M. D., Pennock, D. J., van Kessel, C. and Elliot, D. K. (1999) Estimation of annual nitrous oxide emissions from a transitional grassland-forest region in Saskatchewan, Canada. *Biogeochemistry* **44**: 29–49.

Dobbie, K.E. and Smith, K.A. (1996) Comparison of CH_4 oxidation rates in woodland, arable and set aside soils. *Soil Biology and Biochemistry* **28**: 1357–1365.

Dong, Y., Scharffe, D., Lobert, J.M., Crutzen, P.J. and Sanhueza, E. (1998) Fluxes of CO_2, CH_4, and N_2O from a temperate forest soil: the effects of leaves and humus layer. *Tellus Series B – Chemical and Physical Meteorology* **50**: 243–252.

Dörr, H., Katruff, L., and Levin, I. (1993) Soil texture parameterization of the methane uptake in aerated soils. *Chemosphere* **26**: 697–713.

dos Santos, M.B.P. (1997) Medidas de fluxo de metano em solos de florestas da mata Atlântica do estado de Rio de Janeiro. Ph.D. dissertation, Universidade Federal Fluminense Rio de Janeiro, Brazil.

Firestone, M.K. and Davidson, E.A. (1989) Microbiological basis of NO and N_2O production and consumption in Soil. In: Andreae, M.O. and Schimel, D.S. (eds) *Exchange of Trace Gases between Terrestrial Ecosystems and the Atmosphere*, pp. 7–21. John Wiley, New York.

Flessa, H., Dörsch, P. and Beese, F. (1995) Seasonal variation of N_2O and CH_4 fluxes in differently managed arable soils in southern Germany. *Journal of Geophysical Research* **100**: 23,115–23,124.

García-Méndez, G., Maas, J.M., Matson, P.A. and Vitousek, P.M. (1991) Nitrogen transformations and nitrous oxide flux in a tropical deciduous forest in México. *Oecologia* **88**: 362–366.

Goldman, M.B., Groffman, P.M., Pouyat, R.V., McDonnell, M.J., and Pickett, S.T.A. (1995) CH_4 uptake and N availability in forest soils along an urban to rural gradient. *Soil Biology and Biochemistry* **27**: 281–286.

Goodroad, L.L. and Keeney, D.R. (1984) Nitrous oxide emission from forest, marsh, and prairie ecosystems. *Journal of Environmental Quality* **13**: 448–452.

Goreau, T.J. and deMello, W.Z. (1987) Tropical deforestation: some effects on atmospheric chemistry. *Ambio* **17**: 275–281.

Granli, T. and Bøckman, O.C. (1994) Nitrous oxide from agriculture. *Norwegian Journal of Agricultural Sciences* **12**: 1–128.

Gulledge, J. and Schimel, J.P. (2000) Controls on soil carbon dioxide and methane fluxes in a variety of taiga forest stands in interior Alaska. *Ecosystems* **3**: 269–282.

Gulledge, J., Doyle, A.P. and Schimel, J.P. (1997) Different NH_4^+ inhibition patterns of soil CH_4 consumption: a result of distinct CH_4 oxidizer populations across sites? *Soil Biology and Biochemistry* **29**: 13–21.

Holdridge, L.R. (1967) *Life Zone Ecology*. Tropical Science Center, San Jose, Costa Rica.

Hudgens, D.E. and Yavitt, J.B. (1997). Land-use effects on soil methane and carbon dioxide fluxes in forests near Ithaca, New York. *Ecoscience* **4**: 214–222.

Huttunen, J., Nykänen, H., Turunen, J., Nenonen, O. and Martikainen, P.J. (2002) Fluxes of nitrous oxide on natural peatlands in Vuotos, an area projected for a hydroelectric reservoir in northern Finland. *Suo* **53 (3–4)**: 87–96.

Huttunen, J.T., Nykänen, H., Turunen, J. and Martikainen, P.J. (2003a) Methane emissions from natural peatlands in the northern boreal zone in Finland, Fennoscandia. *Atmosperic Environment* **37**: 147–151.

Huttunen, J.T., Nykänen, H., Martikainen, P.J. and Nieminen, M. (2003b). Fluxes of nitrous oxide and methane from drained peatlands following forest clear-felling in southern Finland. *Plant and Soil* **255**: 457–462.

IPCC (2001) *Climate Change 2001: The Scientific Basis*, Houghton, J.T., Ding, Y., Griggs, D.J., Noguer, M., van der Linden, P.J., Dai, X., Maskell, K. and Johnson, C. A. (eds). Cambridge University Press, Cambridge and New York. 881 pp.

Ishizuka, S., Tsuruta, H. and Murdiyarso, D. (2002) An intensive field study on CO_2, CH_4, and N_2O emission from soils at four land-use types in Sumatra, Indonesia. *Global Biogeochemical Cycles* **16**: 1049, doi:10.1029/2001GB001614.

Jarvis, P.J., Saugier, B. and Schulze, E.D. (2001) Productivity of boreal forests. In: Roy, J., Saugier, B. and Mooney, H.A. (eds) *Terrestrial Global Productivity*, pp. 211–244. Academic Press, San Diego.

Kasimir-Klemedtsson, A. and Klemedtsson, L. (1997) Methane uptake in Swedish forest soil in relation to liming and extra N-deposition. *Biology and Fertility of Soils* **25**: 296–301.

Kauppi, P.E., Mielikäinen, K. and Kuusela, K. (1992) Biomass and carbon budget of European forests, 1971 to 1990. *Science* **256**: 70–74.

Keller, M. and Reiners, W.A. (1994) Soil–atmosphere exchange of nitrous oxide, nitric oxide and methane under secondary succession of pasture to forest in the Atlantic lowlands of Costa Rica. *Global Biogeochemical Cycles* **8**: 399–409.

Keller, M., Goreau, J., Wofsy, S.C., Kaplan, W.A. and McElroy, M.B. (1983) Production of nitrous oxide and consumption of methane by forest soils. *Geophysical Research Letters* **10**: 1156–1159.

Keller, M., Kaplan, W.A. and Wofsy, S.C. (1986) Emissions of N_2O, CH_4 and CO_2 from tropical forest soils. *Journal of Geophysical Research* **91**: 11791–11802.

Keller, M., Kaplan, W.A., Wofsy, S.C. and DaCosta, J.M. (1988) Emissions of N_2O from tropical forest soils: Response to fertilization with NH_4^+, NO_3^-, PO_4^{3-}. *Journal of Geophysical Research* **93**: 1600–1604.

Keller, M., Mitre, M.E. and Stallard, R.F. (1990) Consumption of atmospheric methane in tropical soils of central Panama: effects of agricultural development. *Global Biogeochemical Cycles* **4**: 21–28.

Keller, M., Veldkamp, E., Weitz, A.M. and Reiners, W.A. (1993) Effect of pasture age on soil trace-gas emissions from a deforested area of Costa Rica. *Nature* **365**: 244–246.

Kiese, R. and Butterbach-Bahl, K. (2002) N_2O and CO_2 emissions from three different tropical forest sites in the wet tropics of Queensland, Australia. *Soil Biology and Biochemistry* **34**: 975–987.

Klemedtsson, L., Kasimir-Klemedtsson A., Moldan, F. and Weslien, P. (1997) Nitrous oxide emission from Swedish forest soils in relation to liming and simulated increased N-deposition. *Biology and Fertility of Soils* **25**: 290–295.

Kosov, V.I. and Kreshtapova, V.N. (1966) The peat resources of Russia and their utilization. In: Lappalainen, E. (ed) *Global Peat Resources*, pp. 127–131. International Peat Society, Jyskä, Finland.

Lappalainen, E. (1996) General review on world peatland and peat resources. In: Lappalainen, E. (ed) *Global Peat Resources*, pp. 53–56. International Peat Society, Jyskä, Finland

Lappalainen, E. and Zurek, S. (1996) Peat in other European countries. In: Lappalainen, E. (ed) *Global Peat Resources*, pp. 153–162. International Peat Society, Jyskä, Finland.

Legates, D. R. and Willmott, C. J. (1990) Mean seasonal and spatial variability in gauge-corrected, global precipitation. *International Journal of Climatology* **10**: 111–127.

Lessard, R., Rochette, P., Topp, E., Pattey, E. and Desjardins, R.L. (1994) Methane and carbon dioxide fluxes from poorly drained adjacent cultivated and forest sites. *Canadian Journal of Soil Science* **74**: 139–146.

Livingston, G.P., Vitousek, P.M. and Matson, P.A. (1988) Nitrous oxide flux and nitrogen transformations across a landscape gradient in Amazonia. *Journal of Geophysical Research* **93**: 1593–1988.

Luízao, F., Matson, P., Livingston, G., Luizao, R. and Vitousek, P.M. (1989) Nitrous oxide flux following tropical land clearing. *Global Biogeochemical Cycles* **3**: 281–285.

Lundgren, B. and Söderström, B. (1983) Bacterial numbers in a pine forest soil in relation to environmental factors. *Soil Biology and Biochemistry* **15**: 625–630.

MacDonald, J.A., Skiba, U., Sheppard, L.J., Ball, B., Roberts, J.D., Smith, K.A. and Fowler, D. (1997) The effect of nitrogen deposition and seasonal variability on methane oxidation and nitrous oxide emission rates in an upland spruce plantation and moorland. *Atmospheric Environment* **31**: 3693–3706.

Malhi, Y., Baldocchi, D.D. and Jarvis, P.G. (1999) The carbon balance of tropical, temperate and boreal forests. *Plant, Cell and Environment* **22**: 715–740.

Maljanen, M., Liikanen, A., Silvola, J. and Martikainen, P.J. (2003) Methane fluxes on agricultural and forested boreal organic soils. *Soil Use and Management* **19**: 73–79.

Maljanen, M., Jokinen, H., Saari, A., Strommer, R. and Martikainen, P.J. Short-term effects of wood ash and nitrogen fertilization on CH_4, N_2O and CO_2 gas fluxes from a Finnish spruce forest. (Manuscript).

Maljanen, M., Nykänen, H., Moilanen, M. and Martikainen, P.J. Effect of ash fertilizaton on CH_4, N_2O and CO_2 gas fluxes from boreal forest soils (Manuscript).

Martikainen, P.J. (1996) Microbial processes in boreal forest soils as affected by forest management practices and atmospheric stress. In: Stotzky, G. and Bollag, J.M. (eds) *Soil Biochemistry*, volume 9, pp. 195–232. Marcel Dekker, New York.

Martikainen, P.J., Nykänen, H., Crill, P. and Silvola, J. (1993). Effect of a lowered water table on nitrous oxide fluxes from northern peatlands. *Nature* **366**: 51–53.

Martikainen, P. J., Nykänen, H., Lång, K. and Ferm, A. (1994) Kasvihuonekaasujen päästöt turkistarhojen lähimetsissä. In: Mälkönen, E. and Sivula, H. (eds) *Suomen Metsien Kunto*, pp. 212–223. (In Finnish.) Metsäntutkimuslaitoksen tiedonantoja 527. Metsien terveydentilan tutkimusohjelman väliraportti. Helsinki, Finland.

Martikainen, P. J., Nykänen, H., Regina, K., Lehtonen, M. and Silvola, J. (1996) Methane fluxes in a drained and forested peatland treated with different nitrogen compounds. In: Laiho, R., Laine, J. and Vasander, H. (eds) *Northern Peatlands in Global Climatic Change*, pp. 105–109. Proceedings of the international workshop held in Hyytiälä, Finland, 8–12 Oct 1995. Publications of the Academy of Finland 1/96.

Matson, P.A. and Vitousek, P.M. (1987) Cross-system comparison of soil nitrogen transformations and nitrous oxide flux in tropical forest ecosystems. *Global Biogeochemical Cycles* **1**: 163–170.

Matson, P.A. and Vitousek, P.M. (1990) Ecosystem approach to a global nitrous oxide budget. *BioScience* **40**: 667–672.

Matson, P.A., Vitousek, P.M. and Schimel, D.S. (1989) Regional extrapolation of trace gas flux based on soils and ecosystems. In: Andreae, M.O. and Schimel, D.S. (eds) *Exchange of Trace Gases Between Terrestrial Ecosystems and the Atmosphere*, pp. 97–108. John Wiley, New York.

Melillo, J.M., Steudler, P.A., Feigl, B.J., Neill, C., Garcia, D., Piccolo, M.C., Cerri, C.C. and Tian H. (2001) Nitrous oxide emissions from forests and pasture of various ages in the Brazilian Amazon. *Journal of Geophysical Research* **106**: 34179–34188.

Mogge, B., Kaiser, E.A. and Munch, J.C. (1998) Nitrous oxide emissions and denitrification N-losses from forest soils in the Bornhöved lake region (Northern Germany). *Soil Biology and Biochemistry* **30**: 703–710.

Mosier, A.R., Klemedtsson, L.K., Sommerfield, R.A. and Musselmann, R.C. (1993). Methane and nitrous oxide flux in a Wyoming subalpine meadow. *Global Biogeochemical Cycles* **7**: 771–784.

Mosier, A.R., Parton, W.J., Valentine, D.W., Ojima, D.S., Schimel D.S. and Delgado, J.A. (1996) CH_4 and N_2O fluxes in Colorado shortgrass steppe: 1. Impact of landscape and nitrogen addition. *Global Biogeochemical Cycles* **10**: 387–399.

Mosier, A.R., Morgan, J.A., King, J.Y., LeCain, D. and Milchunas, D.G. (2002) Soil-atmosphere exchange of CH_4, CO_2, NO_x, and N_2O in the Colorado shortgrass steppe under elevated CO_2. *Plant and Soil* **240**: 201–211.

Nevison, C.D., Esser, G. and Holland, E.A. (1996) A global model of changing N_2O emissions from natural and perturbed soils. *Climate Change* **32**: 327–378.

Paavolainen, L., Fox, M. and Smolander, A. (2000) Nitrification and denitrification in forest soil subjected to sprinkling infiltration. *Soil Biology and Biochemistry* **32**: 669–678.

Papen, H. and Butterbach-Bahl, K. (1999) A 3-year continuous record of nitrogen trace gas fluxes from untreated and limed soil of a N-saturated spruce and beech forest ecosystem in Germany 1. N_2O emissions. *Journal of Geophysical Research* **104**: 18487–18503.

Phillips, R.L., Whalen, S.C. and Schlessinger, W.H. (2001) Influence of atmospheric CO_2 enrichment on methane consumption in a temperate forest soil. *Global Change Biology* **7**: 557–563.

Potter, C. S., Matson, P. A., Vitousek, P. M. and Davidson, E. A. (1996a) Process modeling of controls on nitrogen trace gas emissions from soils world-wide. *Journal of Geophysical Research* **101**: 1361–1377.

Potter, C.S., Davidson, E.A. and Verchot, L.V. (1996b) Estimation of global biogeochemical controls and seasonality in soil methane consumption. *Chemosphere* **32**: 2219–2246.

Priemé, A. and Christensen, S. (1997) Seasonal and spatial variation of methane oxidation in a Danish spruce forest. *Soil Biology and Biochemistry* **29**: 1665–1172.

Regina, K., Nykänen, H., Maljanen, M., Silvola, J. and Martikainen, P.J. (1998) Emissions of N_2O and NO and net nitrogen mineralization in a boreal forested peatland treated with different nitrogen compounds. *Canadian Journal of Forest Research* **28**: 132–140.

Rubec, C. (1996) The status of peatland resources in Canada. In: Lappalainen, E. (ed) *Global Peat Resources*, pp. 243–260. International Peat Society, Jyskä, Finland.

Saari, A., Heiskanen, J. and Martikainen, P.J. (1998) Effect of the organic horizon on methane oxidation and uptake in soil of a boreal Scots pine forest. *FEMS Microbiology Ecology* **26**: 245–255.

SAS Institute (1992) *SAS/STAT Users Guide*. SAS Institute, Cary, NC.

Savage, K., Moore, T.R. and Crill, P.M. (1997) Methane and carbon dioxide exchanges between the atmosphere and northern boreal forest soils. *Journal of Geophysical Research* **102 (D24)**: 29279–29288.

Schiller, C.L. and Hastie, D.R. (1996) Nitrous oxide and methane fluxes from perturbed and unperturbed boreal forest sites in northern Ontario. *Journal of Geophysical Research* **101 (D17)**: 22767–22774.

Schmidt, J., Seiler, W. and Conrad, R. (1988) Emission of nitrous oxide from temperate forest soils into the atmosphere. *Journal of Atmospheric Chemistry* **1:** 95–115.

Schulte-Bisping, H., Brumme, R. and Priesack, E. (2003) Nitrous oxide emission inventory of German forest soils. *Journal of Geophysical Research* **108 (D4):** 4132. doi: 10.1029/JD002292. ACH 2-1, ACH 2-9.

Simpson, I.J., Edwards, G.C., Thurtell, G.W., den Hartog, G., Neumann, H.H. and Staebler, R.M. (1997) Micrometeorological measurements of methane and nitrous oxide exchange above a boreal aspen forest. *Journal of Geophysical Research* **102 (D24):** 29331–29341.

Smith, K.A., Dobbie, K.E., Ball, B.C., Bakken, L.R., Sitaula, B.K., Hansen, S. *et al.* (2000) Oxidation of atmospheric methane in Northern European soils, comparison with other ecosystems, and uncertainties in the global terrestrial sink. *Global Change Biology* **6:** 791–803.

Sommerfeld, R.A., Mosier, A.R., Musselman, R.C. (1993) CO_2, CH_4 and N_2O flux through a Wyoming snowpack and implications for global budgets. *Nature* **361:** 140–140.

Steudler, P.A., Bowden, R.D., Melillo, J.M., and Aber, J.D. (1989) Influence if nitrogen fertilization on methane uptake in temperate forest soils. *Nature* **341:** 314–316.

Steudler, P.A., Jones, R.D., Castro, M.S., Mellilo, J.M. and Lewis, D.L. (1996a) Microbial controls of methane oxidation in temperate forest and agricultural soils. In: Murrell, J.C. and Kelly, D.P. (eds) *Microbiology of Atmospheric Trace Gases,* pp. 69–84. NATO ASI Ser. G vol. 39. Springer-Verlag, New York.

Steudler, P.A., Melillo, J.M., Feigl, B.J., Neill, C., Piccolo, M.C. and Cerri, C.C. (1996b) Consequence of forest-to-pasture conversion on CH_4 fluxes in the Brazilian Amazon. *Journal of Geophysical Research* **101:** 18547–18554.

Striegl, R.G. (1993) Diffusional limits to the consumption of atmospheric methane by soils. *Chemosphere* **26:** 715–720.

Teepe, R., Brumme, R. and Beese, F. (2000) Nitrous oxide emissions from frozen soils under agricultural, fallow and forest land. *Soil Biology and Biochemistry* **32:** 1807–1810.

Teepe, R., Brumme, R. Beese, F. and Ludwig, B. (2004a) Nitrous oxide emissions and methane consumption following compaction of forest soils. *Soil Science Society of America Journal* **68:** 605–611.

Teepe, R., Vor, A., Beese, F. and Ludwig, B. (2004b) Emissions of N_2O from soils during cycles of freezing and thawing and the effects of soil water, texture and duration of freezing. *European Journal of Soil Science* **55:** 347–367.

van Ham, J., Baede, A.P.M., Guicherit, R. and Williams-Jacobse, J.G.F.M. (eds) (2002) *Non-CO_2 Greenhouse Gases: Scientific Understanding, Control Options and Policy Aspects.* Proceedings of the third international symposium, Maastrich, The Netherlands. Millpress, Rotterdam, The Netherlands.

Verchot, L.V., Davidson, E.A., Cattânio, J.H., Ackerman, I.L., Erickson, H.E. and Keller, M. (1999) Land use change and biogeochemical controls of nitrogen oxide emissions from soils in eastern Amazonia. *Global Biogeochemical Cycles* **13:** 31–46.

Verchot, L.V., Davidson, E.A., Cattânio, J.H. and Ackerman, I.L. (2000) Land-use change and biogeochemical controls of methane fluxes in soils of eastern Amazonia. *Ecosystems* **3:** 41–56.

Weitz, A.M., Veldkamp, E., Keller, M., Neff, J. and Crill, P.M. (1998) Nitrous oxide, nitric oxide, and methane fluxes from soils following clearing and burning of tropical secondary forest. *Journal of Geophysical Research* **103**: 28047–28058.

Whalen, S.C., Reeburgh, W.S. and Kizer, K.S. (1991) Methane consumption by taiga. *Global Biogeochemical Cycles* **5**: 241–265.

Yavitt, J.B., Downey, D.M., Lang, G.E. and Sexstone, A.J. (1990) Methane consumption in two temperate forest soils. *Biochemistry* **9**: 39–52.

Yavitt, J.B., Simmons, J.A. and Fahey, T.J. (1993) Methane fluxes in a northern hardwood forest ecosystem in relation to acid precipitation. *Chemosphere* **26**: 721–730

Yavitt, J.B., Fahey, T.J. and Simmons, J.A. (1995) Methane and carbon dioxide dynamics in a northern hardwood ecosystem. *Soil Science Society of America Journal* **59**: 796–804.

Zechmeister-Boltenstern, S. and Meger, S. (1997) Nitrous oxide emissions from two beech forests near Vienna, Austria. In: Becker, K.H. and Wiesen, P. (eds) *Proceedings of the 7th International Workshop on Nitrous Oxide Emissions in Cologne, Germany*, pp. 429–432. Bergische University, Gesamthochschule Wuppertal.

Zechmeister-Boltenstern, S., Hahn, M., Meger, S. and Jandl, R. (2002) Nitrous oxide emissions and nitrate leaching in relation to microbial biomass dynamics in a beech forest soil. *Soil Biology and Biochemistry* **34**: 823–832.

15

Effects of reforestation, deforestation, and afforestation on carbon storage in soils

Claudia I. Czimczik, Martina Mund, Ernst-Detlef Schulze and Christian Wirth

1. Introduction

Land-use change, especially in potentially forested areas, often leads to changes of soil organic carbon (SOC) pools in the mineral topsoil. Thus, there is scope for land-use and land-management activities to increase or decrease CO_2 concentrations in the atmosphere over this century. The cumulative worldwide historical carbon losses from forest vegetation and soils to the atmosphere resulting from land-use change up to the year 2000 have been estimated as 200–220 Pg (C) (House *et al.*, 2002). This is of the same order of magnitude as the estimated cumulative fossil fuel emissions from pre-industrial times to the year 2000 of about 280 Pg (C) (IPCC, 2001). Two-thirds to three-quarters of the total carbon losses probably resulted from conversion of forest land to cropland or other land-use types (DeFries *et al.*, 1999). In the 1990s, deforestation remained the most important land-use change in the tropical regions (see *Table 1*). By contrast, in the non-tropical regions of the developed countries, deforestation has been reversed by natural reforestation of former croplands and pastures.

However, a modelling experiment has shown that even very extensive reforestation and afforestation activities over the next 50 years would result in reduction of the atmospheric CO_2 concentration of only about 40–70 parts per million by volume (p.p.m.v.) by 2100 (House *et al.*, 2002). Among various options for changes to land use and land management for temperate regions, afforestation or reforestation of former cropland has the highest carbon mitigation potential. This potential could largely

The Carbon Balance of Forest Biomes, edited by H. Griffiths and P.G. Jarvis.
© 2005 Taylor & Francis Group

Table 1. Forest area change in the 1990s (based on FAO, 2001).

Change in forest area	Tropical regions	Non-tropical regions
	10^6 hectares per year	
Losses of natural forests by:		
deforestation	142	4
reforestation	10	5
Gains of forest land by:		
afforestation	8	7
natural regeneration	10	26

be realized by using wood and other forest products for bio-energy production, as a substitute for fossil fuels (Smith *et al.*, 2000). An even stronger option would be to change forest management altogether from conventional timber production to carbon management (WBGU, 2003).

We refer to *afforestation* as the establishment of forest on land that has carried forest a long time ago, and on land that has never carried forest (e.g., peat-land, natural grassland), and where afforestation is thus ecologically not acceptable. We will refer to *reforestation* only as the establishment of forest on former forested land (but excluding forested land that has recently been clear-felled as part of the normal management cycle). Large uncertainties in constraining potential sink capacities for atmospheric CO_2 through afforestation and reforestation arise from a lack of understanding of the response of SOC pools in the mineral topsoil to land-use and land-management change activities (see *Figure 1*).

Globally, forest soils store 57–69% of the carbon present in forest ecosystems (704–787 Pg (C) or 516–563 Mg (C) ha^{-1} (1 hectare (ha) = 10^4 m^2)) (IPCC, 2001), and yet in many regional carbon budget studies of forests, SOC pools are not considered (e.g., Harmon *et al.*, 1990). Therefore, in this study we first review how much SOC is lost from the mineral topsoil by forest management activities through the regular silvicultural cycle or by land-use change (i.e., harvest and subsequent reforestation, or deforestation and conversion to cropland or pasture). Secondly, we compare rates of SOC losses after deforestation to rates of SOC accumulation after afforestation.

2. Carbon storage in forest soils

To assess the effects of changes in land-use and land-management activities in forested areas on atmospheric CO_2 concentrations, it is important to consider the net balance between carbon input and decomposition of SOC pools. Carbon enters soils mainly as above-ground or below-ground plant detritus (or litter), and thus input of carbon to soils is highly correlated with net primary productivity (NPP) of the vegetation. However, an increase in NPP does not necessarily lead to a proportionately larger accumulation of carbon in soils because the input may be counteracted by higher respiration losses (Finzi and Schlesinger, 2002; Schlesinger and Lichter, 2001). Decomposition and mineralization of the SOC is primarily affected by accessibility of the organic carbon to micro-organisms.

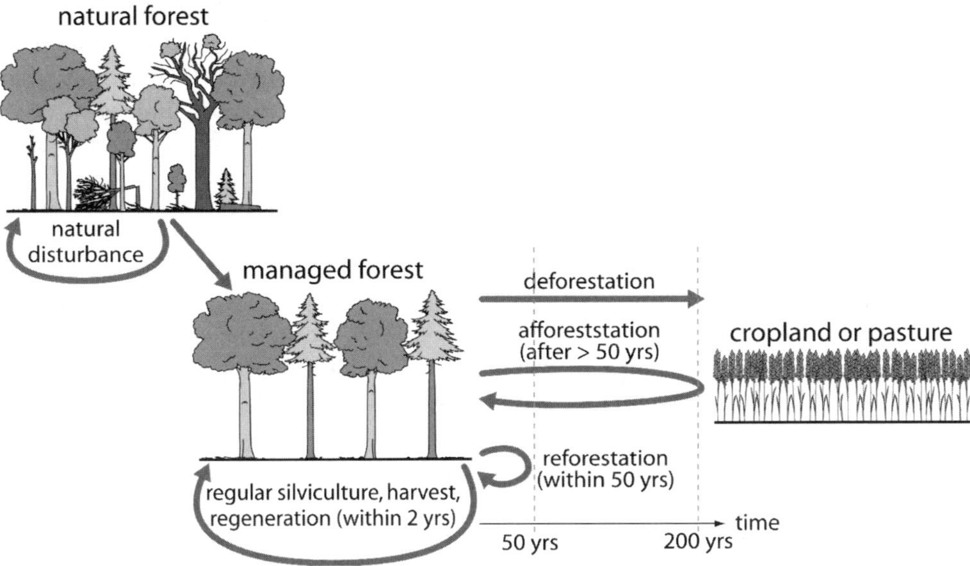

Figure 1. *Land-management change activities in forest land according to Article 3.4 of the Kyoto Protocol refer to stand-replacing activities. These are the conversion of natural forests into managed forests or regular silviculture where harvest is followed by artificial or natural regeneration within two years. Land-use change activities according to Article 3.3 involve the conversion of forest land into non-forest land or vice versa. The term reforestation applies where forest land is used as non-forest land for up to 50 years after felling and before planting or natural regeneration. Deforestation includes harvesting followed by conversion to cropland or pasture. The term afforestation applies to areas which have been non-forest land for at least 50 years, but is also used for establishing forest on land that has never carried forest previously, e.g., peat land, natural grassland.*

Using a simplified model, one can assume that the mineral soil SOC pool in forests consists of three fractions that can be separated by density fractionation into the following components: SOC that is not bound to the mineral phase (i.e., the 'free-light fraction', density about 1.6 g cm^{-3}); SOC that is occluded into mineral soil aggregates (i.e., the 'occluded-light fraction', density less than 1.6 g cm^{-3}); and SOC that is bound to clay particles or to aluminium or iron hydrous oxides (i.e., the 'heavy fraction', density greater than 1.6 g cm^{-3}). SOC in the free-light fraction is relatively easily accessible to micro-organisms. SOC in the occluded-light fraction becomes accessible only if soil aggregates are fragmented by moisture-induced or temperature-induced shrinking and swelling processes, caused by bioturbation, or by cultivation. SOC in the heavy fraction is much less accessible to micro-organisms (Golchin *et al.*, 1994; Six *et al.*, 1998). Thus, the mean residence time of SOC increases from the organic layer to the free-light fraction to the heavy fraction. For a temperate forest soil, Gaudinski *et al.* (2000) calculated mean residence times in the top mineral soil layer of 73–100+ years for the free-light plus occluded-light fraction, and 130–200+ years for the heavy fraction. These residence times can be compared with four years for recognizable plant material and 40 years for more humified material in the organic layer.

Furthermore, decomposition of SOC is a function of climate, hydrological conditions, and litter quality (Saleska *et al.*, 2002). Decomposition is retarded at low temperatures, and at high or low soil moisture contents. Relatively recalcitrant litter components are black carbon (e.g., charcoal) (Glaser *et al.*, 2001) and lipids (e.g., cutin, suberin), whereas lignin is less recalcitrant than has been assumed earlier (Gleixner *et al.*, 2002).

In many soils, only a small fraction of litter-derived SOC accumulates, whereas the major fraction is lost by heterotrophic respiration. For European forests in 1990, carbon inputs into soils accounted for 76% of NPP (0.596 Pg (C) per year) (Nabuurs *et al.*, 2003). Of this input, 87% was lost by heterotrophic respiration, whereas 13.5% (0.061 Pg (C) per year) accumulated. The accumulating SOC corresponded to 10% of NPP. This fraction is lower for Australian eucalypt or Monterey pine plantations, in which less than 3% of NPP has been estimated to accumulate as SOC (Paul *et al.*, 2003).

3. Effects of harvest and natural regeneration or reforestation on soil organic carbon pools

A harvest, followed by natural regeneration or replanting of similar productive managed stands or tree plantations, results in SOC losses in the mineral topsoil of 6–13% when it is accompanied by intensive site preparation activities (Guo and Gifford, 2002; Johnson and Curtis, 2001). This is because activities such as ploughing, mounding, or harrowing, lead to SOC in the free-light and occluded-light fractions becoming more easily accessible to micro-organisms. Furthermore, mineralization of SOC is increased when the residues (slash) and the soil organic layer are mixed into the mineral soil, removed, or burned.

However, in the absence of intensive site preparation, harvest has little or no impact on SOC storage in the mineral topsoil. By comparing stands of different age class in a temperate forest of beech (*Fagus sylvatica* L.), managed with conventional silviculture as a regular shelterwood system with rotation length of 140 years, we found that thinning and final harvest of the trees had, on average, no effects on the amounts of SOC in the mineral topsoil (0–0.3 m depth) (see *Figure 2*). Within each chronosequence studied, SOC pools did not change significantly over the period of one rotation and thus did not increase with increasing stand age (i.e., there was no so-called 'age effect'). Among the chronosequences, SOC pools mainly differed because of site effects. With an increase in altitude, SOC pools increased with decreasing temperature and increasing precipitation, and with decreasing clay concentrations (M. Mund, unpublished results). The absence of an 'age effect' on SOC after harvest has also been found in Boreal forests under conventional silviculture (i.e., clear-cut followed by natural regeneration without intensive site preparation) (Preston *et al.*, 2002; Trofymow and Blackwell, 1998). This has been confirmed by meta-analyses of other studies of natural regeneration and reforestation in which changes in SOC have been compared (Guo and Gifford, 2002; Johnson and Curtis, 2001). Leaving the slash on site after a harvest of saw-logs had no effect on SOC pools in the mineral topsoil of deciduous forests, but increased SOC pools in coniferous forests by 18%, probably only for the period until the slash had decomposed (Johnson and Curtis, 2001). However, removing the slash, for example in whole tree harvesting, not only reduces the potential for carbon input to the SOC pool during the next rotation, but could also seriously interrupt the nutritional cycle and affect tree growth, and in this way also reduce carbon inputs to the SOC pool. (See Chapter 16, this volume.)

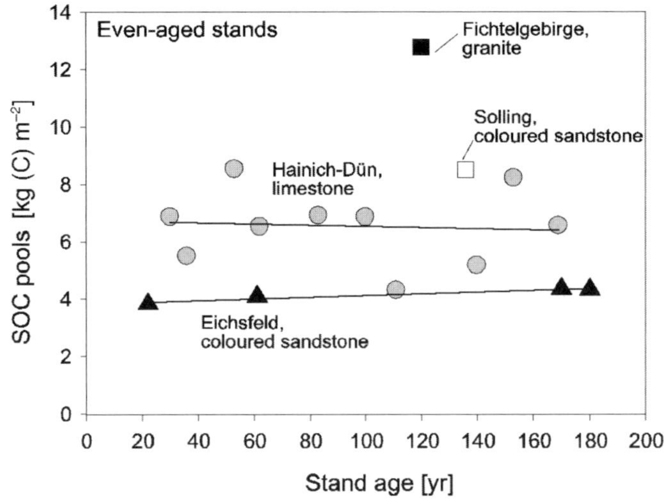

Figure 2. *Organic carbon pools in the mineral topsoil (0–0.3 m) of different age classes of similarly managed, even-aged beech forest stands (conventional shelter-wood systems of* Fagus sylvatica *L., rotation length 140 years) in Germany.*

Thus, an 'age effect' after harvest does not occur when the slash is left on site, and the organic layer is not removed or mixed into the mineral soil, provided that a similar productive managed forest or tree plantation replaces the harvested stand. In such cases, a harvest followed by natural regeneration or reforestation is quite comparable to natural disturbances caused by insects and pathogens or windthrow, which lead to stand replacement by natural regeneration (Boone *et al.*, 1988). Thus we conclude that to understand the consequences for atmospheric CO_2 concentrations of natural or managerial disturbances, followed by either natural regeneration or reforestation, it is useful to differentiate between events according to their impacts on the SOC pools in the mineral soil (see *Table 2*).

4. Effects of deforestation on soil organic carbon pools

Deforestation followed by conversion into cropland or pasture significantly changes SOC pools in the mineral topsoil (plough horizon to 0.3 or 0.6 m) (Guo and Gifford, 2002; Paul *et al.*, 2003). In warm-temperate regions of the USA, over 100 years of agriculture on former deciduous forest land reduced SOC pools in the mineral topsoil (0–0.3 m layer) by 1.3 kg m⁻² or 40% (Richter *et al.*, 1999), and on former pine (*Pinus palustris* Mill.) forest land by 1.5 kg m⁻² or 58% in the 0–0.1 m layer (Markewitz *et al.*, 2002). In tropical regions, losses of 14–56% from the initial SOC pools have been measured in the 0–0.5 m layer (Brown and Lugo, 1990). A meta-analysis of such data on the global scale indicates that the conversion of forests into cropland has, on

Table 2. *Losses of SOC from forest soils after disturbances, assuming that the stand is replaced by similar productive vegetation. Large losses result in increase of the SOC pools with stand age. In chronosequence studies this becomes apparent as an 'age effect' (or 'time since disturbance' effect).*

Stand-replacing disturbance		
	Small	Large
Natural	Insect or pathogen attack Windthrow	Stand-replacing fires
Anthropogenic	Saw-log harvest with residues left on site and without site preparation[1]	Saw-log harvest with residues removed from site and/or prescribed burning
	Herbicide treatment	Clear cut with subsequent intensive site preparation[1]

[1]For example, cutting, prescribed burning, harrowing, scarification, ploughing, or mounding.

average, resulted in losses of 42% of the initial SOC pools (Guo and Gifford, 2002). In contrast, the conversion of forests into pastures can result in SOC gains. Data are scarce, but a meta-analysis has indicated that SOC pools in the mineral topsoil increase on average by approximately 8% of the initial pool size as a result of conversion of forest into grassland (Guo and Gifford, 2002).

The estimates of SOC losses after conversion to cropland are conservative in most cases, while the estimates of SOC gains after conversion to pasture may be too high. This is because SOC pools of agricultural land are often compared to SOC pools of managed forests (including also prescribed surface fires) and not to SOC pools of old-growth, natural forests. Thus, the interpretation depends on the assumed base line. An objective quantification is further complicated because natural forests simply do not exist any more in regions that have been inhabited by people for a long time, and in areas with intensive agricultural land use.

Losses of SOC from mineral topsoil occur within one or two decades after deforestation (Mann, 1986). Most SOC is lost from the free-light fraction or the occluded-light fraction. The amount of SOC in the free-light fraction mainly depends on the litter input, and this is strongly reduced in cropland because residues are repeatedly removed (Six *et al.*, 1998). The SOC in the occluded-light fraction becomes subject to mineralization when the mineral soil is mechanically disturbed (e.g., by tillage) and losses of SOC from the mineral topsoil increase with repeated mechanical disturbance (Matson *et al.*, 1997). The amount of SOC in the free-light plus the occluded-light fraction can be reduced by agricultural activities to less than 40% of its initial amount (Golchin *et al.*, 1994).

5. Restoring cultivation-induced soil organic carbon losses by afforestation

We have shown that the largest SOC losses from mineral topsoil occur within two decades after deforestation. In contrast, the time needed for recovery of the mineral soil SOC pools is likely to be considerably longer. Support for this assumption is based on a recent study of the regional forest carbon budget in Thuringia (southeast Germany). This study showed that the smallest SOC pools in the mineral topsoil occurred in reforested areas. The largest SOC pools occurred in forest stands that formerly were deciduous or mixed coniferous/deciduous forests (see *Figure 3*). In these stands, SOC pools were approximately 0.5 kg (C) m^{-2} larger than the SOC pools in stands on other former land uses. The effect of the former land-use (i.e., the land use 'memory effect') was independent of the soil substrate type. However, to understand the effects of land-use history on SOC pools at a large spatial scale, it remains a challenge to distinguish the variation amongst mineral soil SOC pools attributable to site factors from the variation resulting from land-use history.

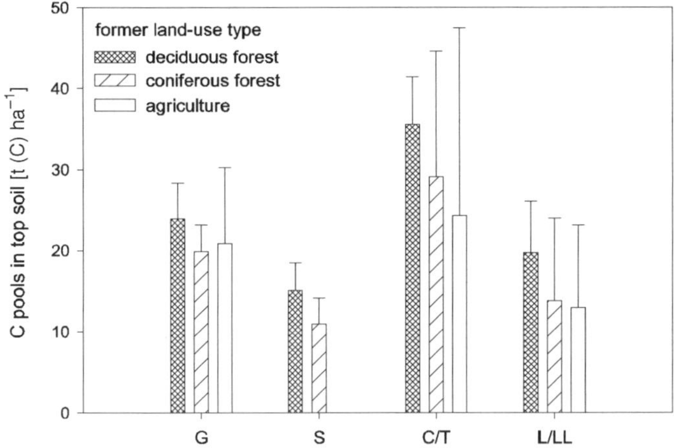

Figure 3. *Organic carbon pools in the top mineral soil layer (0–0.1 m depth) of forest stands in Thuringia, southeast Germany. Forest stands were sorted by former land use and soil substrate group (G, silicate; S, sand/sandstone; C/T, clay/carbonate; L/LL, loam/silty loam). Effects of climate on the SOC pools were eliminated by introducing altitude as a covariate in an analysis of covariance. The columns do not show mean values but predicted values for SOC pools corrected for climate. The error bars indicate the residual error at the mean of the covariate 'climate'.*

Long-term (0–10 000 years) chronosequences on mineral soil formations provide hints for a baseline of SOC accumulation in soils under natural vegetation. However, such studies integrate over different stages of vegetation succession, generally starting on shallow under-developed soils with low productivity, non-forest vegetation that differs in floristic and structural composition, and thus in NPP, from the climax forest. Syers *et al.* (1970) calculated SOC storage in sand dunes in New Zealand of 8 g (C) m^{-3} per year within the first 500 years of soil development, and 0.2 g (C) m^{-3} per year within the next 99 500 years. Slightly higher values were found for volcanic soils in Hawaii with 0.4 g (C) m^{-3} per year for the first 150 000 years (Torn *et al.*, 1997). After afforestation of former croplands or pastures, rates of SOC accumulation by the mineral soil are generally higher than the figures above because of higher productivity of the vegetation (see *Figure 4*) and the initial restoration of the free-light fraction of SOC (which might, of course, be lost again later as a result of intermittent disturbances).

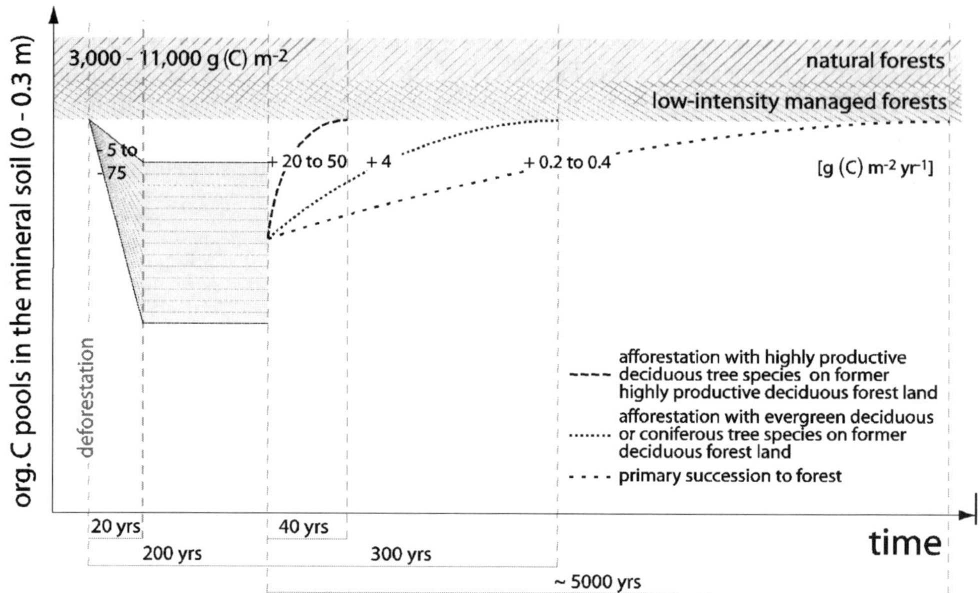

Figure 4. Scenarios of SOC loss after deforestation and of SOC gain after afforestation in the mineral topsoil (0–0.3 m) (based on Brown and Lugo, 1990; Mann, 1986; Markewitz et al., 2002; Paul et al., 2002; Richter et al., 1999; Syers et al., 1970).

Changes of the SOC pools after afforestation depend on the former land use (cropland or pasture), type of forest established, and climate and soil texture (Paul *et al.*, 2002). Afforestation of former cropland generally results in carbon accumulation in the SOC pools in the mineral topsoil, whereas afforestation of pastures often results in SOC losses. Afforestation with hardwood broadleaves (except eucalypts) often results in higher SOC accumulation rates in the mineral topsoil than afforestations with softwood conifers (see *Table 3*). Furthermore, SOC accumulation is usually found to be largest in clay-rich soils and warm moist climates, but these variables are often covariant factors of the former land use, previous vegetation and kind of forest established.

These relative differences (cropland versus grassland; broadleaves versus conifers) in accumulation of carbon in the mineral soil SOC pools may be reversed if the superficial soil organic layer is included in the SOC accumulation rates. In these cases, afforestation generally results in an increase in the total (organic layer plus mineral topsoil) SOC pool size compared with the former land use (Paul *et al.*, 2002), because afforestation with conifers results in thick organic layers on top of the mineral soil, whereas the mineral soil SOC pools are hardly affected (Markewitz *et al.*, 2002; Richter *et al.*, 1999).

However, including the litter layer in SOC budgets for afforestation and reforestation may be misleading in the longer term because SOC in the organic layer can

Table 3. *Changes in SOC pools in the mineral topsoil after afforestation (based on Paul et al., 2002).*

Site management	SOC accumulation (0–0.3 m mineral soil) Per cent per year of initial SOC	g (C) m^{-2} per year
Former land use		
Cropland or cropland/pasture-rotation	0.56–1.51	20–50
Pasture	–0.2	–10
Type of forest established		
Hardwood other than eucalypts	1.04	40
Eucalypts	0.16	<10
Softwood other than Monterey pine	0.87	20
Monterey pine	–0.40	–20

be rapidly lost by subsequent intensive management after harvest, such as prescribed burning, or during site preparation for the next rotation.

The recovery of SOC pools in the mineral topsoil is fastest under productive deciduous vegetation on pH neutral or slightly acidic soils with a loamy texture. There, litter production is high and litter-derived carbon is rapidly mixed into the mineral soil by bioturbation. If forests on these soils are converted to agricultural land and later afforested to become managed deciduous forests with similar litter production rates as the natural forest, SOC pools can recover within 30–50 years (Brown and Lugo, 1990; Paul *et al.*, 2002) (see *Figure 4*). In contrast, if natural deciduous forests are converted into agricultural land and later afforested to become coniferous or evergreen deciduous managed forests, SOC accumulation can be very small (see *Figure 4*). However, if warm–temperate deciduous forests are converted to agricultural land and subsequently afforested to coniferous forest, Richter *et al.* (1999) found that complete recovery of the initial SOC pools in the mineral topsoil would occur only within 320 years. Although, the SOC accumulation rate in the litter layers was 94.5 g (C) m^{-2} per year, only 4.1 g (C) m^{-2} per year accumulated in the mineral soil (0–0.3 m, over 35 years after planting). Clearly, considerable differences exist in the length of the time during which SOC is lost from the mineral topsoil after deforestation and the period required for SOC to accumulate to the original amount after afforestation or reforestation.

The situation is further complicated in developed countries because devegetation, rather than deforestation, frequently takes place for the development of new infrastructure (mainly roads and highways). Taking Germany as an example, we may assume that 1000 ha were devegetated between 1991 and 1999 to provide new land for highways, and, as compensation, 2000 ha were reforested (by law the area of reforestation must be larger than the area of devegetation by a factor of 2). If we take as a basis the average tree biomass of managed forest (80 Mg (C) ha^{-1}) and the average SOC content (110 Mg (C) ha^{-1}) and assume a 70% loss of SOC, the total carbon loss resulting from devegetation would be about 155 000 Mg. Let us assume also that the compensating reforestation took place mainly on grassland, because land for arable crops has priority. The grasslands will lose about 10% of their SOC on reforestation, but the sites will then gain carbon from accumulating litter and from the new tree biomass. Thus the total carbon balance is slightly negative for about 20 years and we are left with an overall carbon loss of about 160 000 Mg for the 1000 ha devegetated. On current assessments, this will need hundreds of years to re-accumulate, despite the fact that the total forest area has been increased.

6. Conclusions

Our analyses have shown that forest harvest combined with intensive site preparation practices results in SOC losses in the first decades after natural regeneration or reforestation (the so called 'age effect'). The losses of SOC are largest when forests are converted to arable cropland and occur mainly from depletion of the free-light and the occluded-light fractions of carbon in the mineral topsoil as a result of reduced carbon inputs and increased mechanical disturbance of the soils. Furthermore, there is large temporal asymmetry between the period of time in which a certain depletion of the mineral SOC pools occurs after deforestation and cultivation, and the length of time needed for recovery of the SOC pools to their initial state after afforestation with its associated disturbances. The resulting land-use 'memory effect' introduces a large, poorly quantified assessment of changes in SOC pools into contemporary carbon budgets. We conclude that besides differences in species composition, climate, soil texture, drainage, and current management, land-use history is a factor that needs to be considered in carbon budget studies of forest ecosystems today.

7. Summary

This study reviews the effects of changes in land use and land management on SOC pools in forest soils. In the 1990s, deforestation remained the most important land-use change in tropical regions (-142×10^6 ha per year). In non-tropical regions the forested area increased in developed countries as a result of natural reforestation ($+26 \times 10^6$ ha per year). Deforestation also continued in under-developed countries in temperate regions.

Without intensive site preparation, harvest followed by natural regeneration or reforestation has little impact on SOC pools in the mineral topsoil (0–0.3 m). Intensive site preparation results in losses of 6–13% of the initial SOC from the topsoil in the first decades. On average, deforestation followed by conversion to cropland results in SOC losses of 42% (or 0.1–1500 g (C) m^{-2}) from the mineral topsoil, whereas conversion to pasture results in gains of 8%. The largest changes in SOC storage occur within the first two decades.

After reforestation, SOC accumulation depends on the kind of managed forest established. Under productive deciduous reforestation (excluding eucalypts), SOC in the mineral topsoil accumulates at a rate of 20–50 g (C) m^{-2} per year, and SOC pools could recover from cultivation-induced losses within 40 years. Under coniferous reforestation, the rate of accumulation of carbon is highest (95 g (C) m^{-2} per year) in the organic layer, which is very susceptible to site preparation practices. In the mineral topsoil, the rate of accumulation is much lower (4 g (C) m^{-2} per year), and recovery of the initial SOC pools might take several hundred years. The resulting land-use 'memory effect' has introduced large variation of the SOC pools in contemporary carbon budget studies. Thus, there seems to be a large temporal asymmetry between the period of time over which depletion of SOC occurs and the time needed for recovery of the SOC pools in the mineral soil. This should be taken into account when considering land-use and land-management activities to decrease atmospheric CO_2 concentrations over this century.

Acknowledgements

We thank A. Boerner for preparing Figures 1 and 4.

References

Boone, R.D., Sollins, P. and Cromack Jr, K. (1988) Stand and soil changes along a mountain hemlock death and regrowth sequence. *Ecology* **69**: 714–722.

Brown, S. and Lugo, A.E. (1990) Effects of forest clearing and succession on the carbon and nitrogen content of soils in Puerto Rico and US Virgin Islands. *Plant and Soil* **124**: 53–64.

De Fries, R.S., Field, C.B., Fung, I., Collatz, G.J. and Bounoua L. (1999) Combining satellite data and biogeochemical models to estimate global effects of human-induced land cover change on carbon emissions and primary productivity. *Global Biogeochemical cycles* **13**: 803–815.

FAO (2001) *State of the World's Forests 2001*. FAO, Rome. 200 pp.

Finzi, A.C. and Schlesinger, W.H. (2002) Species control variation in litter decomposition in a pine forest exposed to elevated CO_2. *Global Change Biology* **8**: 1217–1229.

Gaudinski, J.B., Trumbore, S.E., Davidson, E.A. and Zheng, S. (2000) Soil carbon cycling in a temperate forest: radiocarbon-based estimates of residence times, sequestration rates and partitioning of fluxes. *Biogeochemistry* **51**: 33–69.

Glaser, B., Haumaier, L., Guggenberger, G. and Zech, W. (2001) The "Terra Preta" phenomenon: a model for sustainable agriculture in the humid tropics. *Naturwissenschaften* **88**: 37–41.

Gleixner, G., Poirier, N., Bol, R. and Balesdent, J. (2002) Molecular dynamics of organic matter in a cultivated soil. *Organic Geochemistry* **33**: 357–366.

Golchin, A., Oades, J.M., Skjemstad, J.O. and Clarke, P. (1994) Study of free and occluded particulate organic matter in soils by solid state 13C CP/MAS NMR spectroscopy and scanning electron microscopy. *Soil Biology and Biochemistry* **32**: 185–309.

Guo, L.B. and Gifford, R.M. (2002) Soil carbon stocks and land use change: a meta analysis. *Global Change Biology* **8**: 345–360.

Harmon, M.E., Ferrell, W.K. and Franklin, J.F. (1990) Effects on carbon storage of conversion of old-growth forests to young forests. *Science* **247**: 699–702.

House, J.I., Prentice, I.C. and Le Quéré, C. (2002) Maximum impacts of future reforestation or deforestation on atmospheric CO_2. *Global Change Biology* **8**: 1047–1052.

IPCC (2001) (Prentice, I.C., Farquhar, G.D., Fasham, M.J.R., Goulden, M.L., Heimann, M., Jaramillo, V.J., Kheshgi, H.S., LeQuéré, C., Scholes, R.J. and Wallace, D.W.R) *The carbon cycle and atmospheric carbon dioxide*. In: Houghton, J.T., Dung, Y., Griggs, D.J., Noguer, M., Vander Linden, P.J., Dai, X., Maskell, K. and Johnson, C.A. (eds) *Climate change 2001: The Scientific Basis. Contribution of Working Group I to the Third Assessment Report of the Intergovernmental Panel on Climate Change*. pp. 183–237. Cambridge Universtiy Press, Cambridge, USA.

Johnson. D.W. and Curtis, P.S. (2001) Effects of forest management on soil C and N storage: meta analysis. *Forest Ecology and Management* **140**: 227–238.

Mann, L.K. (1986) Changes in soil carbon storage after cultivation. *Soil Science* **142**: 279–288.

Markewitz, D., Sartori, F. and Craft, C. (2002) Soil change and carbon storage in longleaf pine stands planted on marginal agricultural lands. *Ecological Applications* **12**: 1276–1285.

Matson, P.A., Parton, W.J., Power, W.J. and Swift, M.J. (1997) Agricultural intensification and ecosystem properties. *Science* **277**: 504–508.

Nabuurs, G.J., Schelhaas, M.J., Mohren, G.M.J. and Field, C.B. (2003) Temporal evolution of the European forest sector carbon sink from 1950 to 1999. *Global Change Biology* **9**: 152–160.

Paul, K.I., Polglase, P.J., Nyakuengama, J.G. and Khanna, P.K. (2002) Change in soil carbon following afforestation. *Forest Ecology and Management* **168**: 241–257.

Paul, K.I., Polglase, P.J. and Richards, G.P. (2003) Predicted change in soil carbon following afforestation or reforestation, and analysis of controlling factors by linking a C accounting model (CAMFor) to models of forest growth (3PG), litter decompostion (GENDEC) and soil C turnover (RtohC). *Forest Ecology and Management* **177**: 485–501.

Preston, C.M., Trofymow, J.A., Niu, J. and Fyfe, C.A. (2002) Harvesting and climate effects on organic matter characteristics in British Columbia coastal forests. *Journal of Environmental Quality* **31**: 402–413.

Richter, D.D., Markewitz, D., Trumbore, S.E. and Wells, C.G. (1999) Rapid accumulation and turnover of soil carbon in a re-establishing forest. *Nature* **400**: 56–58.

Saleska, S.R., Shaw, M.R., Fischer, M.L., Dunn, J.A., Still, C.J., Holman, M.L. and Harte, J. (2002) Plant community composition mediates both large transient decline and predicted long-term recovery of soil carbon under climate warming. *Global Biogeochemical Cycles* **16**: 1055 doi: 10.1029/2001GB001573.

Schlesinger, W.H. and Lichter, J. (2001) Limited carbon storage in soil and litter of experimental forest plots under increased atmospheric CO_2. *Nature* **411**: 466–469.

Six, J., Elliott, E.T., Paustian, K. and Doran, J.W. (1998) Aggregation and soil organic matter accumulation in cultivated and native grassland soils. *Soil Science of America Journal* **62**: 1367–1377.

Smith, P., Powlson, D.S., Smith, J.U., Falloon, P. and Coleman, K. (2000) Meeting Europe's climate change commitments: quantitative estimates of the potential for carbon mitigation by agriculture. *Global Change Biology* **6**: 525–539.

Syers, J.K., Adams, J.A. and Walker, T.W. (1970) Accumulation of organic matter in a chronosequence of soils developed on wind-blown sand in New Zealand. *Journal of Soil Science* **21**: 146–153.

Torn, M.S., Trumbore, S.E., Chadwick, O.A., Vitousek, P.M. and Hendricks, D.M. (1997) Mineral control of soil organic carbon storage and turnover. *Nature* **389**: 170–173.

Trofymow, J.A. and Blackwell, B.A. (1998) Changes in ecosystem mass and carbon distribution in coastal forest chronosequences. *Northwest Science* **72**: 40–42.

WBGU (2003) *Welt Im Wandel-Energlewende zur Nachhaltigkeit.* Wissenschaftlicher Beirat der Bundersregierung Globale Umweltveranderungen, 254 pp. Springer-Verlag, Berlin-Heidelberg.

16

'Carbon forestry': managing forests to conserve carbon

Paul G. Jarvis, Andreas Ibrom and Sune Linder

1. Multiple purpose forestry

Forests currently provide services to communities for conservation and biodiversity of wildlife, watershed and soil protection, for recreation and amenity, for sport and for landscape, in addition to timber production. To these goals should now be added 'carbon forestry', the direct role of forest management in protecting forest carbon stocks and enhancing forest sink capacity for carbon. Best estimates indicate that over the past decade terrestrial vegetation, particularly woods and forests, was responsible for the removal of about 40% of the carbon dioxide (CO_2) being put into the atmosphere by the burning of fossil fuels and land-use change (Royal Society, 2001). Now that we appreciate that forests have a vital role in the global carbon budget, forest management should take on responsibility to protect existing carbon stocks and to increase them through enhancement of the forest sink capacity. The recent developing emphasis on continuous cover forestry coupled with selective harvesting goes some way in this direction, but not far enough. Positive measures are particularly required to conserve and enhance soil carbon stocks in forests. A key question is whether we can hope to maintain or even enhance removal of CO_2 from the atmosphere in the immediate future by planting trees, and get credit for so doing. In our view, we are really only concerned with the next 50–100 years. Within that period, emissions of CO_2 are likely to increase to such an extent that our problems will become insuperable, unless a practical source of alternative energy is developed before then.

Historically much of the land available for tree planting in Europe has been marginal agricultural land, often with a large soil organic matter content. Although agricultural land is becoming available for tree planting in increasing amounts today, large areas of marginal land with podsols, peaty gleys, and peats continue to provide major opportunities for reforestation and afforestation. It is likely that the operations associated with the establishment and management of forests on such soils will exacerbate return of CO_2 to the atmosphere, thereby largely negating the removal of CO_2 from the atmosphere

(Jarvis, 2003). We must, therefore, be selective in the sites where we can expect forests to make an effective contribution to net removal of CO_2 from the atmosphere.

Two other land-based trace gases that contribute to global warming, methane (CH_4) and nitrous oxide (N_2O), must also be considered. Both are produced from organic soils and influenced by forest management, emissions of methane by drainage and nitrous oxide by application of fertilizer. Although emissions are small compared with those of CO_2, methane has a global warming potential (GWP) of about 23 times and nitrous oxide of about 296 times that of CO_2, on a 100 year time-scale (IPCC, 2001), so that even small rates of emission may be relevant to the combined trace-gas budget of a forest. Another key question then is whether emissions of these gases are large in forests and can be moderated by forest management practices (see Chapters 13 and 14, this volume).

2. Kyoto and two kinds of forest

The Kyoto Protocol (Article 3.3) is concerned with new plantations after initiation of afforestation or reforestation, and focuses on only a part of the stand life-cycle, at present no more than a few decades. During the early years of this period while the trees are small, a forest stand is likely to be a source of carbon, becoming a sink for carbon only when the trees are sufficiently large that net primary production by the trees exceeds heterotrophic respiration within the soil (*Figure 1*). Because of relatively short rotations and regular management actions, there is considerable potential for carbon losses, but there is also scope for conservation and enhancement of carbon stocks through a positive management approach.

Later Articles in the Protocol (3.4, 6 and 12) (see Chapter 1, this volume) have a focus on semi-natural and old growth forest that is largely unmanaged or lightly managed and intermittently exploited. These forests may be significant carbon sinks in the tropics and sub-tropics (e.g., Arain *et al*, 2002; see Chapter 10, this volume) but are closer to carbon neutrality in the Boreal region, where a few stands have been identified as carbon sources rather than sinks (see Chapter 8, this volume). There is considerable scope for management actions to increase their carbon sequestration (Bergh *et al*., 1999; Linder, 1995) but the small current returns and low level of current management are unlikely to encourage this very widely, particularly in regions where forests are marginally economic. The introduction of new and more intense management methods will, however, require differentiated regional planning, taking into account environmental, economic and technical constraints.

3. What a forest does

A sustainably managed forest landscape, comprising stands representing all stages in the life cycle, operates as a functional system that maintains an overall carbon balance, taking in carbon (as CO_2) from the atmosphere, retaining a part in the growing trees, transferring another part into the soils and exporting carbon as forest products. On a time-scale of tens of years, most forests accumulate carbon through growth of the trees and increase in the soil carbon reservoir, until major disturbance occurs. Recently disturbed and newly regenerating areas lose carbon, young stands gain carbon rapidly, mature stands gain carbon at a lesser rate, and over-mature stands may lose carbon (*Figure 1*).

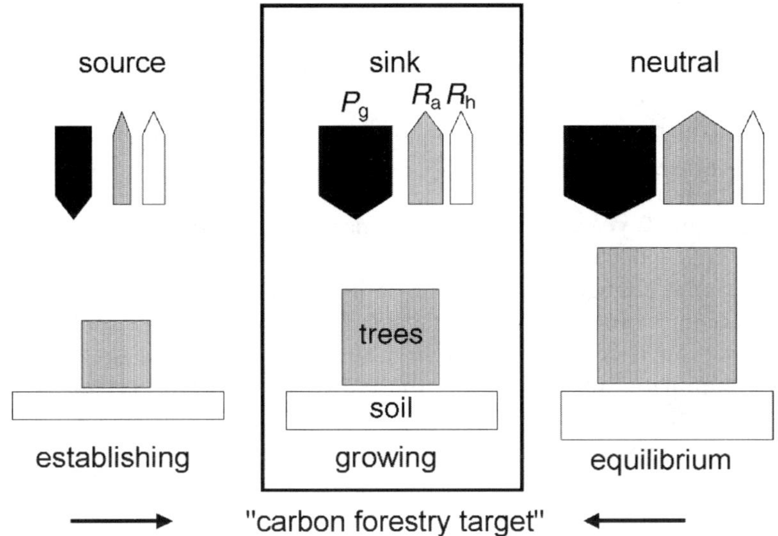

Figure 1. *Schematic diagram of carbon stocks in forest trees (squares) and soil (rectangles). The arrows represent the CO_2 fluxes of gross photosynthesis (black P_g), autotrophic respiration (grey R_a), and heterotrophic respiration (white R_h) during stand development. Net CO_2 uptake is largest in the central panel ('growing'), which is the developmental stage targeted by 'carbon forestry'. It can be attained from the left and from the right, through appropriate forestry measures applied to site preparation, planting, thinning, harvest, irrigation, compensatory liming, and fertilization.*

An extensive forest being managed *sustainably* for timber production would comprise compartments representing all age classes from regeneration or planting through to harvest. There would be a net transfer of carbon from the atmosphere into such a forest and carbon would be removed from the forest in wood products (see Chapter 4, this volume). Most compartments would accumulate carbon through growth of the trees unless major disturbance intervenes. On a time-scale of years, the standing stock of carbon within the trees in such a forest would be constant, if management is sustainable in the sense that the harvest cut equals the growth. However, the forest as an ecosystem might be gaining or losing carbon, depending on whether the soil carbon reservoir is increasing or decreasing, and could be functioning unsustainably from a carbon perspective. Similarly, if a full range of tree ages were present within unmanaged old-growth, or lightly managed semi-natural forest, the standing mass of wood might well be constant (or increasing as recent evidence indicates: see Chapter 10, this volume), but the soil carbon stock could be increasing *or* decreasing.

These are ideal management scenarios: there are only few forests in the world being managed for sustainable timber production such that all age classes are equally represented. In most forests the age structure is heavily biased as a result of past management practices, as a result of exploitation during wars, or natural disturbances, such as fire and windthrow. In other cases, local and global economics determine whether the harvest cut currently exceeds or is less than the growth.

4. Trees and soil

Forests worldwide contain about 45% of the stock of carbon on the surface of the globe, the larger part of which is to be found in the forest soils. The global pool of carbon in soil exceeds that in vegetation by 4:1. The average carbon stock in soils relative to vegetation ranges from a ratio of about 1:1 in tropical forests, 2:1 in temperate forests, to 5:1 in Boreal forests, and to 17:1 in deserts and wetlands, with residence times ranging from 10 years in tropical savannahs to more than 200 years in deserts and tundra (IPCC, 2000; Schlesinger, 1997). Thus changes in soil carbon stocks are at least as important as changes in vegetation carbon stocks, if not more so. For lack of good data on changes in soil carbon stocks it is often assumed, for simplicity, that the soil carbon reservoir is constant in amount, but it is clear that in northern latitudes carbon has accumulated in soils during the Holocene as the ice retreated and trees moved northwards.

Carbon is added to the soil as debris from the vegetation above ground (leaves, twigs, branches and sometimes trunks of trees) and below ground as the result of turnover of fine roots and sometimes also coarse roots. Thus, to some extent, total carbon additions to the soil are coupled to net primary production of the trees on the site, whereas losses of carbon result from heterotrophic respiration of the soil fauna and microflora, associated with decomposition of the soil organic matter (SOM). Carbon losses are also coupled to the vegetation present to some extent, but somewhat less directly because some fractions of SOM have very long turnover times, ranging from years to centuries (Schlesinger, 1997).

5. Carbon budgets of forests

Change in the carbon stocks of stands of forest trees over a five year period can be assessed with good precision by standard forestry mensuration methods coupled with predetermined allometric relationships between stem biomass and other parts of the trees (see Chapter 15, this volume). Stocks of carbon in soils can also be determined by standard sampling techniques, but because of heterogeneity in carbon content in both the horizontal and vertical dimensions, particularly in disturbed forest soils, very many samples are required for good precision. There are considerable doubts as to whether changes in soil carbon stocks can be assessed with adequate precision over a period as short as five years (see Chapters 11 and 15, this volume). On the other hand, such periods are too long to be informative about the fast microbiological processes that determine the long-term changes in soil carbon stocks.

As an alternative to repeated stock-taking, measuring additions to and removals from the stock of carbon as CO_2 can assess changes in carbon stocks. The eddy covariance methodology enables direct measurement of the *net* change in carbon stock (i.e., the net ecosystem production, NEP) of a forest stand, trees and soil together, over periods from hours to years. Over the past 10 years, this approach has been applied worldwide to over 100 predominantly mature forest stands, of a range of species and history, to provide direct measures of their carbon balance. Broadly speaking, the annual NEP of Boreal forests lies in the range 1–2.5 Mg (C) ha^{-1}, of temperate forests 2.5–7 Mg (C) ha^{-1}, and of humid tropical forests 4–6 Mg (C) ha^{-1} (Malhi *et al.*, 1999). Annual NEP of forest stands can be large (up to 7 Mg (C) ha^{-1}) in tropical and temperate forests but is constrained in Boreal forests to less than *ca* 2.5 Mg (C) ha^{-1} by timing of the spring thaw, short growing season, low temperatures and low nitrogen availability (Arain *et al.*, 2002; Jarvis and Linder, 2000; Jarvis *et al.*, 2001b). The carbon stock in forest soils is constrained in the humid tropics by comparatively high rates of carbon circulation but is large in soils of the Boreal forest because of slow decomposition.

To assess the sensitivity of these net carbon sequestration rates to management actions, the gain and loss components can be evaluated in several ways. Most simply, NEP measured in this way can be partitioned directly into gain and loss components by using daytime and night-time data (see, for example, Valentini *et al.*, 2000). In addition, the same approach can be used to estimate the net efflux of CO_2 from the soil and ground vegetation by positioning an eddy flux system in the trunk space. More explicitly, the component fluxes of leaves, branches, stems and soil can be individually evaluated by using chambers of a range of dimensions and operating principles (see, for example, Rayment and Jarvis, 1997, 1999) together with models to upscale the CO_2 flux rates to the entire ecosystem.

Complete, quantitative carbon balances for forest stands demonstrate that annual NEP is the net result of two large, opposed terms within the overall carbon balance: the gain of CO_2 by assimilation by leaves in the crowns of trees; and the loss of CO_2 by autotrophic respiration of tree structures and heterotrophic respiration by micro-organisms comprising the soil fauna and flora (Malhi *et al.*, 1999). Thus carbon sequestration is particularly sensitive to managerial actions that affect amount and structure of the tree crowns, above and below ground microclimate, and structure and chemical composition (i.e., C/N ratio) of the forest floor and soil beneath. Measurements of the scale and activities of these terms separately provide a basis for evaluation of the likely impacts of management actions.

The current range of measurements of NEP of forest stands is insufficient to provide estimates of the net carbon budget of a forest comprising all stages in the management cycle (Gower, 2003). Although there have been a few research studies of chronosequences recently, in general we lack data covering stands of all species at all stages in the life cycle from regeneration to harvest, including impacts of disturbances such as fire, wind-throw, drought, pollution, pests and diseases, etc. Alternative methods are required to estimate spatially integrated carbon sequestration at the biome or regional scale (see Chapter 7, this volume).

6. Natural disturbance

Disturbance occurs in forests as a result of natural causes: trees are blown down in gales, fires start in thunderstorms, beetles kill the tops of trees, fungi and microfauna kill the roots! Some natural disturbances can contribute substantially to carbon sequestration. Whereas fire is a disaster from perspectives of both timber production and carbon stocks, windthrow is a disaster only from a timber production perspective and can be beneficial from a carbon sequestration viewpoint. So long as no attempt is made to harvest the trees, windthrow results in an immediate large transfer of living carbon from above-ground vegetation onto the forest floor and ultimately into pools of soil carbon that are largely immobilized with long turnover times. This is a major feature of woodlands in the Boreal zone, where such *coarse woody debris* provides a ready environment through which natural regeneration of young, fast-growing trees rapidly occurs, i.e., there can be a double benefit of windthrow to carbon sequestration.

7. Managerial disturbance

Current forest management is oriented towards increasing timber production and thus the stocks of carbon stored within trees, but takes little account of the larger stocks of carbon in forest soils (Jarvis 2003; Johnson and Curtis, 2001). Operations that are disruptive of the soil are likely to lead to a stand becoming a source of CO_2, rather than a sink, at least temporarily. Practically all forest operations today make use of heavy machines that disrupt the forest floor, particularly, the organic soil horizons, and stimulate return of carbon to the atmosphere through enhanced oxidation of the soil organic matter. A key question is how can current practices be modified to minimise loss of carbon through disturbance resulting from forest management operations.

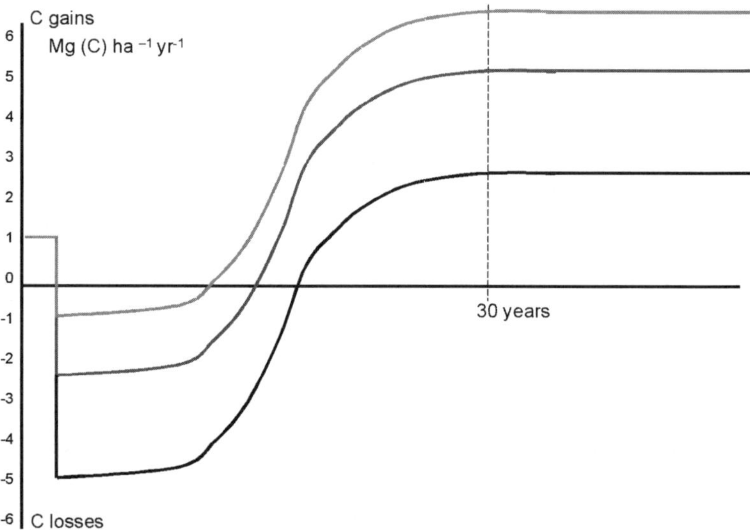

Figure 2. *The time taken for the carbon budget of a newly afforested site to break even depends on the amount and potential to decompose the SOM in the original heathland vegetation. With a large amount of SOM present initially and typical site preparation by mounding, it is likely that it will take about 15 years for the break-even point to be reached, and the plantation to become a net carbon sink. The curves are all of the same shape, simply displaced.*

Such operations include *site preparation*, such as burning, soil scarification, ploughing, mounding, and draining before tree planting, all of which lead to considerable disruption of the uppermost soil horizons that are rich in soil organic matter. Afforestation of carbon-rich soils inevitably leads to loss of carbon with a variable payback time that depends on the prior carbon content of the soil and subsequent rates of carbon acquisition (*Figure 2*). None the less, these operations are usually necessary to provide a seedbed or a microsite for successful planting. They result in stimulation of mineralization, and so facilitate reasonably rapid re-establishment of a forest canopy, and thus regeneration of sink capacity, but at the same time lead to immediate loss of carbon to the atmosphere. After mounds have settled, *planting* itself results in additional disturbance that may stimulate further decomposition and carbon loss. After afforestation of carbon-rich heathland with Sitka spruce, for example, 3.5 Mg (C) ha^{-1} per year was lost from the soil to the atmosphere during the first rotation, but this was recovered during the second rotation (*Figure 3*). *Thinning* (i.e., removing a proportion of the trees and leaving the residues on the ground), or *respacing to waste* (i.e., cutting a proportion of the trees and leaving them on the ground) and *selection harvesting* all temporarily reduce uptake of CO_2 from the atmosphere until the canopy has re-grown. *Clear felling* instantly converts a carbon sink into a carbon source that persists until a vegetation cover is re-established with full sink capacity.

Such management operations lead to immediate loss of capacity to take carbon from the atmosphere until the canopy has filled in again. In addition, the large machines used seriously churn up the ground, leading to accelerated loss of soil carbon, although measurements of this loss have rarely been made. Currently, we do not have adequate information to provide full carbon budgets, embracing the gains and losses of carbon resulting from these operations through a complete management cycle. Recent investigations using chronosequences of stands are, however, beginning to provide some answers at the scale of forest stands. (At the larger scale of the forest and region, we would need to include the use of fossil fuels for management, processing, transport and trade.)

A contentious issue today is whether the debris that is left behind after thinning and harvesting operations (branches, tops, and too-small logs) should be baled up and removed from the forest to be used as biofuel for heat and/or power generation. On the face of it, use of the debris for fossil fuel substitution would seem to be advantageous, but removal of the debris could be at the subsequent expense of the soil carbon stock. Using a chronsequence of stands of Sitka spruce, Zerva (2004) showed that after the significant loss of soil carbon during the first rotation after afforestation, referred to above, the soil stock was replenished during the second rotation from the debris left at harvest at approximately the same rate of about 3.5 Mg (C) ha^{-1} per year (*Figure 3*). That is, retention of the debris in the forest after harvest led to effective transfer of significant amounts of carbon from the debris into the soil carbon reservoir. Consequently, it would seem undesirable to remove this debris from the forest, because that would very likely lead to further reduction of the soil carbon stock, rather than replenishment, during the second rotation.

8. A case study

Since 1997 the net exchange of CO_2 has been measured continuously by eddy covariance at a compartment in Griffin Forest, near Aberfeldy, Perthshire, Scotland. The forest covers about 4000 ha (1 hectare (ha) = 10^4 m^2) and was planted with 80% Sitka spruce (*Picea sitchensis* (Bong.) Carr.) on heathland dominated by heather (*Calluna vulgaris* L.) in 1981–1983, with standard site preparation practice at that time. This consisted of deep ploughing at about 4 m intervals to suppress the heather and improve drainage. Regrettably, we do not have any information about the carbon

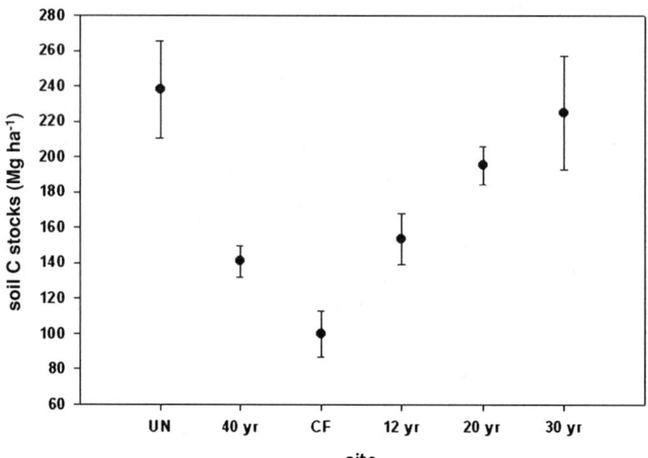

Figure 3. *Reduction in soil carbon content during the first rotation after afforestation of heathland by planting Sitka spruce, followed by the increase in soil carbon content during the second rotation, after clear fell with the lop and top brash left on the site. Data from a Sitka spruce chronosequence at Harwood Forest as part of the CARBOAGE project (Zerva, 2004).*

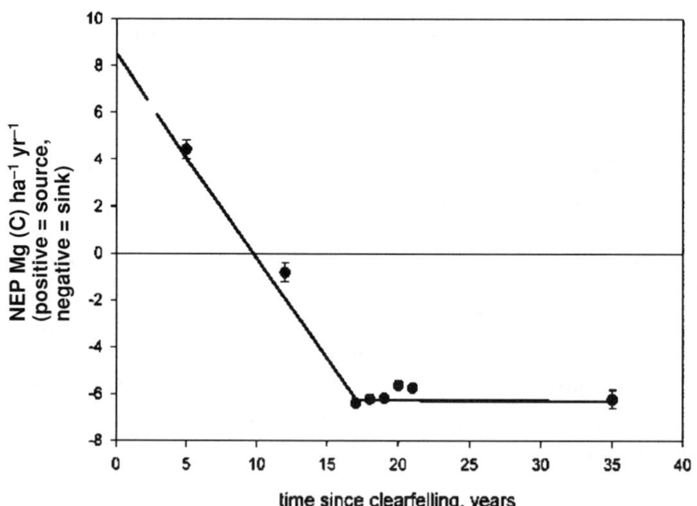

Figure 4. *Age-related changes in the net ecosystem exchange of Sitka spruce. Data points compiled from the time sequence of data from Griffin Forest (Clement et al., 2003) and the chronosequence of data from Harwood Forest (Grace, 2004).*

budget of the heathland before ploughing and planting or of the forest during the period immediately after planting. However, on the basis of a recent study of an age series of stands (of the same species, of similar yield class, on similar sites) at Harwood

Forest in northern England, we surmise that the stand passed from carbon source to carbon sink about 12 years after planting (*Figure 4*).

The research site is in a basin at about 330 m above sea level (a.s.l.), facing north with a slope of about 5°. Mean annual temperature is about 8 °C and precipitation 1200 mm. The soil is largely peaty gley, in part overlying glacial deposits. The fetch (about 500 m radius) around the measurement mast is almost entirely Sitka spruce. The whole forest has been fertilized at least once during the past 10 years by helicopter, with about 350 kg ha^{-1} urea, 43% nitrogen. Growth of the trees in the footprint is 'good', with general yield class (YC) 16 (i.e., mean annual increment (MAI) over the rotation of 16 m^3 ha^{-1} per year). There are about 2200 trees ha^{-1}, and across the footprint the canopy has been closing over the past four years, so that it is now closed throughout (Clement *et al.*, 2003).

In such a maritime climate, in general lacking extremes of temperature, some carbon assimilation has occurred on almost all days of the year. On a daily basis, there has been a net loss of carbon on some days throughout the year when it has been exceptionally cloudy. In the winter, net carbon assimilation is limited by the length of day, so that on a weekly or monthly basis there has been a net loss of carbon over the three winter months. Over the past five years of measurement, the annual net uptake of CO_2 from the atmosphere has been in the range 6.2–7.1 Mg (C) ha^{-1}, with no evident time trend. This is a large uptake of CO_2 from the atmosphere compared with comparable measurements throughout Europe and North America in more continental climates at similar latitudes (Jarvis *et al.*, 2001a). Measurements of some of the partial processes within the stand enable a preliminary carbon balance to be drawn up (*Table 1*).

These data provide an analogue today for sites planted in the early 1990s in relation to Article 3.3 of the Kyoto Protocol, which will be assessed in the first 'Commitment Period', 2008–2012. Clearly, young Sitka spruce on a reasonable site in a maritime climate sequesters carbon at a high rate; not all young forests can be expected to take up carbon at equivalent rates. Thinning, following current industry practice, took place in the somewhat better grown stands (YC 18) at lower elevations

Table 1. *Preliminary carbon budgets for the Griffin flux site. Based on data from R. Clement, F. Conen, Jarvis (1981), and L. Wingate. Quantities all in units of Mg (C) ha^{-1} yr^{-1} rounded to nearest 0.5.*

(1)	GPP (2 + 3)	20.0
(2)	NEP measured	6.5
(3)	Ecosystem respiration	13.5
(4)	Above-ground tree growth (YC 16)	4.0
(5)	Above-ground detritus	2.0
(6)	Above-ground autotrophic respiration	7.0
(7)	Translocation from above to below ground	8.0
(8)	Below-ground autotrophic respiration	3.5
(9)	Below-ground heterotrophic respiration	3.0
(10)	Below-ground tree growth	1.5
(11)	Below-ground detritus	2.0
(12)	Increase in soil carbon	1.0
(13)	NPP (1 - 6 - 8)	9.5

in winter 2001–2, and the footprint of the experiment site was thinned in spring 2004 with measurements re-commencing in summer 2004. A key question now is what will happen to the carbon balance as a result of this management intervention. Measurements of NEP are being restarted. We have an excellent set of pre-thinning NEP calibration data and similar, overlapping, measurements have continued in a comparable stand at Harwood Forest and serve as a control. Before the forest experiment, we have addressed the question, at least in part, through the model analysis that follows. Our simulations suggest that the consequences of the disturbance will be very short-lived; the practical forest experiment is designed to test this.

9. A model investigation of thinning

Thinning is done for long-term silvicultural goals to improve the quality of the final harvested product, but the thinnings are themselves a product equivalent to the removal of a significant quantity of CO_2 from the atmosphere. There are several potential impacts of thinning on the carbon budget: the amount of timber removed, fate of the woody debris, change in the soil physical conditions, and the capacity of the canopy to acquire carbon as CO_2 from the atmosphere. We are far from being able to forecast the consequences of all of these changes and possible associated feedbacks affecting the biophysical process rates. Here, we focus on the consequences of thinning-induced reduction in leaf area index (LAI), gross primary productivity (GPP), and energy exchange with the atmosphere. The key question addressed by this analysis is whether the disturbance caused by current thinning practice leads to a greatly reduced annual carbon gain, or even a temporary net carbon loss, and how long recovery from thinning may take.

In a numerical experiment, we simulated instantaneous photosynthesis in the Griffin Forest Sitka spruce plantation after thinning according to current commercial practice and made comparisons with some alternative spatial variants. We used the process-based three-dimensional canopy model MAESTRA (Medlyn, 2004; Wang and Jarvis, 1990). With this model we calculated light distribution, energy exchange, transpiration, gross photosynthesis (P_g), and respiration of components (R) on a tree basis, taking into account tree crown and stand structure, which, together with the spatial tree distribution, were mimicked for the plantation (Wingate, 2003). Physiological parameters for Sitka spruce were taken from the abundant literature (e.g., Leverenz and Jarvis, 1980; Ludlow and Jarvis, 1971; Meir *et al.*, 2002; for details see Ibrom *et al.*, 2004). Driving meteorological variables, namely photosynthetic photon flux density (PPFD), air temperature, humidity, and horizontal wind speed, were measured at the site (Clement *et al.*, 2003). We input the undisturbed variant at a stocking density of 2500 trees ha^{-1} before any thinning (S_1) and tested the following three thinning scenarios: a one in three line thinning (S_2), a selection thinning (S_3), and current practice, a one in five line thinning with selection thinning in the intervening four rows (S_4). The simulations were run using weather data for the three months, May–July, 1998. The proportion of trees extracted was 1/3 in S_2 and S_3, and a little more, 11/25, in S_4 (*Figure 5*). LAI was reduced from an assumed initial LAI of 8.7, in relation to the proportion of trees extracted. Scenarios S_2 and S_3 were used to analyse the effects of the resulting heterogeneity in tree distribution imposed on the stand remaining by the pattern of thinning.

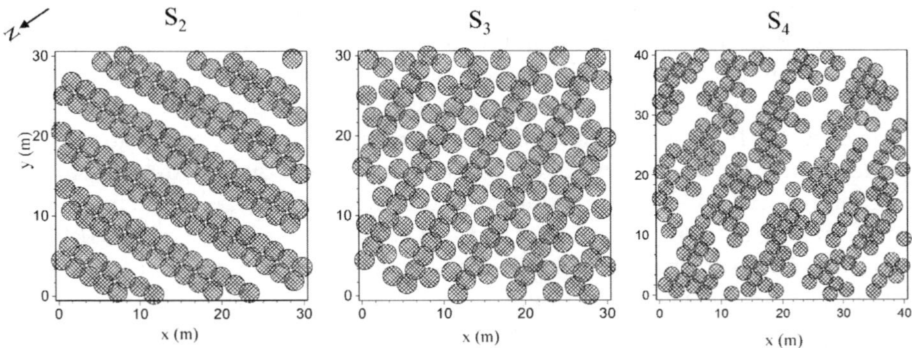

Figure 5. *Stand structure in the three thinning scenarios. S_2, a traditional line thinning, 1/3 trees removed; S_3, selection thinning, 1/3 trees removed; S_4, current line thinning with selection thinning in between, 11/25 trees removed.*

The simulation results were compared with the eddy covariance flux measurements and were found to be in surprisingly tight agreement. Values of P_g derived from the flux measurements were only 5% higher than the simulated rates for the unthinned scenario, S_1 (slope 1.05, offset 0.03 μmol (CO_2) m^{-2} s^{-1}, R^{-2} = 0.84) (Ibrom *et al.*, 2004).

The three thinning scenarios resulted in different reductions of absorbed PPFD, of P_g and, to a lesser extent, of net photosynthesis of the leaves, P_n (*Table 2*). The last is a consequence of taking into account leaf respiration, which scales linearly with LAI. The effect would be even less if total tree respiration were to be considered (i.e., NPP). Comparison of the results for the two scenarios S_2 and S_3 shows that thinning pattern does have an effect. The relative effects of thinning are smaller than the relative proportion of leaf area reduction, because of the non-linear processes that are involved. The more homogeneous tree extraction pattern (S_3) results in a smaller change of photosynthesis and PPFD absorption than the other variants in which the same amount of tree extraction and leaf area reduction occurs. Given an optimal extraction pattern, the resulting GPP would be reduced by only 4%, for a tree extraction of 33%. Assuming a negligible difference in stand albedo between thinning patterns S_2 and S_3, the potential effect of thinning pattern on transmission of PPFD to the soil would be 35% less in the case of the more uniform selection thinning.

Table 2. *Results of the simulations of the thinning scenarios (S_2–S_4) relative to values before the thinning (scenario S_1). S_2, a traditional line thinning, 1/3 trees removed; S_3, selection thinning, 1/3 trees removed; S_4, current line thinning with selection thinning in between, 11/25 trees removed.*

Scenario	removed LAI	absorbed PPFD	P_g	P_n
2	1/3	0.83	0.81	0.88
3	1/3	0.89	0.87	0.96
4	11/25	0.77	0.75	0.86

Current thinning practice (S_4) takes a higher proportion of trees out of the stand, and is, therefore, not directly comparable to S_2 and S_3. The resulting canopy pattern is even more irregular than in S_2, with larger canopy gaps and parts of the canopy almost unchanged. This results in sub-optimal P_g and a larger heterogeneity of PPFD transmission to the soil. The 44% reduction in LAI resulted in a 14% reduction of P_g.

As far as they go, the quantitative estimations presented on how thinning pattern is likely to influence PPFD absorption and photosynthesis show a clear picture: given an optimal thinning pattern, a substantial leaf area reduction would leave the GPP almost unchanged. However, these simulations did not consider the stand dynamics that are likely to be a consequence of the thinning. So far, any kind of physiological and structural acclimation of the leaves to their post-thinning local environment has been left out of consideration. Shade-acclimated needles will most likely gain photosynthetic capacity soon after opening up of the canopy, and canopy gaps will fairly quickly be closed by crown growth, over three years for the selection thinning, over five years for the line thinnings. Both acclimation and re-growth responses will likely lead to an increase of GPP compared with the simulations. Furthermore, a complete assessment of the impacts of thinning on carbon sequestration must take into account the consequent effects on soil and woody debris respiration. Additional, new, empirical data are needed to support description and parameterization of these processes: these data will be obtained during the post-thinning period of measurements now starting.

10. What can we do?

Having regard to the importance of the carbon in forests, how can current practice be modified to minimize loss of carbon through the disturbance resulting from management operations. We propose that forest management needs to consider: (a) how to *preserve* the existing stock of carbon in our forests, much of which may have been accumulated before recent afforestation by the vegetation that developed in the past; and (b) how to *increase* the acquisition of CO_2, and hence the stock of carbon in our forests in both trees and soil. A new style of forest management is needed to conserve and enhance the carbon stocks in forests, rather than to disrupt them, particularly the stocks in forest soils. In essence, the net carbon sink strength can be increased by operations that both minimize soil disturbance and increase gross primary productivity.

10.1 *Minimize soil disturbance*

For socio-economic reasons, the forest operations enumerated in Section 7 are today very largely accomplished in the developed world using large machines, frequently track mounted, that seriously disturb the forest floor and churn up the soil organic matter. The consequences of this for CO_2 emissions have not been systematically evaluated, but it seems likely that emissions from forest soils are considerably enhanced. Without significant economic incentives, there is little realistic prospect of this changing, but some amelioration is possible. For example, the brash can be used to provide trackways through stands, thereby reducing mechanical disturbance of the forest floor. What is needed at the present time is heightened awareness of the problem and its quantitative evaluation in terms of enhanced carbon losses. The machines themselves contribute emissions and these should be evaluated and factored into the budget.

In developing countries, traditional practices may still pertain and forest operations may be less mechanized. The carbon benefits of this should be quantified and the practices not lightly changed.

10.2 *Increase gross primary productivity*

Many forests are chronically nutrient deficient, particularly in nitrogen and phosphorus, or lack a balanced nutrient supply in areas of considerable nitrogen deposition (Tamm, 1991; Vitousek and Howarth, 1991). Raising fertility is possibly the most effective means of rapidly increasing uptake of CO_2 from the atmosphere and transferring carbon as detritus to the soil reservoir for storage and is, of course, compatible with enhancing timber production. Many experiments worldwide have demonstrated that the growth of temperate forests is very responsive to the application of fertilizers, particularly of nitrogen (see, for example, Albaugh *et al.*, 1997; Linder *et al.*, 1987; Pereira *et al.*, 1989; Tamm, 1991). Enhancement of nitrogen fixation by free-living soil microorganisms and by leguminous plants, or applications of wood ash, can also increase productivity, growth, and detritus production.

In forests at high latitudes in particular, the small stock of nutrients is locked up in wood and soil organic matter and turnover is slow; CO_2 uptake from the atmosphere, tree growth, and carbon transfer to the soil go on at low rates. Applications of fertilizer coupled with extension of length of the rotation can partly reverse the age-related decline in growth and productivity and restore the declining sink capacity of aging forests, and is possibly the most effective way of rapidly increasing carbon sink capacity. Experimental, annual, low-intensity applications of fertilizer have been shown to be particularly effective in these respects in northern Europe (Bergh *et al.*, 1999; Tamm, 1991).

The potential to increase biomass production by nutrient optimization is illustrated by two ongoing long-term experiments in northern and southern Sweden (*Figure 6*). The treatment at the northern site, Flakaliden (64°07″ N, 19°27″ E, 310–320 m a.s.l.), started in 1987 and treatments at the southern site, Asa (57°08″ N, 14°45″ E) started one year later (Bergh *et al.*, 1999). The northern site has a low natural fertility, but the southern site is among the top site indices for Norway spruce in Sweden. Two different nutrient optimization treatments were included. The first treatment was a complete nutrient solution, injected into the irrigation water and supplied every second day during the growing season, and the second an annual application of a solid fertilizer mix. Non-treated controls and plots with irrigation were also included. The annual rates of nutrients supplied were estimated to maintain non-limiting foliar concentrations of all essential macro- and micronutrients, at the same time as nutrient leakage to the groundwater was avoided.

As expected, the largest *relative* response to fertilization was found at the low fertility site. At Flakaliden there was a spectacular response, with more than a tripling of annual volume growth, but with no additional effect of irrigation (*Figure 6*). At the southern site, Asa, there was a doubling of yield when fertilization was combined with irrigation. Fertilization without irrigation also resulted in a substantial yield increase, but was on average, over the period 1988–2001, 18% lower than in the combined treatment. During the same period, irrigation alone increased yield by 12% compared with non-treated control plots. If the response to treatment is compared in *absolute* terms, the highest increase (11.6 m³ ha⁻¹ yr⁻¹) was found in the irrigation–fertilization treatment in Asa, which was almost 50% higher than the extra yield in Flakaliden (7.8 m³ ha⁻¹ yr⁻¹). Without irrigation, however, the absolute increase in yield was as high in Flakaliden as in Asa (7.7 and 7.3 m³ ha⁻¹ yr⁻¹, respectively).

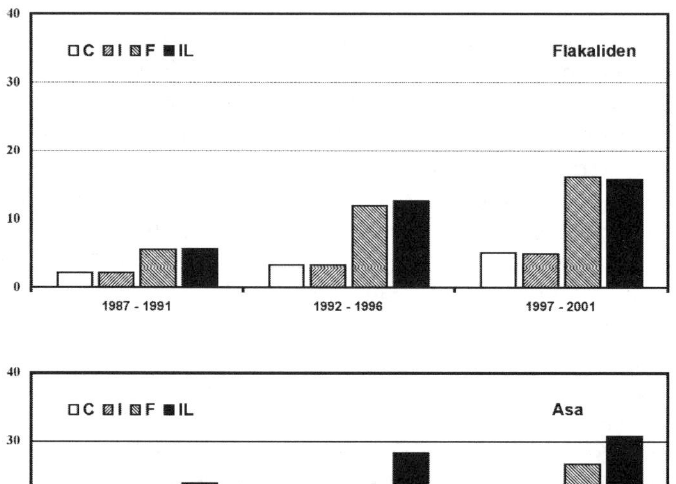

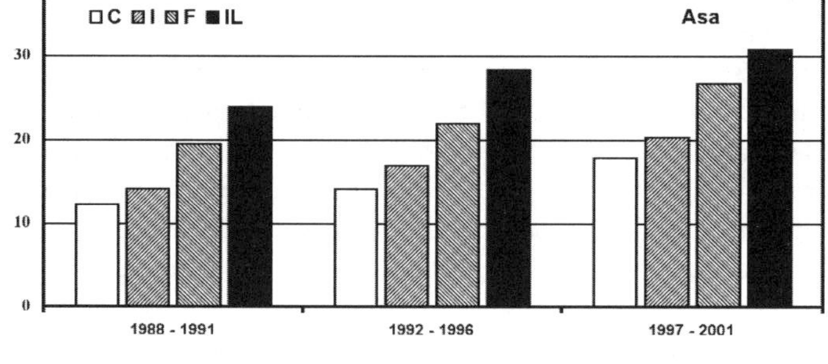

Figure 6. *Periodic annual increments of standing volume (m³ ha⁻¹ yr⁻¹) in two long-term fertilization experiments. The experiments were done in Norway spruce* (Picea abies *(L.) Karst.) plantations in the north (Flakaliden, 64°07″ N, 19°27″ E) and south (Asa, 57°08″ N, 14°45″ E) of Sweden. Treatments were: non-treated controls (C), irrigation (I), annual supply of a complete mix of fertilisers (F), and a complete balanced nutrient solution, injected into the irrigation water and supplied every second day during the growing season (IL) (see Bergh et al., 1999; Linder, 1995).*

Fertilization not only increases biomass production, but was in a recent meta-analysis found to be the only forest management practice, which had a clear positive effect on the soil carbon pool (Johnson and Curtis, 2001). The total soil-surface CO_2 flux is also reduced by fertilization, but the processes behind this are not fully understood (cf., Ågren *et al.*, 2001). None-the-less, it can be concluded that the carbon and nitrogen cycles are very closely linked, and that nitrogen fertilization has a positive effect on the carbon balance and carbon sequestration in most Boreal forest ecosystems and in many temperate forest ecosystems too (Jarvis *et al.*, 2001b).

In northern temperate and Boreal regions, where atmospheric nitrogen deposition is of the order of 10 kg (N) ha⁻¹ yr⁻¹, enhanced tree growth may result from the widespread distribution of nitrogen of industrial origin by atmospheric transport

(Spiecker *et al.*, 1996). However, in heavily industrialized areas, where atmospheric nitrogen deposition is excessive (more than 50 kg (N) ha^{-1} yr^{-1}) (Ibrom *et al.*, 1995), such as in central Europe, fertilization does carry the risk of eutrophication of ecosystems and consequent damage to forests. Large-scale modelling has shown that the so-called critical nitrogen load to forests has already been exceeded in Germany and the Benelux countries (see, for example, Posch *et al.*, 2003). In such conditions, compensation liming, rather than nitrogen fertilization, is needed to amend functioning of the forest soils and to stabilize tree growth.

A further potential problem arising from the use of nitrogen-containing fertilizers is the risk of stimulating nitrogen emissions, but quantitative data on N_2O emissions in fertilized forests are scarce at present (see Chapter 14, this volume).

10.3 *Lock the gate*

A third way by which we can increase the uptake of CO_2 from the atmosphere by our forests, favoured by some conservation organizations, is 'zero management', with the emphasis on restoration of natural forest and endorsement of conservation goals.

Starting with afforestation or reforestation, one can stimulate natural regeneration or plant trees with minimum disturbance and then shut the gate and leave the ecosystem to equilibrate! Over time, semi-natural forest may develop that will both accumulate carbon and be a recreational, educational and biodiversity resource of substantial benefit to local communities.

Similarly, one can leave mature degraded forest to revert to a natural condition, like that before the human-initiated disturbance it has previously experienced. This may lead quickly to increase and retention of carbon on the forest floor as coarse woody detritus, and will likely have immediate conservation benefits for wildlife biodiversity. Although not particularly attractive to foresters, this is also not without risk of carbon loss from disturbance. For example, the risk of eventual fire as coarse woody debris accumulates can be very high in eucalypt forest in Australia.

11. The wider perspective

We have considered thinning in some detail because forests planted for Kyoto goals will be coming up for thinning within the next ten years. Thinning is done for long-term silvicultural goals to improve the quality of the final harvested product, but the thinnings are themselves a product equivalent to the removal of a significant quantity of CO_2 from the atmosphere. However, if the market is such that the price for thinnings is not realistic, or the nearest site of utilization is some hundreds of kilometers away, it would be advantageous from a carbon perspective for the thinnings to be used *locally* to substitute for fossil fuels in one of several ways. They could, for example, be used as a bio-fuel to provide combined heat and power for a school or community centre or be used within the forest to generate electricity for local community use. Local power generation is commonplace in developing countries but has been superseded by the demanding, inefficient, ubiquitous grid in developed countries (Anon., 2004).

In this paper we have gone beyond the carbon economy of trees alone and endeavoured to raise awareness of some aspects of the larger-scale carbon budget arising from disturbance and management of the forest ecosystem. We have not, however, included the carbon cost of the use of the machines used in management: the 'harvesters' that cut the trees, the 'forwarders' that pick up the logs, the 'trucks' that take them out of the forest to the processing plant, or the cars that bring the foresters and machine operators in and out of the forest. One could, and should, eventually go further and include a full quantitative carbon analysis of the 'wood chain' through the management cycle (Gower, 2003).

12. Conclusions

- We need a new approach to forestry to embrace the importance of forests in carbon sequestration, *carbon forestry*!
- Forest management operations may decrease or increase carbon stocks and CO_2 fluxes. Operations that increase soil disturbance and reduce productivity lead to a decrease in carbon stocks, whereas operations that both minimize soil disturbance and increase GPP of a stand increase the sink strength.
- Application of balanced fertilizer can effect a large increase in annual carbon uptake and increase in carbon stocks, particularly in north temperate and sub-Boreal regions.
- The impact of management operations on the instantaneous carbon balance of a *forest* should take into account the spatial representation of all stands present at whatever stages they are in their life and management cycles.
- The impact of management operations on the carbon balance of a *stand* should be calculated over all stages in its life cycle.
- Even if assessments of the impacts of management operations are restricted to the forest within the forest gate, they should take into account the carbon cost of the operators and their machines within the forest.
- Assessments of the effects of management operations on forest sinks are in their infancy, and we cannot currently provide quantitative estimates of the likely consequences for forest carbon stocks of most management operations.
- We consider it likely that forest management oriented towards *protecting* and *enhancing* carbon stocks could lead to enhancement of the global forest carbon sink, perhaps up to 0.3 Pg per year, but at present we need many more data to support this suggestion.

Acknowledgements

We thank the many friends and colleagues who have stimulated the arguments and ideas put forward here.

References

Ågren, G.I., Bosatta, E. and Magill, A.M. (2001) Combining theory and experiment to understand effects of inorganic nitrogen on litter decomposition. *Oecologia* **128**: 94–98.

Albaugh, T.J., Allen, H.L., Dougherty, P.M., Kress, L.W. and King, J.S. (1997) Leaf area and above- and belowground growth responses of loblolly pine to nutrient and water additions. *Forest Science* **44**: 317–327.

Anon. (2004) [Editorial.] *Nature* **427**: 661.

Arain, M.A., Black, T.A., Barr, A.G., Jarvis, P.G., Massheder, J.M., Verseghy, D.L. and Nesic, Z. (2002).Effects of seasonal and interannual climate variability on net ecosystem productivity of boreal deciduous and coniferous forests. *Canadian Journal of Forest Research* **32**: 878–891.

Bergh, J., Linder, S., Lundmark, T. and Elfving, B. (1999) The effect of water and nutrient availability on the productivity of Norway spruce in northern and southern Sweden. *Forest Ecology and Management* **119**: 51–62.

Clement, R., Moncrieff, J.B. and Jarvis, P.G. (2003) Net carbon productivity of Sitka spruce forest in Scotland. *Scottish Forestry* **57**: 5–10.

Gower, S.T. (2003) Patterns and mechanisms of the forest carbon cycle. *Annual Review of Environmental Resources* **28**: 169–204.

Grace, J. (2004) Age-related Dynamics of Carbon Exhange in European forests. Final Report and Technological Implementation Plan: The CARBO-AGE project. www.ierm.ed.ac.uk/CARBO-AGE/HOME.htm. 170 pp. University of Edinburgh.

Ibrom, A., Oltchev, A., Constantin, J., Marques, M. and Gravenhorst, G. (1995) Die Stickstoffimmision und -deposition in Wäldern. *UBA Texte* **28**: 20–29.

Ibrom, A., Jarvis, P.G., Clement, R.B., Morgenstern, K., Oltchev, A., Medlyn, B., Wang, Y.P., Wingate, L., Moncrieff, J. and Gravenhorst, G. (2004) A comparative analysis of simulated and observed photosynthetic CO_2 uptake in two coniferous forest canopies. Submitted to *Tree Physiology*.

IPCC (2000) *Land Use, Land-Use Change, and Forestry*. 377 pp. Cambridge University Press, Cambridge.

IPCC (2001) *Climate Change 2001: The Scientific Basis*. 881 pp. Cambridge University Press, Cambridge.

Jarvis, P.G. (1981) Production efficiency of coniferous forest in the UK. In: *Physiological Processes Limiting Plant Productivity* (ed. C.B. Johnson) pp. 81–107. Butterworth Scientific Publications, London.

Jarvis, P. (2003) Trees and peat: carbon sources or sinks? Summary report of a seminar. *Scottish Forestry* **57**: 81–90.

Jarvis, P.G. and Linder, S. (2000) Constraints to growth of boreal forests. *Nature* **405**: 904–905.

Jarvis, P.G., Dolman, A.J., Schulze, E.-D., Matteucci, G., Kowalski, A.S., Ceulemans, R. *et al.* (2001a) Carbon balance gradient in European forests: should we doubt 'surprising' results? A reply to Piovesan & Adams. *Journal of Vegetation Science* **12L**: 145–150.

Jarvis, P.G., Saugier, B. and Schulze, E.-D. (2001b) Productivity of boreal forests. In: Roy, R., Saugier, B. and Mooney, H.A. (eds) *Terrestrial Global Productivity*, pp. 211–244. Academic Press, San Diego and London.

Johnson, D.W. and Curtis, P.S. (2001) Effects of forest management on soil C and N storage: meta analysis. *Forest Ecology and Management* **140**: 227–238.

Leverenz, J. and Jarvis, P.G. (1980) Photosynthesis in Sitka spruce (*Picea sitchensis* (Bong.) Carr.). IX: The relative contribution made by needles at various positions on the shoot. *Journal of Applied Ecology* **17**: 59–68.

Linder, S. (1995) Foliar analysis for detecting and correcting nutrient imbalances in Norway spruce. *Ecological Bulletins* **44**: 178–190.

Linder, S., Benson, M.L., Myers, B.J. and Raison, R.J. (1987) Canopy dynamics and growth of *Pinus radiata*. I. Effects of irrigation and fertilisation during a drought. *Canadian Journal of Forest Research* **10**: 1157–1165.

Ludlow, M.M. and Jarvis, P.G. (1971) Photosynthesis in Sitka spruce (*Picea sitchensis* (Bong.) Carr.) I. General characteristics. *Journal of Applied Ecology* **8**: 925–953.

Malhi, Y., Baldocchi, D.D. and Jarvis, P.G. (1999) The carbon balance of tropical, temperate and boreal forests. *Plant, Cell and Environment* **22**: 715–740.

Medlyn, B. (2004) A MAESTRO retrospective. In: Mencuccini, M., Grace, J., Moncrieff, J. and McNaughton, K.G. *Forests at the Land–Atmosphere Interface*, pp. 105–121. CAB International, Wallingford, UK.

Meir, P., Kruijt, B., Broadmeadow, M., Barbosa, E., Kull, O., Carswell, F., Nobre, A. and Jarvis, P.G. (2002) Acclimation of photosynthetic capacity to irradiance in tree canopies in relation to leaf nitrogen concentration and leaf mass per unit area. *Plant, Cell and Environment* **25**: 343–357.

Pereira, J.S., Linder, S., Araújo, M.C., Pereira, H., Ericsson, T., Borallho, N. and Leal, L. (1989) Optimization of biomass production in *Eucalyptus globulus* – A case study. In: Pereira, J.S. and Landsberg, J.J. (eds.) *Biomass Production by Fast-growing Trees*, pp. 101–121. Kluwer. Dordrecht.

Posch, M., Hettelingh, J.-P. and Downing, R.J. (2003) Modelling and Mapping of Critical Thresholds in Europe Status Report 2003. RIVM Report No. 259101013/2003, National Institute for Public Health and the Environment, Bilthoven, The Netherlands.

Rayment, M. and Jarvis, P.G. (1997) An improved open chamber system for measuring soil CO_2 effluxes in the field. *Journal of Geophysical Research* **102**: 28779–28784.

Rayment, M.B. and Jarvis, P.G. (1999) Seasonal gas exchange of black spruce using an automatic branch bag system. *Canadian Journal of Forest Research* **29L**: 1528–1538.

Royal Society (2001) *The Role of Land Carbon Sinks in Mitigating Global Change.* 27 pp. The Royal Society, London.

Schlesinger, W.H. (1997) *Biogeochemistry, an Analysis of Global Climate Change.* 588 pp. Academic Press, San Diego and London.

Spiecker, H., Mielikäinen, K., Köhl, M. and Skovsgaard, J.P. (1996) *Growth Trends in European Forests – Studies from 12 Countries.* 354 pp. Springer-Verlag, Heidelberg.

Tamm, C.O. (1991) *Nitrogen in Terrestrial Ecosystems. Questions of Productivity, Vegetational Changes and Ecosystem Stability. Ecological Studies 81.* 115 pp. Springer-Verlag, Berlin.

Valentini, R., Matteucci, G., Dolman, A.J., Schulze, E.-D., Rebmann, C., Moors, E.J. *et al.* (2000). Respiration as the main determinant of carbon balance in European forests. *Nature* **404**: 861–865.

Vitousek, P.M. and Howarth, R.W. (1991) Nitrogen limitation on land and in the sea: How can it occur? *Biogeochemistry* **13**: 87–115.

Wang, Y.P. and Jarvis, P.G. (1990) Description and validation of an array model MAESTRO. *Agricultural and Forest Meteorology* **51**: 257–280.

Wingate, L. (2003) The contribution of photosynthesis and respiration to the net ecosystem exchange and ecosystem ^{13}C discrimination of a Sitka spruce plantation. PhD thesis, University of Edinburgh, UK.

Zerva, A. (2004) Effects of afforestation and forest management on soil carbon dynamics and trace gas emissions in a Sitka spruce (*Picea sitchensis* (Bong.) Carr.) forest. Ph.D. thesis, University of Edinburgh.

Index

Note: Page references in **bold** refer to Tables; those in *italics* refer to Figures